ENZYME ENGINEERING

Volume 6

ENZYME ENGINEERING
Volume 6

Edited by

Ichiro Chibata

Research Laboratory of Applied Biochemistry
Tanabe Seiyaku Company
Osaka, Japan

Saburo Fukui

Department of Industrial Chemistry
Faculty of Engineering
Kyoto University
Kyoto, Japan

and

Lemuel B. Wingard, Jr.

Department of Pharmacology
School of Medicine
University of Pittsburgh
Pittsburgh, Pennsylvania

PLENUM PRESS · NEW YORK AND LONDON

The Library of Congress cataloged the second volume of this title as follows:

Engineering Foundation Conference on Enzyme Engineering, 2d, Henniker, N. H., 1973.
   Enzyme engineering; [papers] Edited by E. Kendall Pye and Lemuel B. Wingard, Jr. New York, Plenum Press [1974]

   Called volume 2 in continuation of a volume with the same title published in 1972, which contains the papers of the 1st Engineering Foundation Conference on Enzyme Engineering.
   1. Enzymes — Industrial applications — Congresses. I. Pye, E. Kendall, ed. II. Wingard, L., ed. III. Title. [DNLM: 1. Biomedical engineering — Congresses. 2. Enzymes — Congresses. W3 EN696]

TP248.E5E53 1973	660'.63	74-13768

ISBN 978-1-4615-9292-1   ISBN 978-1-4615-9290-7 (eBook)
DOI 10.1007/978-1-4615-9290-7

LC 74-13768
ISBN 978-1-4615-9292-1

Proceedings of the Sixth International Enzyme Engineering
Conference, held September 20 – 25, 1981, in Kashikojima, Japan

©1982 Plenum Press, New York
Softcover reprint of the hardcover 1st edition 1982

A Division of Plenum Publishing Corporation
233 Spring Street, New York, N.Y. 10013

ORGANIZATION OF THE CONFERENCE

COSPONSORS:

The Engineering Foundation
345 East 47th Street
New York, NY   10017
U.S.A.

Japanese Society of Enzyme
   Engineering
c/o Tanabe Seiyaku Co.
Osaka, Japan

EXECUTIVE COMMITTEE:

Saburo Fukui . . . .   Executive Chairman
Ichiro Chibata . . . .   Program Chairman
David Fink . . . .   U.S.A. Member
Jan Konecny. . . . .   European Member
Garfield Royer . . . .   U.S.A. Member
Howard Weetall . . . .   U.S.A. Member
Michael Weibel . . . .   U.S.A. Member
Lemuel B. Wingard Jr.. . . . .   Permanent Member
Sandford Cole. . . . .   Conference Director

PROGRAM COMMITTEE:

I. Chibata, chm.
C. Horvath
H. Inoue
A. I. Laskin
K. Mosbach
T. Murachi

D. Ryu
H. Samejima
S. Suzuki
D. Thomas
F. Wagner
H. Yamada

ADVISORY BOARD:

F. Bartoli
I. V. Berezin
G. Broun
T. Cayle
T. M. S. Chang
F. F. Davis
H. Filippusson
R. M. Flora
F. Garcia-Hernandez
S. Gestrelius
H. U. Geyer
C. Horvath
J. V. Hupkes
H. Inoue
E. Katchalski-Katzir
J. Klein
M-R. Kula
A. I. Laskin
P. Linko
G. Manecke

B. Mattiasson
A. S. Michaels
K. Mosbach
T. Murachi
A. H. Nishikawa
H. Okada
D. Ryu
H. Samejima
G. Schmidt-Kastner
M. Seidman
B. P. Sharma
S. Suzuki
D. Thomas
W. R. Vieth
F. Wagner
J. C. Weaver
M. Wilchek
H. Yamada
F. Widmer

FINANCIAL SUPPORT:

103 Japanese companies contributed through the Japanese Society
    of Enzyme Engineering
Engineering Foundation
    New York, NY, U.S.A.
Assoreni
    Milan, Italy
BASF Aktiengesellschaft
    Ludwigshafen, F. R. Germany
Bayer AG
    Leverkusen, F. R. Germany
Behringwerke AG
    Marburg (Lahn), F.R. Germany
Boehringer Mannheim GmbH.
    Mannheim, F. R. Germany
British Petroleum Co.
    Sunbury-on-Thames, England
Cetus Corp.
    Berkeley, CA, U.S.A.
Ciba-Geigy AG
    Basel, Switzerland
Corning Glass Works
    Corning, NY, U.S.A.

Cutter Laboratories, for Bayer/Cutter/Miles
    Elkhart, IN, U.S.A.
Dow Chemical Co.
    Midland, MI, U.S.A.
DSM, N.V.
    Heerlen, Netherlands
Exxon Research and Engineering Co.
    Florham Park, NJ, U.S.A.
FMC Corp.
    Princeton, NJ, U.S.A.
Genex Corp.
    Rockville, MD, U.S.A.
Gist-Brocades, N.V.
    Delft, Netherlands
Chr. Hansen's Laboratorium
    Copenhagen, Denmark
Henkel & Cie, GmbH.
    Dusseldorf, F. R. Germany
Hoechst, AG
    Frankfurt (Main), F. R. Germany
International Minerals & Chemicals Inc.
    Terre Haute, IN, U.S.A.
Kraft, Inc.
    Glenview, IL, U.S.A.
Miles Kali-Chemie, GmbH.
    Hannover, F. R. Germany
Monsanto Co.
    St. Louis, MO, U.S.A.
Novo Industries
    Bagsvaerd, Denmark
Pfizer, Inc.
    Groton, CT, U.S.A.
Pharmacia Inc.
    Piscataway, NJ, U.S.A.
G. D. Searle & Co.
    Skokie, IL, U.S.A.
A. E. Staley Manufacturing Co.
    Decatur, IL, U.S.A.
National Science Foundation
    Washington, DC, U.S.A.

NEXT ENZYME ENGINEERING CONFERENCE

TIME: September 25-30, 1983

LOCATION: Pocono Hershey Resort Hotel
White Haven, PA, U.S.A.
(on highway Interstate 80 near
Scranton, Pennsylvania)

EXECUTIVE COMMITTEE:

Allen I. Laskin . . . .   Executive Chairman

George T. Tsao. . . . .   Program Chairman

Gunter Schmidt-Kastner. . . . . .   European Member

Shuichi Suzuki. . . . .   Japan Member

Daniel Thomas . . . .   European Member

Howard H. Weetall . . . .   U.S.A. Member

Michael H. Weibel . . . .   U.S.A. Member

Lemuel B. Wingard Jr. . . . . . .   Permanent Member

Sandford S. Cole . . . .   Conference Director

PREFACE

Presently, intensive and global attention is being devoted to "biotechnology"--the technology utilizing marvelous capacities of living things for human welfare. Each country is strongly promoting its development. In particular, enzyme engineering, whose purpose is to utilize efficiently enzymes, microorganisms, and cultured plant as well as animal cells as organic catalysts, is one of the main themes in the field of biotechnology.

Under these circumstances, the Sixth Enzyme Engineering Conference was held at Kashikojima, Mie Prefecture, Japan from September 20 to 25, 1981, under the joint auspices of the Engineering Foundation of New York and the Japanese Society of Enzyme Engineering. This series of international conferences has been held biannually since 1971. The first three and the fifth conferences were held in the United States and the fourth one was in the Federal Republic of Germany. This sixth conference was the first to be held in Asia; and it was significant that a number of participants could visit Japany, which has produced successful achievements in the field of biotechnology.

This conference had 203 participants from 23 countries. There were participants from China for the first time, thus enlarging the international scope of the conference series. Besides the opening and closing lectures, 36 plenary lectures and 96 poster papers were presented; and most of them are included in this volume. It is worthy of notice that some topics of genetic engineering now share common interests with enzyme engineering, so that scientists and engineers in both fields should work together in the future. If these two areas can be combined successfully, it will surely lead to greater contributions and further attainments of biotechnology.

This successful conference was aided markedly by the continued efforts of Dr. Sandford S. Cole of the Engineering Foundation, members of the Executive Committee, and the General Secretaries of the Japanese Society of Enzyme Engineering. Also, we are deeply indebted to the many industrial firms in Europe, the United

States, and Japan that contributed financial support. We heartily
express our gratitude to all of them. We also express our appreciation to Ms. Hall of the University of Pittsburgh for her fine
efforts in retyping all of the edited manuscripts and to the production staff at Plenum Press for their assistance in the publi-
cation of this volume.

Ichiro Chibata
Saburo Fukui
Lemuel B. Wingard, Jr.

April, 1982

CONTENTS

CONFERENCE HISTORY   1

    The First Ten Years   3
        L.B. Wingard Jr.

KEYNOTE PAPER   9

    Recent Studies on Antibiotics and Low Molecular
    Weight Enzyme Inhibitors   11
        H. Umezawa

SESSION I.   LARGE SCALE PRODUCTION AND PURIFICATION OF
            BIOMOLECULES   35
            Chairmen: *H. Weetall and H. Yamada*

    Integration of Large Scale Production and Purifi-
    cation of Biomolecules   37
        K. Venkatasubramanian

    Role of Active Transport of Inducer in Enzyme
    Biosynthesis and Process Scale-Up   45
        *W.R. Vieth, K. Kaushik and K. Venkatasubraman-
        ian*

    Multi Enzyme Systems in Membrane Reactors   61
        C. Wandrey, R. Wichmann and A.S. Jandel

    Scale-Up of Protein Purification by Liquid-Liquid
    Extraction   69
        *M.-B. Kula, K.H. Kroner, H. Hustedt and H.
        Schutte*

    Rapid and Large-Scale Purification of Angiotensin-I
    Forming Enzymes and Milk Clotting Enzymes by
    Affinity Chromatography   75
        K. Murakami, S. Hirose and H. Kobayashi

Separation and Purification of Enzymes Via Continuous Parametric Pumping   77
    S.Y. Huang, C.K. Lin and L.Y. Juang

SESSION II.   APPLICATIONS OF BIOCATALYSTS FOR NEW
               REACTIONS AND ORGANIC SYNTHESIS   79
               Chairmen: A.L. Laskin and I. Chibata

Enzymatic Synthesis of Penicillins and Cephalosporins by Penicillin Acylase   81
    R. Okachi, Y. Hashimoto, M. Kawamori, R.
    Katsumata, K. Takayama and T. Nara

Enzymatic Acyl Transfer in Penicillin and Cephalosporin Chemistry.   91
    J. Konecny, A. Schneider and M. Sieber

Enzymatic Processes for the Synthesis of Optically
Active Amino Acids   97
    H. Yamada

Horse Liver Alcohol Dehydrogenase: An Illustrative
Example of the Potential of Enzymes in Organic
Synthesis.   107
    J.B. Jones

Immobilized Enzymes and Synzymes: Applications in
Organic Synthesis   117
    G.P. Royer

Immobilization of Living Microbial Cells and Transformation of Steroids   123
    K.A. Koshcheyenko and G.K. Skryabin

Steroid Bioconversion by Immobilized Cells   125
    C. Glomon, P. Germain, A. Miclo and J.M.
    Engasser

Steroid Δ^1-Dehydrogenase Isoenzymes in *Corynebacterium* Species Chol 73 T 191   127
    W.-R. Mueller, W. Preuss and R.D. Schmid

Use of Immobilized Enzyme Requiring Cofactor Regeneration and of Immobilized Mycelium for Steroid
Modification   129
    M.D. Legoy, F.Ergan, P. Dhulster, M.N. Kim and
    G. Gellf

Hydroxylation of Steroids by Immobilized Microbial
Cells   131
    A. Tanaka, K. Sonomoto, M. Hoq, N. Usui,
    K. Nomura and S. Fukui

Enzymatic Synthesis of Sulfur and Selenium Amino
Acids   135
    K. Soda, N. Esaki and H. Tanaka

Production of Alkaloids with Immobilized Cells of
Catharanthus Roseus   137
    C.A. Lambe, A. Reading, S. Roe, A. Rosevear,
    and A.R. Thomson

Propyleneoxide Production by Immobilized
Nocardia Corallina B-276   139
    K. Furuhashi, S. Uchida, I. Karube and S.
    Suzuki

Air-Oxidation of Linoleic Acid by Lipoxygenase-
Containing Particles Suspended in Water-Insoluble
Organic Solvent   141
    T. Yamane

Progress Toward Artificial Photosynthesis   143
    D.J. Graves

Light-Sensitization of a Microbial Protease   145
    Y.Y. Lee, K.N. Kuan, and P. Melius

Enzymatic Photosensitive Materials   149
    N.F. Kazanskaya

Ester Exchange of Triglyceride by Entrapped Lipase
in Organic Solvent   151
    K. Yokozeki, T. Tanaka, S. Yamanaka, T.
    Takinami, Y. Hirose, K. Sonomoto, A. Tanaka and
    S. Fukui

Solvent Production by Clostridium Acetobutylicum in
Aqueous Two-Phase Systems   153
    B. Mattiasson, M. Suominen, E. Andersson, L.
    Haggstrom, P.-A. Albertsson and B. Hahn-Hagerdal

Enzyme Kinetics and Mass-Transfer on Two Liquid Phase
Heterogeneous Systems   157
    J.M.C. Duarte

Physico—Chemical Means of Increasing the Yield of End
Products in Biocatalysis   159
   I.V. Berezin

Heat-Stable L-Lactate Dehydrogenase and its Application to an Enzyme Reactor   161
   T. Ohta, H. Taguchi and H. Matsuzaswa

Multienzymes and Cofactors Immobilized within Lipid-
Polyamide Membrane Microcapsules for Sequential
Substrate Conversion   163
   Y.T. Yu and T.M.S. Chang

Tyrosine and 3,4-Dihydroxyphenyl-Alanine Synthesis
by Citrobacter Freundii   165
   N.S. Egorov, M.B. Koupletskaja, and E.N.
   Kondratieva

Enzyme Catalysts for the Stereo and Regio Selective
Oxyfunctionalization of Organic Substrates   167
   S.W. May

SESSION III.   NEW IMMOBILIZATION TECHNIQUES OF BIO-
            MATERIALS AND THEIR APPLICATIONS   171
           Chairmen: D. Thomas and N. Ise

High-Rate, Continuous Waste Processor for the Production of High BTU Gas Using Immobilized Microbes . . .   173
   R.A. Messing

New Developments in the Preparation and Characterization of Polymerbound Biocatalysts   181
   J. Klein and G. Manecke

Bioconversion of Lipophilic or Water Insoluble Com-
pounds by Immobilized Biocatalysts in Organic Solvent
Systems   191
   S. Fukui and A. Tanaka

Multiple Carrier Regeneration and Enzyme Immobilization In Situ   201
   J.E. Prenosil and E. Stuker

Viability and Biosynthetic Capacity of Immobilized
Plant Cells   203
   P. Brodelius, F. Constabel and W.G.W. Kurz

Immobilized Enzymes of Heterogenous Structure   205
    K. Nakamura, K. Hibino and Y. Yano

Fibrous Support for Immobilization of Enzymes   207
    H. Ichijo, T. Suehiro, A. Yamauchi, S. Ogawa,
    M. Sakurai and N. Fujii

High Temperature Cell-Trapped Ultrafiltration Membranes   209
    E. Drioli, G. Iorio, M. DeRosa, A. Gambacorta,
    and B. Nicolaus

Immobilized Modifiers for Proteins (IMPs)   211
    W.H. Scouten, C. Lewis, A. Barnett, R. Haller and
    and W. Iobst

Enzyme Immobilization in Porous Supports   213
    J.E. Bailey

Immobilization of Microbial Cells Using Gelatin and
Glutaraldehyde   215
    Q. Wang, X. Ji and Z. Yuan

Immobilization of β-Galactosidase from E. Coli CSH36
and Its Microbial Cells Using Cellulose Beads   217
    Y. Hong, S. Kwon, M. Chun and M. Sernetz

Hydrophobic Immobilization of Enzymes and Polynucleotides on Trityl Agarose   219
    P. Cashion, A. Javed, V. Lentini, D. Harrison,
    J. Seeley and G. Sathe

Influence of the Activation Degree of the Support on
the Properties of Agarose-Nuclease   223
    A. Ballesteros, J.M. Guisan, and J. Serrano

Immobilization of Biofunctional Components by Radiation Polymerization and Applications   225
    I. Kaetsu, M. Kumakura, T. Fujimura, M. Yoshida,
    F. Yoshii, M. Asano, M. Tamada, and N. Kasai

Immobilization of Whole Cell Glucose Isomerase within
Soybean Protein   227
    C.L. Lai

Enzyme Stabilization in High Macromolecular Concentration Environments   229
    G. Greco Jr., G. Marrucci, and L. Gianfreda

Intrinsic Stability of Thermophilic Enzymes: 6-Phosphogluconate Dehydrogenase from *Bacillus Stearothermophilus* and Yeast 233
F.M. Veronese, E. Boccu and A. Fontana

Stabilization of Penicillin Amidohydrolase Immobilized on Eupergit C 235
K. Sauber and D.M. Kramer

Stabilization of Fumarase Activity of *Brevibacterium Flavum* Cells by Immobilization with κ Carrageenan and Polyethyleneimine 237
T. Tosa, I. Takata, and I. Chibata

Application of Polyethylene Glycol-Bound NAD Derivative and Thermostable Dehydrogenase in a Model Enzyme Reactor 239
I. Urabe, N. Katayama, and H. Okada

Use of the Porous Mineral Spherosil as a Carrier for Enzymes: Fixation and Purification 241
B. Mirabel

SESSION IV. INDUSTRIAL APPLICATIONS OF IMMOBILIZED BIOMATERIALS 243
Chairmen: A. Michaels and H. Samajima

Potential Application of Immobilized Viable Cells in the Food Industry: Malolactic Fermentation of Wine . 245
S. Gestrelius

Production of L-Tryptophan with Immobilized Cells . . 251
F. Wagner, S. Lang, W.-G. Bang, K.D. Vorlop, and J. Klein

Applications of Immobilized Tannin for Protein and Metal Adsorption 259
I. Chibata, T. Tosa, T. Mori, T. Watanabe, K. Yamashita and N. Sakata

Industrial Applications of Immobilized Biomaterials in China 265
S. Zhang

Production of L-Alanine from Ammonium Fumarate Using Two Types of Immobilized Microbial Cells 271
T. Sato, S. Takamatsu, K. Yamamoto, I. Umemura, T. Tosa, and I. Chibata

New Process for Production of High Fructose Corn
Syrup Using Combined Adsorption and an Enzyme Reactor 273
    K. Hashimoto, S. Adachi and H. Noujima

Immobilized Multienzyme Systems for Starch Processing
    A. Lindroos, Y.Y. Linko, and P. Linko   275

Glucoamylase Covalently Coupled to Porous Glass . . .   279
    G. Li, J. Huang, X. Kou and S. Zhang

Rotary Multidisc Reactor of Collagen Supported Immobilized Glucoamylase 281
    S. Gondo, H. Koya and M. Morishita

Kinetics for the Hydrolysis of Soluble Starch by
Glucoamylase and Application to an Immobilized Enzyme
System 283
    K. Kusunoki and K. Kawakami

Reduced Thermostability of Modified *Mucor Miehei*
Rennet 287
    S. Branner-Jorgensen, P. Eigtved and P.
    Schneider

Application of Immobilized Enzymes to Milk Curdling . 289
    S. Shimizu, K. Ohmiya, S.-E. Yun and T.
    Kobayashi

Production of 7-Aminodesacetoxycephalosporanic Acid
by Immobilized *E. coli* Cells 291
    Z. Wang, H. Yuo, M. Wang, Q. Jiao, W. Han, W.
    Sun and Q. Zhang

Technical Applications of Lactase and Aminoacid
Acylase Immobilized to Form Plexazym 293
    H. Plainer, B.G. Spörssler and H. Uhlig

Continuous Hydrolysis of Lactose in Skim Milk and
Acid Whey by Immobilized Lactase of *Aspergillus
Oryzae* 295
    H. Hirohara, H. Yamamoto, E. Kawano and T.
    Nagase

Lactic Acid Fermentation with Immobilized *Lactobacillus* Sp. 299
    S.L. Stenroos, Y.Y. Linko, P. Linko, M. Harju,
    and M. Heikonen

Production of l-Malic Acid with Immobilized Thermophilic Bacterium, *Thermus Rubens* Nov. sp.   303
    Y. Ado, T. Kawamoto, I. Masunaga, K. Takayama,
    S. Takasawa, and K. Kimura

Continuous Hydrolysis of Concentrated Sucrose Solutions by Immobilized Invertase   305
    D. Combes and P. Monsan

A Series of Covalently Bonded Enzymes and Their
Applications   307
    S. Liu, Z. Yuan, Q. Wang, J. Wang and Y. Zeng

The Application of Fluidized Beds for Improved
Enzyme Reactor Performance   309
    A. Renken, E. Flaschel and P.-F. Fauquex

Control of Continuous Coenzyme Regeneration   311
    R. Wichmann and C. Wandrey

A New Approach to Membrane Reactor Design and Operation   313
    E. Flaschel, E. Raetz and A. Renken

Influence of Compaction in Gel-Immobilized Enzyme
Packed Bed Reactors   315
    S. Furusaki, Y. Okamura and T. Miyauchi

SESSION V.   BIOMASS CONVERSION WITH ENERGY PRODUCTION . .   317
    Chairmen: *W. Vieth and S. Suzuki*

Enzymatic Removal of Hazardous Pollutants from Industrial Aqueous Effluents   319
    A.M. Klibanov

Enzymatic Hydrolysis of Cellulose: Effects of Structural Properties of Cellulose on Hydrolysis Kinetics   325
    D.D.Y. Ryu and S.B. Lee

Continuous Ethanol Fermentation by Immobilized
Biocatalysts   335
    P. Linko and Y.Y. Linko

An Immobilized Yeast Cell Column for the Fermentation
of Molasses   343
    D.F. Day and D. Sarkar

Pilot Operation for Continuous Alcohol Fermentation
of Molasses in an Immobilized Bioreactor   347
   S. Fukushima and S. Hanai

Enzymatic Hydrolysis of Cellulose   349
   Y. Harano, H. Ooshima, K. Ohmine and M. Sakata

ATP Regeneration by Enzymes of Alcohol Fermentation
and Kinases of Yeast and Its Computer Simulation . .   351
   R. Matsuno, M. Asada, K. Nakanishi and T.
   Kamikubo

Extracellular Cellulases Produced by a Yeast-Like
Fungus   353
   G. Larios, A. Gilbon, Y. Lara and C. Huitron

Relationship Between Extracellular Proteases and the
Cellulase Complex of Trichoderma Reesei   355
   C.P. Dunne

SESSION VI.   ANALYTICAL APPLICATIONS OF IMMOBILIZED
            BIOMATERIALS   357
         Chairmen: L.B. Wingard Jr. and H. Okada

Binding Assays Involving Separation in Aqueous Two-
Phase Systems: Partition Affinity Ligand Assay
(PALA)   359
   B. Mattiasson, T.G.I. Ling, and M. Ramstorp

Use of Immobilized Enzyme Reactors in Automated
Clinical Analyses   369
   T. Murachi

Immobilized Enzymes in Analysis: Applications and
Economic Aspects   377
   M. Gloger, M. Nelboeck, D. Doring, and S. Klose

Microbial Sensors for Gas Analysis   387
   S. Suzuki and I. Karube

Analytical Uses of Immobilized Enzymes   395
   G.G. Guilbault

Cost Analysis and Viability of Immobilized Enzymes in
Routine Analysis   405
   P.V. Sundaram

Multipurpose Enzyme-Collagen Membrane Electrodes   409
   D.C. Gautheron, P.R. Coulet and C. Bertrand

Long-Term Stability of Air-Dried Enzyme Electrodes
with Selective Enzymic Collagen Membranes   411
    P.R. Coulet, D.C. Gautheron, and G. Bardeletti

Enzyme Electrodes Based on Insolubilized Enzyme Membranes Coupled with an Electrochemical Transducer . .   413
    *J.L. Romette, N.D. Tran, P. Durand, J.L.
    Boitieux, and J.L. Navarro*

Potentiometric Glucose Sensor: Enzymatic Activity and
Potentiometric Measurements   415
    L.B. Wingard Jr. and J.F. Castner

Microbial Sensor for Preliminary Screening of
Mutagens   417
    I. Karube and S. Suzuki

Application of Microbial Electrode to Analysis of
Waste Water   419
    *M. Hikuma, H. Suzuki, T. Yasuda, I. Karube and
    S. Suzuki*

Biosensors Based on Enzyme Amplification and Immunochemical Selectivity   421
    M. Aizawa and S. Suzuki

Application of Chemiluminescence of *Cypridina* Luciferin Analog to Immobilized Enzyme Sensors   423
    T. Kobayashi, K. Saga, S. Shimizu, and T. Goto

Use of Hydrogen Sensitive Pd-MOS Components in Biochemical Analysis   425
    *B. Danielsson, F. Winquist, K. Mosbach and I.
    Lundstrom*

Computer Controlled Mass Spectrometer Monitoring of
Fermentations   429
    E. Pungor Jr., C.L. Cooney and J.C. Weaver

SESSION VII.   MEDICAL APPLICATIONS OF ENZYME TECHNOLOGY .   431
        Chairmen: *K. Mosbach and T. Murachi*

Immobilized Heparinase: Production, Purification,
and Application in Extracorporeal Therapy   433
    *R. Langer, R.J. Linhardt, C.C. Cooney and D.
    Tapper*

Clinical Utility of Urokinase-Treated Polymer for
Antithrombogenic Material   443
   T. Ohshiro

Artificial Cell Immobilized Multienzyme Systems
and Cofactors   451
   T.M.S. Chang, Y.T. Yu and J. Grunwald

Antithrombogenic Activity of Artificial Medical
Materials Improved by Enzyme Immobilization
Techniques   457
   Y. Miura, S. Aoyagi and K. Miyamoto

Application of Immobilized Enzymes for Biomaterials
Used in the Field of Thoracic Surgery   459
   S. Watanabe and T. Teramatsu

Use of Immobilization Principles for the Construction
of Drug Targeting Systems   461
   V.P. Torchilin, A.L. Klibanov, V.R. Berdichevsky, V.G. Omelyanenko and V.N. Smirnov

Application of Bioreactors with Immobilized
L-Asparaginase   465
   G. Mazzola, C. Giordano, R. Longhi, G. Vecchio
   and R. Esposito

Acyl-CoA Synthetase and Acyl-CoA Oxidase for Determination of Serum Free Fatty Acids   467
   S. Shimizu, Y. Tani and H. Yamada

Enzyme Immunoassay for Free Thyroxine   469
   H.H. Weetall, W. Hertl, F.B. Ward and L.S. Hersh

Routine Determination of Hydrogen Peroxide in
Clinical Chemistry with Immobilized Aldehyde Dehydro-
genase   473
   P.V. Sundaram

Enzyme Electrodes for Simultaneous Determination of
Creatinine and Creatine in Serum or Whole Blood . .   475
   T. Tsuchida and K. Yoda

SESSION VIII.   GENETIC ENGINEERING FOR ENZYME (OR
            IMPORTANT BIOLOGICAL SUBSTANCES) PRODUCTION   477
            Chairmen: D. Fink and K. Sakaguchi

Construction of Various Host Vector Systems and the
Variation of Enzyme Levels   479
   K. Sakaguchi

Enzymes Active on Unnatural Synthetic Compounds:
Nylon Oligomer Hydrolases Controlled by a Plasmid and
Their Cloning   491
    H. Okada, S. Negoro and S. Kinoshita

CLOSING SESSION   501
            Chairman: *G. Manecke*

Recent Developments and Future Aspects of Enzyme
Engineering   503
    E. Katchalski-Katzir

LIST OF PARTICIPANTS   511

SUBJECT INDEX   525

CONFERENCE HISTORY

THE FIRST TEN YEARS

L. B. Wingard Jr.
Department of Pharmacology
University of Pittsburgh
Pittsburgh, Pennsylvania, USA

The 1981 Kashikojima conference marks the 10th year of the Engineering Foundation series of biannual conferences on Enzyme Engineering. The *Proceedings* of this sixth conference seems like an appropriate place to review the highlights of the first ten years and to answer the often asked question of how this series got started.

The idea for a conference on applied enzymology, i.e. enzyme engineering, originated in 1969 when I saw some literature describing the applied science-engineering series of conferences sponsored by the Engineering Foundation. I discussed the idea with Professor Arthur Humphrey; and we readily agreed that the five topics to be covered should be the production, separation and purification, immobilization, reactor design and performance, and applications of enzymes. I was most fortunate to get the following people to serve with me as the organizing committee for the first conference: Daniel Wang of MIT, Donald Brusca of Exxon, Richard Falb of Battelle, Arthur Humphrey of the University of Pennsylvania, and Lewis Mayfield of the National Science Foundation. The name Enzyme Engineering was the suggestion of Lewis Mayfield.

The first conference was held in August 1971 at Henniker, NH in the U.S.A. The American Institute of Chemical Engineers served as cosponsor, primarily to help in publicity and generation of interest in this new topic. The timing turned out to be perfect since government funded research and development programs in applied enzymology were underway in the U.S.A., England, the Federal Republic of Germany, France, Sweden, and the U.S.S.R., and new industrial applications especially with immobilized enzymes were in operation or under intense development in Japan, the Federal Republic of Germany, the U.S.A., and Italy. The first conference was an exciting event because most of the people in the five different topical areas had not met before. There was much discussion and a free flow of information and "education" among the biochemists/biophysicists, chemical engineers, micro- biologists, polymer and support chemists, and the food and medical scientists that attended. Industry, academic, and government lab- oratories all were well represented. The novelty of the confer- ence subject matter was a major factor in attracting the partici- pants; this was evident since the modest support of $5,800, pro- vided by eight U.S.A. companies and the Engineering Foundation, was sufficient to get participation by nearly all of the key workers in the field.

Much of the success of the conference series is attributed to participation by the vast majority of the recognized leaders in the field from around the world. This international scope is pointed up in Table 1. Note especially the increasing total number of participants that come from outside the host geographical region. The dollars listed in Table 1 are the funds raised by each conference Executive Committee; these funds are used to off- set partially the travel or conference fee for many of the aca- demic participants from outside the geographical region where the conference is held. The funds usually come from industrial com- panies; although the government of the Federal Republic of Germany was a major contributor to the 4th conference. I purposedly did not solicit funds for Conferences 1 and 2 from the U.S. government in order to establish the conferences as an entirely independent format for discussion and exchange of ideas.

Some of the highlights and programming areas of emphasis for the first six conferences are shown in Table 2.

The preparation of a *Proceedings* for the first conference was an afterthought that seemed appropriate, considering the many people that applied to attend but could not be accommodated. A *Proceedings* gave them a chance to have a summary of the main topics that were discussed. The switch from Wiley to Plenum with Volume 2 was done simply because the latter publisher showed more enthusiasm for the topic. The volumes are not intended to be a verbatim record of the conference, since this would put a damper on open discussion and dialogue. Most of the papers given at the conference appear in the *Proceedings*; however, the written ver- sions often differ from the oral versions to protect confidential information. From a cost standpoint it is desirable to limit the size of the *Proceedings* to 500 pages; this necessitates allowing only a few pages per author, since the total number of papers has been 30-40 plenary plus 80-90 poster papers.

The simple organizational structure of a Conference Chairman
and a few volunteers to help set up the program, as used for the
first Conference, has given way to an Executive Committee plus
an Advisory Board. This change was instituted because of the
large amount of work brought about by 1) fund raising, 2) setting
up the program, 3) review of applications, 4) seeing that the international scope of the conference series is maintained, and 5)
formulating plans for the locations and financing of future conferences. Obviously, many people have contributed considerable
time and effort to make these first six conferences so successful.
I had planned to mention their names here; but the list turned
out to be too long. Special mention, though, is due Sandford Cole
and Harold Comerer of the Engineering Foundation for accommodating
our numerous special requests for this series of conferences.

As we move into the second ten years, we need to remind our-
selves that the main reason we travel thousands of miles to participate in these conferences is to have an opportunity to exchange ideas and to be stimulated to come up with an answer or
an approach to a problem. Most of us cannot afford the time or
money to go to conferences to hear a presentation that we could
easily read about in a journal. Therefore, let us look forward
to opening the second ten years of the Enzyme Engineering Conferences with an in-depth discussion of the problems, possible approaches for achieving solutions, and novel ideas that could lead
to more innovation in the future.

TABLE 1

PARTICIPANTS AND FUNDING FOR CONFERENCES
DURING FIRST TEN YEARS

Conferences	Where Held	Participants or Attendees		Support for Overseas Participants
		Number*	Countries**	($)
1 (1971)	USA (East Coast)[+]	155 (230)	13 (20%)	5,800
2 (1973)	USA (East Coast)	190 (350)	18 (25%)	10,000
3 (1975)	USA (West Coast)	190 (380)	14 (31%)	15,000
4 (1977)	F.R. Germany[++]	240	23 (35%)	35,000
5 (1979)	USA (East Coast)	183	22 (51%)	15,000
6 (1981)	Japan[+++]	203	23 (61%)	55,000

  * ( ) indicates total persons applying to participate.

 ** ( ) indicates % of participants from outside host region,
i.e. outside the U.S.A., central Europe, or Japan (incl.
Korea & China)

  +   Cosponsored by the American Institute of Chemical
Engineers

 ++   Cosponsored by DECHEMA

+++   Cosponsored by the Japanese Society of Enzyme Engineering

TABLE 2

HIGHLIGHTS AND SPECIAL EMPHASIS

Conference	Highlights and Emphasis
1	Highlights: first time key people in field got together. Emphasis on putting the five topics together (production, separations, immobilization, reactors, applications), diffusion restrictions, and on experimental data versus mathematical models.
2	Highlights: high degree of industrial interest in immobilized enzymes. Emphasis on applications and on cofactor requiring systems.
3	Highlights: more realistic attitude about industrial prospects. Emphasis on industrial prospects and problems. Major use of Workshop sessions.
4	Highlights: growing commitment to basic studies, also seeing encouraging results on analytical and medical applications. Emphasis on immobilized whole cells and organelles, on industrial prospects, and on affinity techniques. First use of Poster Session (very successful); also included Workshops.
5	Highlights: continued need for basic studies still highly evident. Emphasis on energy transduction, cells and organelles, biomass conversion, and organic syntheses with enzymes. Continued highly successful Poster sessions.
6	Highlights: high degree of success of Japanese industry in applied microbiology and applied enzymology very evident; also noted international resurgence of interest in enzyme engineering due to potential of recombinant DNA techniques for production of enzymes. Emphasis on production and applications of enzymes. Poster sessions again very successful.

KEYNOTE PAPER

RECENT STUDIES ON ANTIBIOTICS AND LOW MOLECULAR WEIGHT

ENZYME INHIBITORS

H. Umezawa

Institute of Microbial Chemistry
Shinagawa-ku, Tokyo, Japan

Continuous progress is being made in antibiotic research in the areas of their genetics, biosyntheses, chemical syntheses, and mechanisms of action, as well as in the development of new chemothera peutic agents. About 12 years ago, the study of low MW inhibitors of various enzymes was started; and the results indicated that microorganisms produced not only antibiotics but also compounds with varying structures which have various biological, pharmacological, or medical activities. In this paper, I will review recent studies on low MW enzyme inhibitors and antibiotics.

LOW MOLECULAR WEIGHT ENZYME INHIBITORS

By 1965, NMR and X-ray crystallography had already been introduced into natural product chemistry; and it became possible to elucidate the structures of low MW microbial products very quickly. Moreover, at about that time, an understanding of the biochemistry of various diseases began to make rapid progress. Therefore, I initiated the screening of culture filtrations for low MW enzyme inhibitors (1).

While testing for plasmin-inhibiting activity of *Streptomyces* culture filtrates, we discovered leupeptin, which inhibits plasmin, trypsin, papain, and cathepsin B (1, 2). Two leupeptins were isolated; and their structures were determined to be acetyl(or propionyl)-L-leucyl-L-leucyl-L-argininal (3) (Fig. 1). Trypsin hydrolyzes an arginyl or lysyl bond in peptides; and leupeptin has an argininal group. This aldehyde structure for the inhibition of serine thiol proteases was first found in leupeptin. We studied the cell-free synthesis of leupeptin (4) and were successful in

isolating a multifunctional enzyme (5) which catalyzes the synthesis of leupeptin acid (acetyl or propionyl-L-leucyl-L-leucyl-L-
arginine) in the reaction mixture containing acetyl-L-leucine,
leucine, L-arginine, and ATP. The sequence of leupeptin synthesis
on this enzyme is as follows: acetylleucine → acetylleucylleucine
→ acetylleucylleucylarginine. Leupeptin acid is reduced by a very
labile enzyme to leupeptin (4). An enzyme which transfers the
acetyl group from acetyl-CoA to leucine is found in leupeptin-
producing strains (6). Leupeptin is produced by more than 50% of
all the soil *Streptomyces* strains which belong to more than 11
species (1). This indicates that the gene which is involved in
the biosynthesis of leupeptin acid synthetase is widely distributed
among various species of *Streptomyces*. The ability to produce
leupeptin was transferred by conjugation from a leupeptin-producing
methionine-requiring mutant to a leupeptin-nonproducing arginine-
requiring mutant in high frequency (7), indicating the involvement
of a plasmid in leupeptin production.

Antibiotics and low MW enzyme inhibitors are secondary metabolites which are not involved in the growth of microbial cells.
Leupeptin-nonproducing mutants grow as well as their leupeptin-
producing parent strains. There was no difference between proteolysis of cells of leupeptin-producing mutants and those of non-
producing mutants; and protein autodegradation in cell homogenates
of leupeptin-producing strains was inhibited much more strongly
by chymostatin (a chymotrypsin inhibitor) or EDTA than by leupeptin (8). The ratio of the intracellular amount of leupeptin to
that in the culture medium was 1:250, whereas this ratio for leupeptin acid was 1:5 (8). This indicates that leupeptin acid, which
has no antiprotease activity, is produced within the cells; and
the leupeptin produced from leupeptin acid is rapidly released
extracellularly. Leupeptin has no function in the growth of leupeptin-producing strains.

In testing *Streptomyces* culture filgrates for inhibition
of papain, chymotrypsin, or pancreatic elastase, we discovered
antipain, chymostatin, and elastatinal (Fig. 1) (1,9). Antipain
inhibits papain, trypsin, and cathepsin B; chymostatin inhibits
chymotrypsin; and elastatinal inhibits pancreatic elastase. All
of these inhibitors have the C-terminal aldehyde structures for
inhibition of serine thiol proteases as does leupeptin. In testing
Streptomyces culture filtrates from inhibition of thermolysin,
a metallo-protease, we found phosphoramidon (Fig. 1) (1, 9).
Phosphoramidon inhibits thermolysin as well as other metallo-pro-
teases. The L-rhamnose moiety of phosphoramidon is not involved
in the inhibition; rather, the N-phosphate of L-leucyl-L-tryptophan
is responsible for the inhibitory activity; and it shows a stronger
action than the whole phosphoramidon molecule.

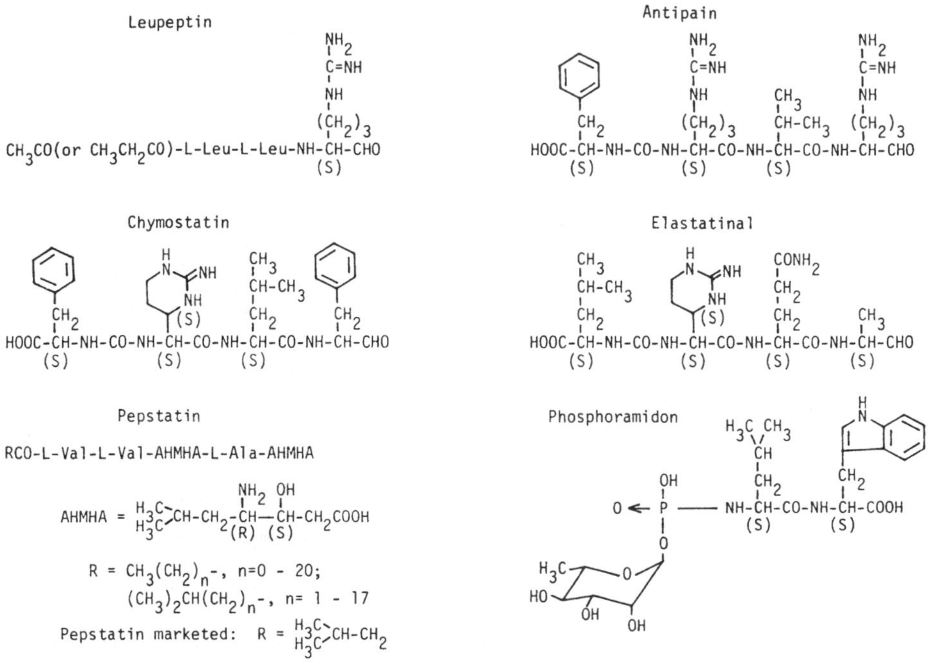

Leupeptin

$$NH_2$$
$$C=NH$$
$$NH$$
$$(CH_2)_3$$
CH₃CO(or CH₃CH₂CO)-L-Leu-L-Leu-NH-CH-CHO
(S)

Antipain

HOOC-CH-NH-CO-NH-CH-CO-NH-CH-CO-NH-CH-CHO
(S)   (S)   (S)

Chymostatin

HOOC-CH-NH-CO-NH-CH-CO-NH-CH-CO-NH-CH-CHO
(S)   (S)   (S)

Elastatinal

HOOC-CH-NH-CO-NH-CH-CO-NH-CH-CO-NH-CH-CHO
(S)   (S)   (S)   (S)

Pepstatin

RCO-L-Val-L-Val-AHMHA-L-Ala-AHMHA

AHMHA = $\mathrm{{}^{H_3C}_{H_3C}}CH-CH_2-\underset{(R)}{CH}-\underset{(S)}{CH}-CH_2COOH$ with NH₂ OH

R = CH₃(CH₂)ₙ-, n=0 – 20;
(CH₃)₂CH(CH₂)ₙ-, n= 1 – 17

Pepstatin marketed: R = $\mathrm{{}^{H_3C}_{H_3C}}CH-CH_2$

Phosphoramidon

O ← P —— NH-CH-CO-NH-CH-COOH
           (S)   (S)

Fig. 1. Inhibitors of proteases.

In the screening for pepsin inhibitors, we discovered pepstatin (1, 9, 10). Pepstatin inhibits pepsin, cathepsin D, and renin. The derivatives which have lactyl or benzoyl groups instead of the isovalerylvalyl group found in commercial pepstatin are more water-soluble than pepstatin and have a similar degree of activity as pepstatin in inhibiting cathepsin D and pepsin (11). However, these water-soluble derivatives have much lower activity against renin. For the preparation of such water-soluble derivatives, we found a bacterial enzyme which hydrolyzes the isovalerylvalyl bond in pepstatin (12). Corvol has reported that other types of derivatives, such as pepstatylaspartic acid, have almost the same activity against renin as pepstatin and are more water-soluble than pepstatin (13).

The protease inhibitors described above have been widely used for the identification of proteases and for analysis of their roles in normal biological and disease processes. Pepstatin was used as the functional group for affinity chromatography of renin; and this enzyme was first purified by such chromatography. Leupeptin inhibits Ca^{2+}-dependent proteases and has therapeutic effects in mouse muscular dystrophy (14). The medicinal applications of pro-

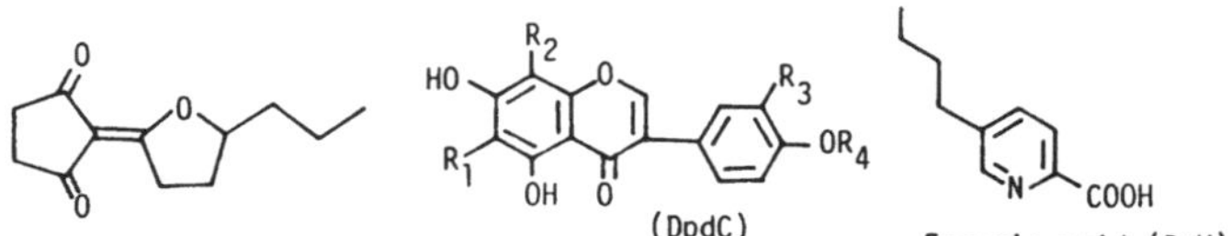

Compound	R_1	R_2	R_3	R_4
I	H	H	OH	H
II	H	OCH_3	OH	H
III	H	OCH_3	H	H
IV	H	OH	H	H
V	OCH_3	H	OH	CH_3
VI	H	OCH_3	OH	CH_3

(I=Orobol)

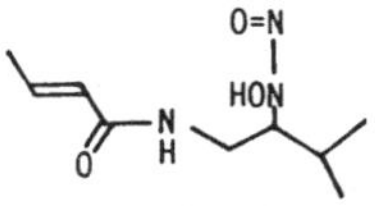

Fig. 2.   Inhibitors of Enzymes Involved in Noradrenaline
Synthesis: tyrosine $\xrightarrow{\text{TH}}$ DOPA $\xrightarrow{\text{DpdC}}$
dopamine $\xrightarrow{\text{D}\beta\text{H}}$ noradrenaline; ( ) means the enzyme
inhibited

tease inhibitors are being studied from various aspects for treatment of thrombosis, muscular dystrophy, and cancer.

We have also found inhibitors of enzymes involved in the biosynthesis of noradrenaline. Oudenone inhibits tyrosine hydroxylase; various isoflavone compounds inhibit dopa decarboxylase, and dopastin inhibits dopamine β-hydroxylase (1, 15) (Fig. 2). During the screening studies, we also found that fusaric (fusarinic) acid (Fig. 2) has a strong inhibiting effect on dopamine β-hydroxylase. All of these inhibitors of enzymes involved in noradrenaline biosynthesis have a hypotensive effect in spontaneously hypertensive rats (1). A clinical study has also confirmed a hypotensive effect of fusaric acid against essential hypertension.

Up to the present, we have found about 50 enzyme inhibitors. All of them have no significant activity in inhibiting microorganisms. The study of enzyme inhibitors is now under way in the laboratories of pharmaceutical companies. For example, an inhibitor (ML-236B, compactins) (Fig. 3) of 3-hydroxy-3-methylglutaryl-CoA reductase involved in mevalonate synthesis was found by Endo et al. (16, 17) and has been reported to be useful in the treatment of cholesteraemina (18).

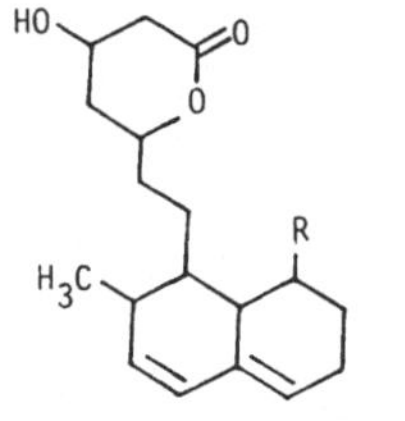

ML-236A: R = OH
     B: R = -OCOCH(CH₃)CH₂CH₃
     C: R = H

Fig. 3.   Inhibitor of 3-hydroxy-3-methylglutaryl-CoA reductase.

PRODUCTS BINDING TO CELL SURFACES OR MEMBRANE IMMUNO-MODIFIERS

     I have further extended the study of enzyme inhibitors to
immuno-modifiers, that is, low MW microbial products which bind
to surfaces or membranes of animal cells (19).

     In 1972, we found that the administration of a very small dose
of diketocoriolin B (Fig. 4) increased the number of mouse spleen
cells producing antibody to sheep red blood cells (20). On the
other hand, we found that diketocoriolin B inhibits Na⁺-K⁺-ATPase,
a membrane-bound enzyme (21). Therefore, I thought that the binding of diketocoriolin B to ATPase in membranes promoted the blasto-
genesis of lymphocytes producing antibodies to sheep red blood
cells. Later, we confirmed that the direct action of diketocoriolin B on B lymphocytes increased the number of antibody-forming
cells (22).

     Since I assumed that the screening of compounds which bind
to cell membranes or surfaces would result in the finding of immuno-
modifiers, I began a search for inhibitors of various enzymes on
cell surfaces or membranes. We first found that all aminopeptidases
are not only in cells but also located on their surfaces (23).
These enzymes are not released extracellularly. Alkaline phosphatase and esterase were also found to be located on cell surfaces.
Searching for inhibitors of these enzymes, we found bestatin (19,

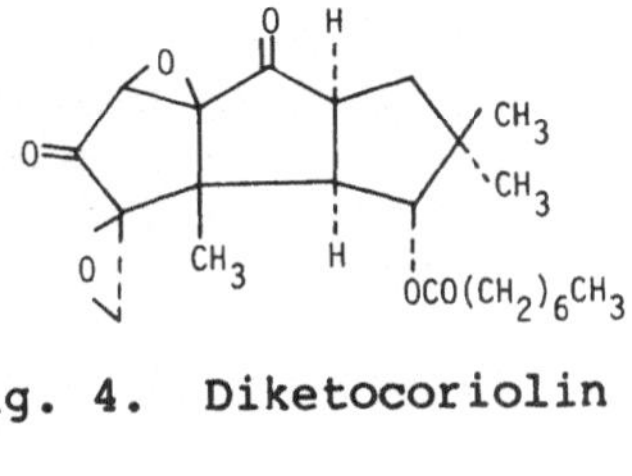

Fig. 4.   Diketocoriolin B.

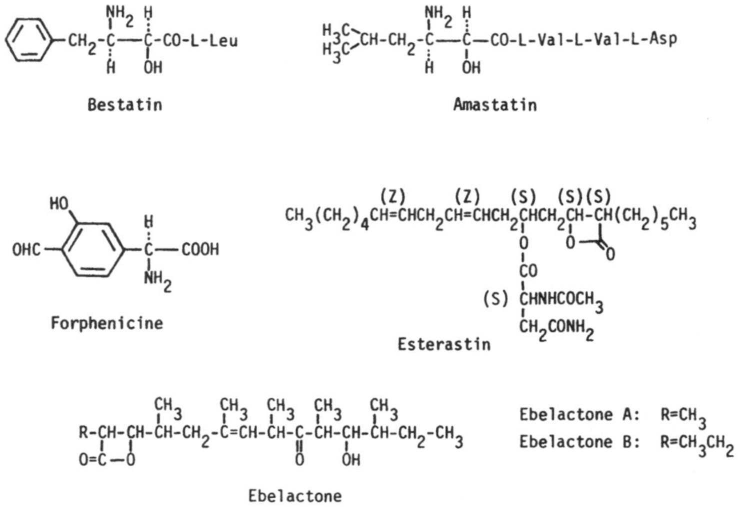

Fig. 5. Bestatin, amastatin, forphenicine, esterastin, and
         ebelactones.

24), amastatin (19, 25), forphenicine (19, 26), esterastin (19,
27), and ebelactone (19, 28), all of which are shown in Fig. 5.
The Ki values of these inhibitors are shown in Table 1. All of
these inhibitors except for esterastin enhanced immune responses
in mice. Bestatin inhibited aminopeptidase B and leucine aminopeptidase. A low dose (1-100 µg/mouse) of bastatin enhanced delayed-
type hypersensitivity (DTH) to sheep red blood cells; and at higher
doses (1 mg/mouse) it increased the number of antibody-forming cells
in spleen. Amastatin, which inhibited aminopeptidase A, increased
the number of antibody-forming cells. Forphenicine enhanced DTH
(1-100 µg/mouse) and increased the number of antibody-forming cells
(10-1,000 µg/mouse). Ebelactone enhanced DTH. But esterastin both
suppressed DTH and reduced the number of antibody-forming cells.
The Ki value for esterastin was about 10^{-10} M, whereas those of
the others ranged from about 10^{-6} to 10^{-8} M.

We have synthesized all stereoisomers of bestatin. Those isomers, which had the same configuration (S) as bestatin at the carbon adjacent to the carbonyl group of the 3-amino-2-hydroxy-4-
phenylbutyryl moiety, showed similar activity as bestatin in inhibiting aminopeptidase B and enhancing DTH; but the other isomers
having the R configuration at this carbon atom had neither activity. The type of inhibition of aminopeptidases by bestatin was
competitive with respect to the substrates.

Forphenicine inhibits chicken intestine alkaline phosphatase;
but its action against other alkaline phosphatases is very weak.

TABLE 1

KINETIC CONSTANTS OF AMASTATIN, BESTATIN, FORPHENICINE, ESTERASTIN, AND EBELACTONES A AND B.

Inhibitor	Enzyme	Substrate	Km $(\times 10^{-4} M)$	Ki $(\times 10^{-8} M)$	Type of Inhibition
amastatin	AP-A[a]	Glu-NA[d]	1	15	competitive
	leu-AP	Leu-NA	37	160	competitive
bestatin	AP-B	Arg-NA	1	6	competitive
	leu-AP	Leu-NA	5.8	2	competitive
forphenicine	alk-phos	PNPP[e]	4.6	16.4	uncompetitive
esterastin	esterase[b]	PNPA[f]	4	0.02	competitive
ebelactone A	esterase[c]	PNPA	6.7	9.2	competitive
ebelactone B	esterase[c]	PNPA	6.7	0.05	competitive

a, aminopeptidase A; b, hog pancreas; c, hog liver; d, L-glutamic acid β-naphthylamide; e, p-nitrophenyl phosphate; f, p-nitrophenyl acetate.

The type of inhibition of chicken intestine alkaline phosphatase
by forphenicine is very interesting; that is, it is uncompetitive
with the substrate. Its derivative, forphenicinol, in which the
aldehyde group of forphenicine is reduced to alcohol, does not inhibit alkaline phosphatase; but it does bind to animal cells, including lymphocytes. Forphenicinol (0.1-100 μg/mouse by intraperi-
toneal injection and 0.1-1,000 μg/mouse by oral administration)
enhanced DTH.

By oral administration, both bestatin and forphenicinol enhanced DTH and showed antitumor effects against mice tumors sensitive to immune-enhancing agents. Therefore, both were selected
for clinical studies. In particular, bestatin has been studied
in detail during the last 4 years.

Bestatin, when given daily in 30 or 60 mg doses, was first
found to increase the T cell percentage which is usually reduced
in cancer patients. Thereafter, the effects of increased daily
doses of 100, 200, 400, 600, and 900 mg were tested. In those
cases in which the percentage of T cells was in the normal range,
200 mg or more decreased T cell numbers in some cases. Therefore,
the proper dosage was estimated to be 30 or 60 mg. The effect of
these doses on minimal residual tumors in cancer patients is being
tested by randomized schedules.

Low MW immuno-modifiers might well become useful in the analysis of immune responses; and some of them may be used to eradicate
minimal residual tumors after surgery, radiation, or chemotherapy.
They would also be useful in that they could avoid the immunosup-
pression which accompanies radiation or chemotherapy.

ANTITUMOR ANTIBIOTICS

The effect of cancer chemotherapy can be enhanced by combination with immunotherapy. Recent studies of derivatives and analogs of effective antitumor antibiotics are contributing greatly
to the development of chemotherapeutic agents against cancer.

Adriamycin (doxorubicin, Fig. 6), discovered by Arcamone
et al. in 1969, exhibits the marked therapeutic effect of rapidly
shrinking the sign of tumors in various kinds of cancer. Derivatives or analogs which have lower cardiac toxicity have a much
stronger therapeutic effect than adriamycin. Therefore, during
the last 10 years, the anthracycline group of antibiotics have been
studied by Arcamone and many other researchers. It is said that
over 600 anthracyclines have been tested at the National Cancer
Institute of the National Institutes of Health, U.S.A., up to
1980.

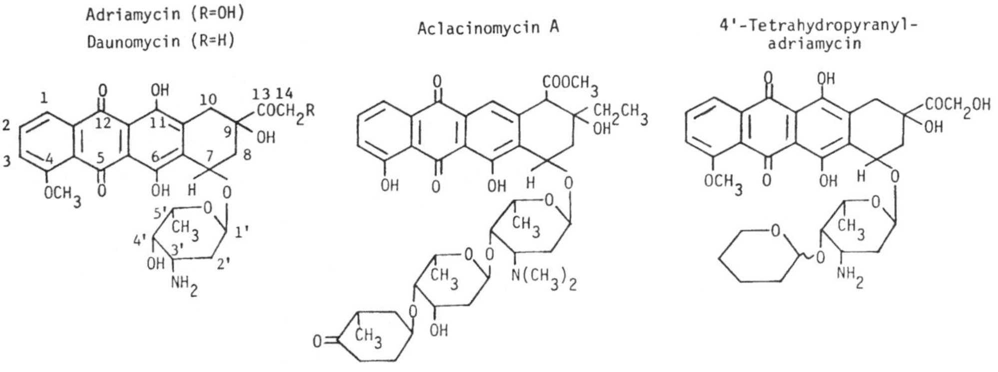

Fig. 6. Structures of anthracyclines clinically used or under clinical test.

    With my colleagues, I searched for new anthracyclines and found the aclacinomycins (Fig. 6) (29). Aclacinomycin A was studied clinically and confirmed to exhibit a therapeutic effect against leukemia and lymphoma. Promisingly, it shows a therapeutic effect against leukemias which are resistant to treatment with other drugs. We also found baumycin Al, which is a 4'-0-glycosidic analog of adriamycin. Since it showed a good effect against L-1210 mouse leukemia, we synthesized the 4'-0-glycosidic derivatives. Among them, one of the isomers of 4'-0-tetrahydropyranyladriamycin (Fig. 6) (30) showed a stronger effect against L-1210 leukemia and had a lower cardiac toxicity than adriamycin itself (31).

    In collaboration with Oki, Yoshimoto, and others of the Research Institute of Sanraku Ocean Co., we studied the biosynthesis of the anthracyclines. The aglycone of aclacinomycin is called aklavinone. Aklavinone (or its 0-demethyl derivative) was found to be an intermediate in the biosynthesis of aclacinomycin and daunomycin (daunorubicin, Fig. 6) (32-34). Daunomycinone is not a biosynthetic intermediate of daunomycin. Before the 4-0-methylation step, the intermediate binds to the sugar moiety (daunosamine) of daunomycin (33). In mutation studies of anthracycline-producing strains, we obtained mutants which produced neither red nor yellow pigments but converted aglycones to their corresponding anthracyclines (32). As shown in Fig. 7, a mutant of an aclacinomycin-producing strain converted various anthracyclinones (aklavinone, 4-0-methylaklavinone, 11-hydroxyaklavinone, 1,11-dihydroxyaklavinone, 10-demethoxycarbonyl-11-hydroxyaklavinone, 10-demethoxycarbonyl-1,11-dihydroxyaklavinone, 10-demethoxycarbonyl-11-hydroxy-7-deoxyaklavinone, and 10-demethoxycarbonyl-6-deoxy-1,11-dihydroxyaklavinone) to anthracyclines consisting of the aglycone added and the trisaccharide contained in aclacinomycin. All aklavinone analogs which contained hydroxyl groups at the C-6 and C-7 posi-

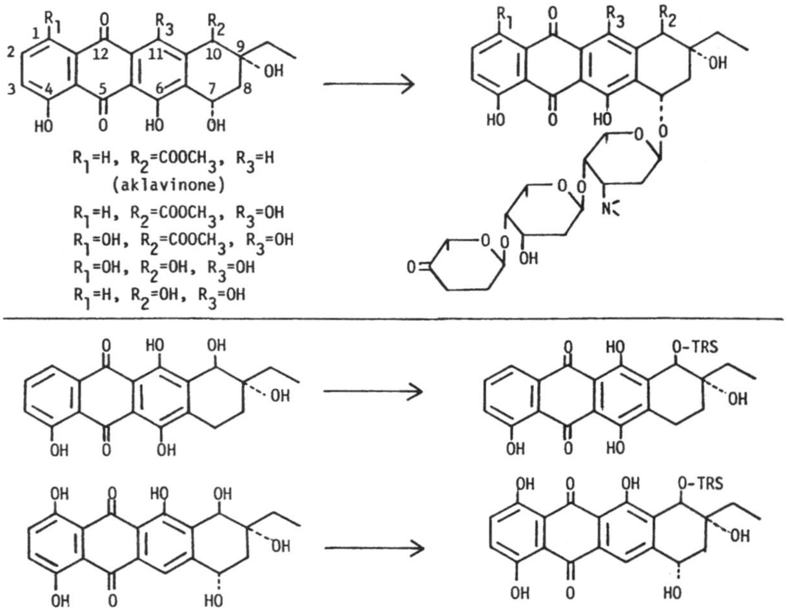

R₁=H, R₂=COOCH₃, R₃=H
(aklavinone)
R₁=H, R₂=COOCH₃, R₃=OH
R₁=OH, R₂=COOCH₃, R₃=OH
R₁=OH, R₂=OH, R₃=OH
R₁=H, R₂=OH, R₃=OH

Fig. 7. Conversion of anthracyclinones to anthracyclines containing the trisaccharide (TRS) of aclacinomycin A by a pigmentless mutant of aclacinomycin-producing stain.

tions were converted to anthracyclines containing the trisaccharide of aclacinomycin A at the hydroxyl group of C-7. In the case of aglycones which lacked the hydroxyl group at C-6 or C-7, but contained a hydroxyl group at C-10 instead of the 10-methoxycarbonyl group, anthracyclines containing the trisaccharide of aclacinomycin A at the hydroxyl group of C-10 were obtained (32).

As shown in Fig. 8, a pigmentless (aglycone-nonproducing) mutant of a daunomycin-producing strain converted aklavinon, 11-hydroxyaklavinone, and 4-0-demethyldaunomycinone (carminomycinone) to daunomycin and 13-dihydrodaunomycin. 1-Hydroxyaklavinone and 1,11-dihydroxyaklavinone were converted to 1-hydroxy-13-dihydro-daunomycin (35). Moreover, by a mutation in an aclacinomycin-producing strain, we obtained a mutant which produced 2-hydroxy-aklavinone. By adding this anthracyclinone to the medium in which was cultivated a aglycone-nonproducing mutant of an aclacinomycin-producing strain, we were successful in the preparation of the 2-hydroxyaclacinomycins, A and B (36). This is the first finding of anthracyclines having the 2-hydroxyl group. We found that both 2-hydroxyaclacinomycin A and B had lower toxicity in mice than aclacinomycins and higher therapeutic indices against mouse L-1210 leukemia.

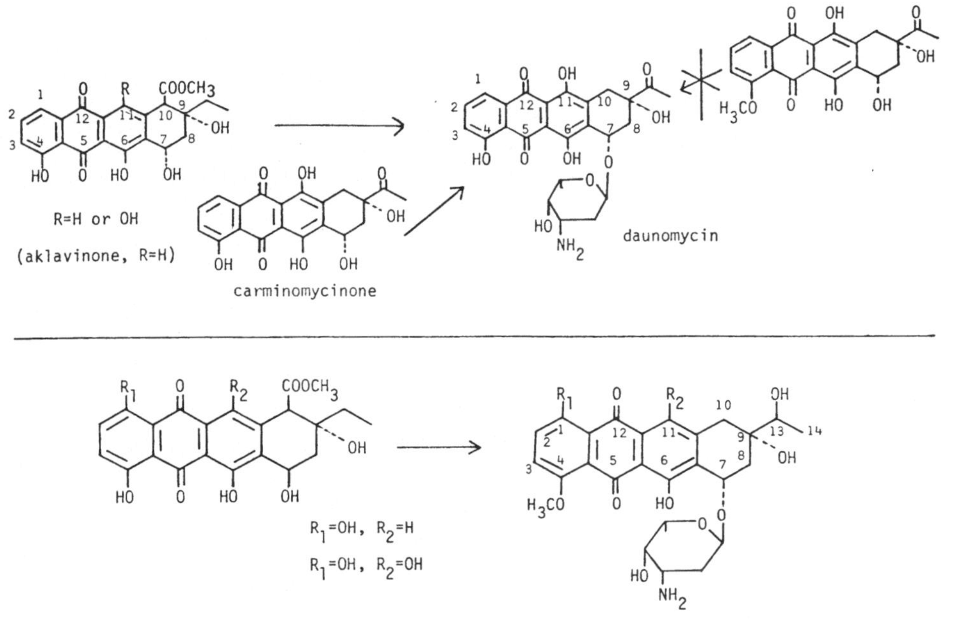

R=H or OH

(aklavinone, R=H)

carminomycinone

daunomycin

R_1=OH, R_2=H

R_1=OH, R_2=OH

Fig. 8.   Conversion of anthracyclinones to anthracyclines containing daunosamine by a pigmentless mutant of a
daunomycin-producing strain.

Bleomycin has contributed to an increased cure rate in patients
having Hodgkin's lymphoma or testicular tumors. It also has a
marked effect in shrinking squamous cell carcinoma of the head,
neck, and skin. In 1972 we clarified its structure, except for
the side chain part of the amino acid moiety containing a pyrimi-
dine ring (37); and in 1978 we proposed the conclusive structure
shown in Fig. 9 (38). The amino acid which consists of an amino-
methylpyrimidine and a side chain is called pyrimidoblamic acid.
The structure of bleomycin shown in Fig. 9 has been confirmed by
chemical syntheses of pyrimidoblamic acid (39) and of the whole
peptide part of bleomycin A 2 (deglycobleomycin A2) (40). Recently,
we were successful in the total synthesis of bleomycin A2 (41).
These structural studies have contributed greatly to the elucida-
tion of the mechanism(s) of cytotoxic and therapeutic actions and
the development of derivatives with improved therapeutic activity.

In collaboration with Tanaka, Fujii, Muraoka, and others of
the Research Institute of Nippon Kayaku Co. over the last 12 years,
we continued studies of the biosynthesis of bleomycins and the development of bleomycin derivatives and analogs which are therapet-
tically more effective than bleomycin itself. From fermentation
broths of a bleomycin-producing strain, we isolated peptides shown

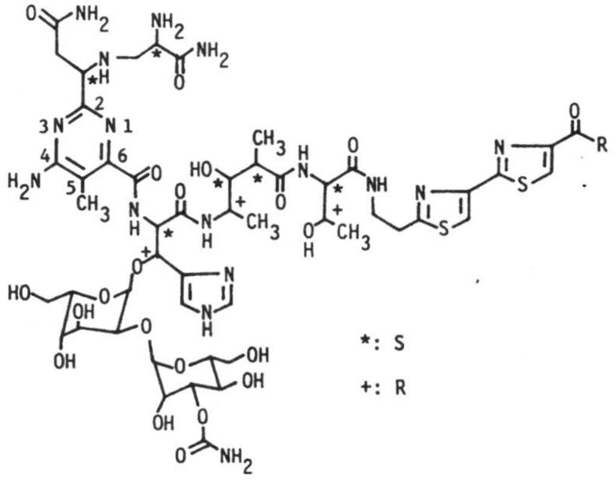

Various bleomycins: A1 ($R=NH-(CH_2)_3-SO-CH_3$),
Demethyl-A2 ($R=NH-(CH_2)_3-S-CH_3$), A2 ($R=NH-(CH_2)_3-S^{+}(CH_3)CH_3$),
A2'-a ($R=NH-(CH_2)_4-NH_2$), A2'-b ($R=NH-(CH_2)_3-NH_2$),
A2'-c ($R=NH-(CH_2)_2$-imidazole ring), A5 ($R=NH-(CH_2)_3-NH-(CH_2)_4-NH_2$),
A6 ($R=NH-(CH_2)_3-NH-(CH_2)_4-NH-(CH_2)_3-NH_2$), B1' ($R=NH_2$),
B2 ($R=NH-(CH_2)_4-NH-C(=NH)-NH_2$), B4 ($R=NH-(CH_2)_4-NH-C(=NH)-NH-(CH_2)_4-NH-C(=NH)-NH_2$)
and Bleomycinic acid (R=OH)
Pepleomycin: $R=NH-(CH_2)_3-NH-C(CH_3)(H)-C_6H_5$
BAPP: $NH-(CH_2)_3-NH-(CH_2)_3-NH-(CH_2)_3-CH_3$

Fig. 9. Natural bleomycins, pepleomycin, BAPP, and bleomycinic acid.

below (42); and their structures suggested the following pathway for the biosynthesis of the peptide portion of bleomycin A2: demethylpyrimidoblamylhistidine → demethylpyrimidoblamylhistidyl-alanine → demethylpyrimidoblamylhistidyl-(4-amino-3-hydroxy-2-methyl)- pentanoic acid [4-amino-3-hydroxy-2-methylpentanoic acid is abbreviated as AHMPA] → demethylpyrimidoblamylhistidyl-AHMPA-threonine → pyrimidoblamylhistidyl-AHMPA-threonyl-[2'-(2-aminoethyl)-2,4'-bithiazole-4-carboxylic acid] (this bithiazole carboxylic acid is abbreviated as BTCA) → pyrimidoblamyl-β-hydroxyhistidyl-AHMPA-Thr-BTCA (β-hydroxyhistidine is abbreviated as β-OH-His) → pyrimidoblamyl-β-OH-His-AHMHA-Thr-BTCA-(3-aminopropyl)dimethylsulfonium.

The methyl group of the pyrimidoblamyl moiety and the 2-methyl group of AHMPA are derived from methionine. BTCA is derived from 3-aminopropionate and two molecules of cysteine. Cysteine is in-

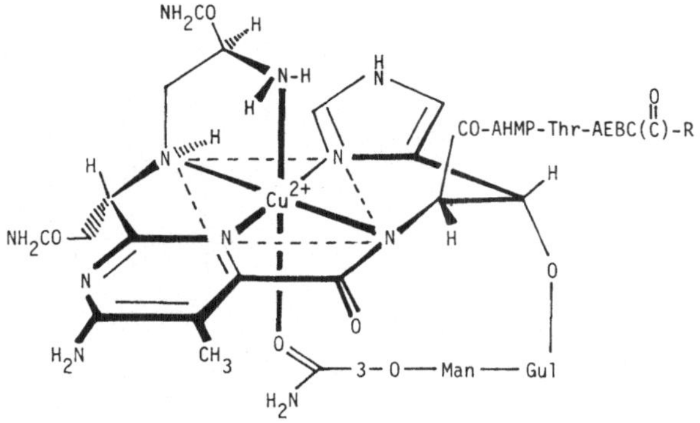

AHMP = 4-amino-3-hydroxy-2-methylpentanoyl

Thr = threonyl

AEBC = 2'-(2-aminoethyl)-2,4'-bithiazole-4-carboxyl

R = terminal amine

Man-Gul = 2-0-(α-D-mannopyranosyl)-α-L-gulopyranosyl

Fig. 10.   Bleomycin copper complex.

corporated into this amino acid moiety; but BTCA added to the medium is not incorporated.

Pyrimidoblamic acid is suggested by incorporation experiments
to be biosynthesized from serine and two asparagine residues (43).

Bleomycin binds strongly to Cu^{2+}; and an equimolar bleomycin-
Cu^{2+} complex is formed (44). The structure of bleomycin-Cu^{2+} complex was also determined, as shown in Fig. 10. The bithiazole
moieties of bleomycin molecules bind to double-stranded DNA by
intercalation with its guanine moieties (45). The terminal amine
also binds to DNA (46). The oxygen molecule of the bleomycin-Fe^{2+}
-O_2 complex which binds to DNA is active; and this oxygen molecule
or a radical produced by it reacts with the C-4 of the deoxyribose
moiety, resulting in strand scission. Grollman and Takeshita (47)
isolated the products of DNA fragmented by this means. On the basis
of these products, the reactions which occur successively after
the reaction of the oxygen molecule of the bleomycin-Fe^{2+}-O_2 complex with the C-4 of the deoxyribose moiety of DNA_2 may be seen in
Fig. 11. The reaction process of the bleomycin-Fe^{2+}-O_2 complex
and the mechanism of double strand scission remain to be solved.
After penetrating into cells, the Cu^{+} of bleomycin-Cu^{2+} complexes
is reduced to Cu^{+} by reducing agents, such as cysteine, in the
cells; and the Cu^{+} is transferred to a cellular protein which can
bind selectively to it (48). Cu^{2+}-free bleomycin thus formed undergoes reaction with bleomycin hydrolase which hydrolyzes the α-

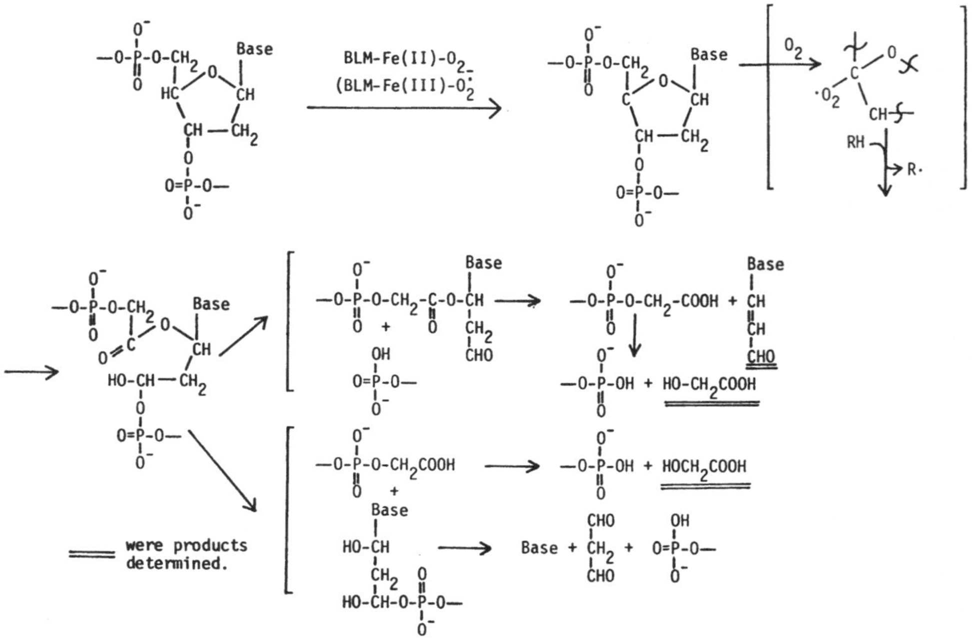

Fig. 11.   Reaction of bleomycin (BLM) with DNA, resulting in
DNA strand scission.

aminocarboxamide bond of the pyrimidoblamyl moiety of bleomycin
(49). This enzyme is widely distributed in human and animal cells.
Cu^{2+}-free bleomycin which escapes this enzymic action reaches
nuclei, binds, and reacts with DNA as described above. Knowledge
of the mechanism of action thus elucidated is very useful for designing studies of derivatives or analogs with improved therapeutic
activities. One of the reasons why bleomycin exhibits a therapeutic effect against squamous cell carcinoma is due to the low content of bleomycin hydrolase in this type of tumor (50).

A side effect caused by bleomycin occurs in the lungs. Therefore, bleomycin derivatives which have lower pulmonary toxicity
than bleomycin can exhibit stronger therapeutic action than the
latter against tumors susceptible to bleomycin treatment. Various
bleomycins are different from one another in the terminal amine
moiety (Fig. 9). The degree of pulmonary of renal toxicity is
different among these bleomycins.

We have established fermentation and chemical processes for
the preparation of various bleomycins (51, 52). By the addition
of many kinds of amines to the fermentation medium, various bleomycins containing the added amines are produced. We found acyl

$$\underset{(S)}{H_2NCNH(CH_2)_4CHCH_2CONHCHCONH(CH_2)_4NH(CH_2)_3NH_2}$$

(with NH above the first C, OH above the second C, and OH above the third CH)

Fig. 12. Spergualin.

agmatine hydrolase in *Fusarium* sp., an enzyme that hydrolyzes
the terminal peptide bond of bleomycin B2, yielded bleomycinic
acid (Fig. 9) (53). From bleomycinic acid various bleomycins can
be synthesized. Pepleomycin (later this name was changed by WHO
and the Japanese Government to peplomycin) shown in Fig. 9 has been
confirmed to be more effective and to have a wider antitumor spectrum than the bleomycin used clinically at the present. It shows
a therapeutic effect not only against bleomycin-sensitive tumors
but also against prostate carcinoma.

Radioactive metal complexes of bleomycin are taken up by
tumors, especially in lung cancer, and are useful for diagnosis.
Therefore, the bleomycins which are resistant to bleomycin hydrolase can be assumed to have a wider anticancer spectrum than bleomycin. Modification of the α-aminocarboxamide part of the bleomycin molecule gives such derivatives resistant to bleomycin hydrolase (55). As already described, bleomycin hydrolase hydrolyzes
the carboxamide bond of the α-aminocarboxamide part; and the products of this hydrolysis have been used as the starting material
for the synthesis of such derivatives.

Chemical derivation studies are also being conducted on other
antibiotics, such as mitomycin (56).

Besides derivative studies of known useful antitumor antibiotics, the continuation of the screening procedure is still pro-
viding interesting antitumor compounds. For instance, recently
we found a new one which we named spergualin (Fig. 12) (57). It
exhibits strong therapeutic effect against L-1210 mouse leukemia
and has no cumulative toxicity. In cancer patients, suppressor
cells which suppress immune responses increase. Therefore, it
would be very beneficial to find some compounds which inhibit the
action or the generation of suppressor cells. We have found that
there are some antitumor compounds which affect the suppressor
system (58).

ANTIBACTERIAL ANTIBIOTICS

As described above, the research area of antibiotics has been
extended from antimicrobial antibiotics to antitumor antibiotics,
enzyme inhibitors, and immuno-modifiers. In the area of antimicrobial antibiotics, during the last 10 years many successful

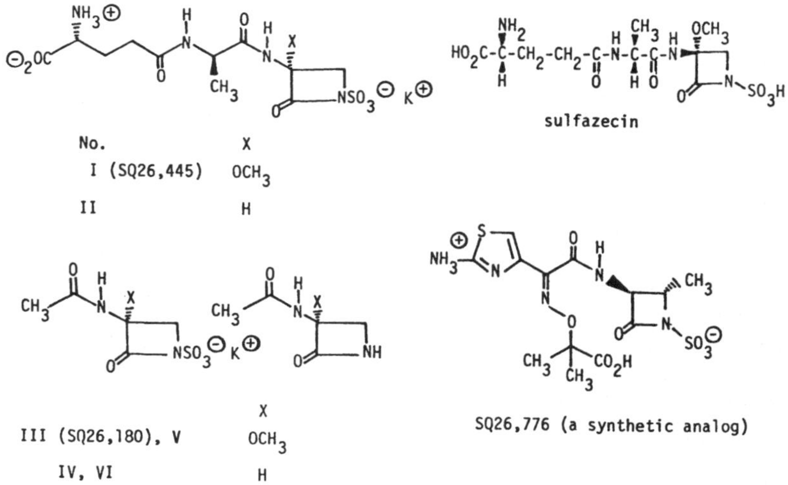

Fig. 13.   β-Lactam antibiotics produced by bacteria and
           synthetic analog.

results have been obtained with β-lactam and aminoglycoside antibiotics.

　　 Imada *et al*. (59) isolated two N-sulfonyl single β-lactam
antibiotics in culture filtrates of *Pseudomonas acidophila* and
named one of them sulfazecin (Fig. 13). Skyes *et al*. (60) found
other 6 single β-lactam antibiotics (Fig. 13) in *Gluconobacter*,
Chromobacterium violaceum, and *Agrobacterium radiobacter*, and
proposed the term monobactam for these β-lactams which are produced by bacteria. Very recently, these authors reported the successful synthesis of a very effective monobactam analog (SQ26, 776,
Fig. 13) at the International Congress of Chemotherapy, Florence,
Italy, July, 1981. This analog both *in vitro* and *in vivo*
strongly inhibits the growth of Gram negative organisms including
Pseudomonas. Ohno *et al*. (61, 62) developed a chemicoenzymatic method of synthesis of (S) and (R)-4-[(methoxycarbonyl)methyl]
-2-azetidione as shown in Fig. 14. This process is useful in the
synthesis of the thienamycin group and monobactam analogs. Depending on the group on the NH of dimethyl β-aminoglutarate (Fig. 14),
the S or R-isomer was obtained; for instance, the chemicoenzymatic
synthesis starting from dimethyl β-(acetylamino)-glutarate gave
(R)-4-[(methoxycarbonyl)methyl]-2-azetidione. Chemicoenzymatic
synthesis will be extended for syntheses of analogs of β-lactam
and other useful antibiotics.

　　 In 1967, we elucidated the mechanism of resistance to aminoglycoside antibiotics (63); and it became possible to predict the
structure of derivatives which would inhibit the growth of resist-

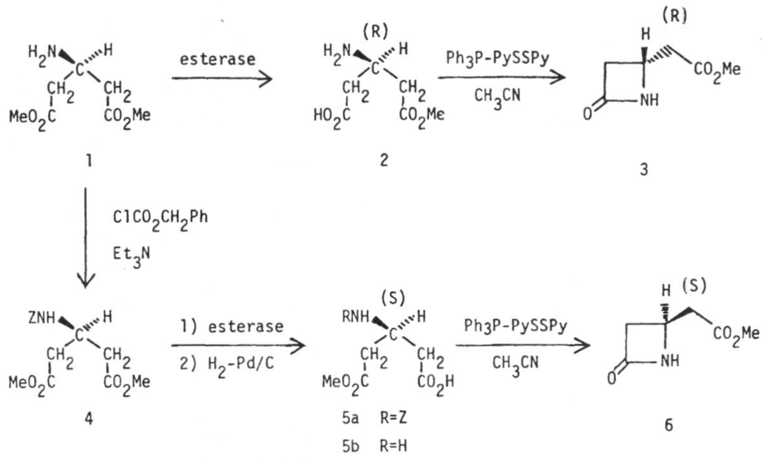

Fig. 14. Chemicoenzymatic synthesis of β-lactam compounds.

ant strains (64). At present, O-phospho or O-adenylyltransferases and N-acetyltransferases shown in Fig. 15 are known to be involved in the resistance mechanism (65). The elimination of the hydroxyl group susceptible to O-phospho or adenylyltransferases or a modification which produces steric hindrance to the enzymes gives derivatives active against resistant strains. Thus, useful derivatives have been developed. I was interested in the antibacterially active structure which had the least number of hydroxyl groups. Therefore, 5,2',3',4',4",6"-deoxy derivatives of kanamycin B were synthesized and confirmed to inhibit the growth of both sensitive and resistant strains (66). Some of these deoxy derivatives may become useful, if resistant strains which have new O-phospho or adenylyltransferases should appear in the future. The antibacterial effect of these deoxy derivatives indicates that only amino groups are involved in the binding of aminoglycosides to bacterial ribosomes.

Microbial transformations of aminoglycosides have been studied in detail. Reinhart et al. (67) have reported on the microbial transformation of various aminocyclitols and pseudodisaccharides to aminoglycoside antibiotics. As shown by our studies in collaboration with Oka, et al. of the Research Institute of the Yamanouchi Pharmaceutical Co., if kanamycin A or B is added to medium in which a gentamicin-nonproducing mutant is cultured, then 3',4'-dideoxygenation and methylation at the 4"-carbon atom occur; and the kanamycins are transformed to combimicins (Fig. 16) (68). 3"-N-Methylation can also occur. Microbial transformation is becoming a useful tool for the development of new aminoglycosides. The cell fusion technique, which has also been applied to develop high-

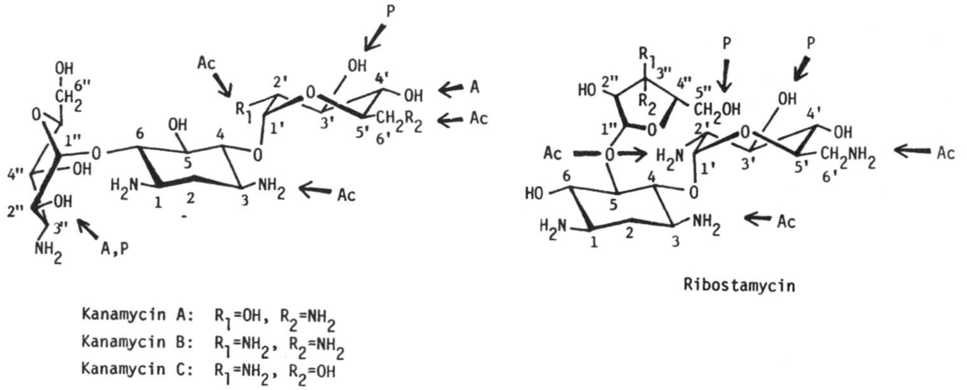

Kanamycin A: R_1=OH, R_2=NH$_2$
Kanamycin B: R_1=NH$_2$, R_2=NH$_2$
Kanamycin C: R_1=NH$_2$, R_2=OH

Fig. 15. O-Phosphotransferases (P), O-adenylyltransferases (A),
          and N-acetyltransferases (Ac) involved in resistance
          mechanism.

yielding strains, is also becoming useful for the development of
new aminoglycosidase (69).

On the other hand, the continuation of the screening of culture filtrates for aminoglycosides may lead to the findings of new
interesting types of antibiotics. The fortimicin group of
antibiotics, shown in Fig. 17, has been found by 4 Japanese research groups (65, 70). Fortimicins are not only useful as them-
selves, but also the study of their derivatives is leading to chemotherapeutic agents which have improved activities. Istamycin B
was found by my group (70); and its 3-O-demethyl derivative has
an improved activity (71).

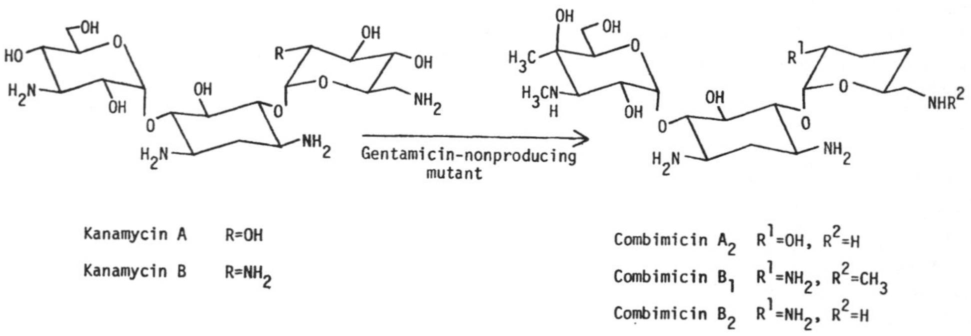

Kanamycin A   R=OH
Kanamycin B   R=NH$_2$

Combimicin A$_2$ R^1=OH, R^2=H
Combimicin B$_1$ R^1=NH$_2$, R^2=CH$_3$
Combimicin B$_2$ R^1=NH$_2$, R^2=H

Fig. 16. Microbial transformation of kanamycin A or B to
          combimicin A$_2$ or B$_1$ and B$_2$.

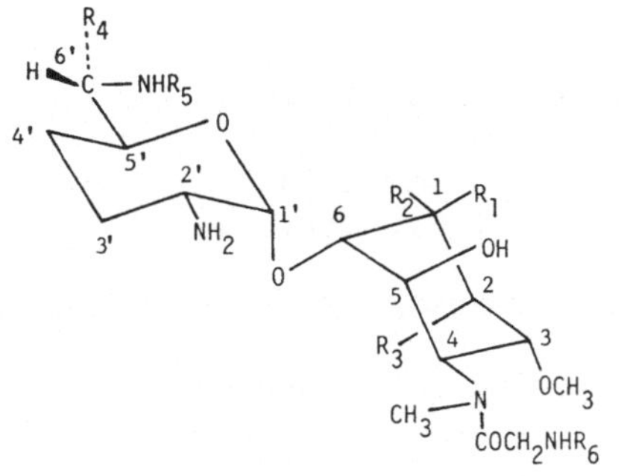

	R_1	R_2	R_3	R_4	R_5	R_6
Fortimicin A	NH_2	H	OH	CH_3	H	H
Fortimicin C	NH_2	H	OH	CH_3	H	$CONH_2$
Fortimicin D	NH_2	H	OH	H	H	H
Sporaricin A	H	NH_2	H	CH_3	H	H
Istamycin A (Sannamycin A)	NH_2	H	H	H	CH_3	H
Istamycin B	H	NH_2	H	H	CH_3	H
Dactimicin	NH_2	H	OH	CH_3	H	CH=NH

Fig. 17. Fortimicin Group Antibiotics

GENETICS OF ANTIBIOTICS AND CONCLUSION

As described above, microorganisms produce an almost infinite member of various compounds which have antibiotic or other pharmacological activities. Antibiotics and other secondary metabolites can be classified according to their structural relationships. Antibiotics within the same group have a common structural part. The production of the same group of antibiotics by different species, genus, and families indicate a wide distribution of a gene or a gene set involved in the biosynthesis of the common structural part. It is thought that the common structural part which has no cytotoxicity is synthesized in cells and transformed ti the final products which are released extracellularly. In different strains, the common structure part is transformed to different final products (15). There are plasmids which increase the production yield. It is thought that during the evolution of microorganisms, the genes for the syntheses of the common structural part of various anti-biotics were formed and transferred to other organisms. It is possible that in some cases the genes reside in plasmids (15).

As shown by the studies of enzyme inhibitors, microorganisms are the treasury of various organic compounds. Parallel to progress in understanding the biochemistry of physiological and abnormal processes, quantitative screening methods will be developed, and useful compounds will be found in microorganisms. Studies of biosyntheses and derivatives of known effect antibiotics and enzyme inhibitors will lead to the development of useful compounds with improved activities.

In order to obtain antibiotics and other useful microbial products, we have been dependent on microorganisms in nature. One day we may be able to produce new genes and establish new methods for the development of additional useful compounds.

REFERENCES

1. UMEZAWA, H., "Enzyme Inhibitors of Microbial Origin," University of Tokyo Press (1972).
2. AOYAGI, T., TAKEUCHI, T., MATSUZAKI, A., KAWAMURA, K., KONDO, S., HAMADA, M., MAEDA, K. & UMEZAWA, H. *J. Antibiotics 22:* 283 (1969).
3. KAWAMURA, K., KONDO, S., MAEDA, K. & UMEZAWA, H. *Chem. Pharm. Bull. 17:* 1902 (1969).
4. HORI, M., HEMMI, H., SUZUKAKE, K., HAYASHI, H., UEHARA, Y., TAKEUCHI, T. & UMEZAWA, H. *J. Antibiotics 31:* 95-98 (1978).
5. SUZUKAKE, K., FUJIYAMA, T., HAYASHI, H., HORI, M. & UMEZAWA, H. *J. Antibiotics 32:* 523 (1979).
6. SUZUKAKE, K., HAYASHI, H., HORI, M. & UMEZAWA, H. *J. Antibiotics 33:* 857 (1980).
7. UMEZAWA, H., OKAMI, Y. & HOTTA, K. *J. Antibiotics 31:* 99 (1978).
8. SUZUKAKE, K., TAKADA, M., HORI, M. & UMEZAWA, H. *J. Antibiotics 33:* 1172 (1980).
9. UMEZAWA, H. *Meth. Enzymol. 45:* 678 (1976).
10. UMEZAWA, H., AOYAGI, T., MORISHIMA, H., MATSUZAKI, M., HAMADA, M. & TAKEUCHI, T. *J. Antibiotics 23:* 259 (1970).
11. MATSUSHITA, Y., TONE, H., HORI, S., YAGI, Y., TAKAMATSU, A., MORISHIMA, H., AOYAGI, T., TAKEUCHI, T. & UMEZAWA, H. *J. Antibiotics 28:* 1016 (1975).
12. TONE, H., MATSUSHITA, Y., YAGI, Y. & TAKAMATSU, A. *J. Antibiotics 28:* 1012 (1975).
13. CORVOL, P., personal information.
14. STRATCHER, H. Muscle Dystrophy Treatment Meeting, February, Tokyo (1981).
15. UMEZAWA, H. *Jap. J. Antibiotics 30 (Suppl.)* S138-S163 (1977).
16. ENDO, A., KURODA, M. & TSUJITA, Y. *J. Antibiotics 29:* 1346 (1976).
17. KANEKO, I., HAZAMA-SHIMADA, Y. & ENDO, A. *Eur. J. Biochem. 87:* 313 (1978).
18. YAMAMOTO, A., SUDO, H. & ENDO, A. *Atherosclerosis 35:* 259 (1980).
19. UMEZAWA, H. (ed.: "Small Molecular Immunomodifiers of Microbial Origin—Fundamental and Clinical Studies of Bestatin," Pergamon Press, Oxford (1981).
20. ISHIZUKA, M., IINUMA, H., TAKEUCHI, T. & UMEZAWA, H. *J. Antibiotics 25:* 320 (1972).
21. KUNIMOTO, T., HORI, M. & UMEZAWA, H. *Biochim. Biophys. Acta 298:* 513 (1973).
22. ISHIZUKA, M., TAKEUCHI, T. & UMEZAWA, H. *J. Antibiotics 34:* 95 (1981).
23. AOYAGI, T., SUDA, H., NAGAI, M., OGAWA, K., SUZUKI, J., TAKEUCHI, T. & UMEZAWA, H. *Biochim. Biophys. Acta 452:* 131 (1976).

24. UMEZAWA, H., AOYAGI, T., SUDA, H., HAMADA, M. & TAKEUCHI, T.
     J. Antibiotics 29: 97 (1976).
25. AOYAGI, T., TOBE, H., KOJIMA, F., HAMADA, M., TAKEUCHI, T.
     & UMEZAWA, H. *J. Antibiotics 31:* 636 (1978).
26. AOYAGI, T., YAMAMOTO, T., KOJIRI, K., KOJIMA, F., HAMADA, M.,
     TAKEUCHI, T. & UMEZAWA, H. *J. Antibiotics 31:* 244 (1978).
27. UMEZAWA, H., AOYAGI, T., HAZATO, T., UOTANI, K., KOJIMA, F.,
     HAMADA, M. & TAKEUCHI, T. *J. Antibiotics 31:* 639 (1978).
28. UMEZAWA, H., AOYAGI, T., UOTANI, K., HAMADA, M., TAKEUCHI,
     T. & TAKAHASHI, S. *J. Antibiotics 33:* 1594 (1980).
29. OKI, T., MATSUZAWA, Y., YOSHIMOTO, A., NUMATA, K., KITAMURA,
     I., HORI, S., TAKAMATSU, A., UMEZAWA, H., ISHIZUKA, M.,
     NAGANAWA, H., SUDA, H., HAMADA, M. & TAKEUCHI, T. *J. Anti-*
     biotics 28: 830 (1975).
30. UMEZAWA, H., TAKAHASHI, T., KINOSHITA, M., NAGANAWA, H.,
     MASUDA, T., ISHIZUKA, M., TATSUTA, K. & TAKEUCHI, T. *J.*
     Antibiotics 32: 1082 (1979).
31. DANTCHEV, D., PAINTRAND, M., HAYAT, M., BOURUT, C. & MATHE,
     G. *J. Antibiotics 32:* 1085 (1979).
32. OKI, T., YOSHIMOTO, A., MATSUZAWA, Y., TAKEUCHI, T. & UMEZAWA,
     H. *J. Antibiotics 33:* 1331 (1980).
33. YOSHIMOTO, A. & OKI, T. *J. Antibiotics 33:* 1199 (1980).
34. MCGUIRE, J. C., THOMAS, M. C., STROSHANE, R. M., HAMILTON,
     B. K. & WHITE, R. J. *Antimicrobial Agents Chemoth. 18:*
     454 (1980).
35. YOSHIMOTO, A., OKI, T., TAKEUCHI, T. & UMEZAWA, H. *J. Anti-*
     biotics 33: 1158 (1980).
36. OKI, T., YOSHIMOTO, A., MATSUZAWA, Y., TAKEUCHI, T. & UMEZAWA,
     H. *J. Antibiotics 34:* 916 (1981).
37. TAKITA, T., MURAOKA, Y., YOSHIOKA, T., FUJII, A., MAEDA, K.
     & UMEZAWA, H. *J. Antibiotics 25:* 755 (1972).
38. TAKITA, T., MURAOKA, Y., NAKATANI, T., FUJII, A., UMEZAWA,
     Y., NAGANAWA,H. & UMEZAWA, H. *J. Antibiotics 31:* 801
     (1978).
39. UMEZAWA, Y., MORISHIMA, H., SAITO, S., TAKITA, T., UMEZAWA,
     H., KOBAYASHI, S., OTSUKA, M., NARITA, M. & OHNO, M. *J.*
     Am. Chem. Soc. 102: 6630 (1980).
40. TAKITA, T., UMEZAWA, Y., SAITO, S., MORISHIMA, H., UMEZAWA,
     H., MURAOKA, Y., SUZUKI, M., OTSUKA, M., KOBAYASHI, S. &
     OHNO, M. *Tetrahedron Lett. 22:* 671 (1981).
41. TAKITA, T., UMEZAWA, Y., SAITO, S., MORISHIMA, H., TSUCHIYA,
     T., MIYAKE, T., UMEZAWA, H., MURAOKA, Y., SUZUKI, M.,
     OTSUKA, M. & OHNO, M. *Tetrahedron Lett.,* in press.
42. FUJII, A. "Bleomycin: Chemical, Biochemical and Biological
     Aspects," Springer-Verlag, New York (1979).
43. NAKATANI, T., NISHIKIORI, T., MURAOKA, Y., FUJII, A., TAKITA,
     T. & UMEZAWA, H. *J. Antibiotics,* in press.
44. TAKITA, T., MURAOKA, Y., NAKATANI, T., FUJII, A., IITAKA, Y.
     & UMEZAWA, H. *J. Antibiotics 31:* 1073 (1978).

45. POVIRK, L. F., HOGAN, M. & DATTAGUPTA, N. *Biochemistry 18:*
     96 (1979).
46. KASAI, H., NAGANAWA, H., TAKITA, T. & UMEZAWA, H. *J. Anti-*
     biotics 31: 1316 (1978).
47. GROLLMAN, A. P. & TAKESHIBA, N., personal information.
48. TAKAHASHI, K., YOSHIOKA, O., MATSUDA, A. & UMEZAWA, H.
     J. Antibiotics 30: 861 (1977).
49. UMEZAWA, H., HORI, S., SAWA, T., YOSHIOKA, T., TAKITA, T. &
     TAKEUCHI, *J. Antibiotics 27:* 419 (1974).
50. UMEZAWA, H., TAKEUCHI, T., HORI, S., SAWA, T., ISHIZUKA, M.,
     ICHIKAWA, T. & KANAI, T. *J. Antibiotics 25:* 483 (1972).
51. UMEZAWA, H. *Pure Appl. Chem. 28:* 665 (1971).
52. UMEZAWA, H. "Monograph on Cancer Research," *Gann* No. 19,
     University of Tokyo Press (1976) p 3.
53. UMEZAWA, H., TAKAHASHI, Y., FUJII, A., SAINO, T., SHIRAI, T.
     & TAKITA, T. *J. Antibiotics 26:* 117 (1973).
54. MATSUDA, A., YOSHIOKA, O., TAKAHASHI, K., YAMASHITA, T.,
     H. in "Bleomycin: Current Status and New Development,"
     (S. K. Carter, S. T. Crook, and H. Umezawa, eds.) Academic
     Press, New York (1978) p. 311.
55. FUKUOKA, T., MURAOKA, Y., FUJII, A., NAGANAWA, H., TAKITA,
     T. & UMEZAWA, H. *J. Antibiotics 33:* 114 (1980).
56. URAKAWA, C., NAKANO, K. & IMAI, R. *J. Antibiotics 33:* 804
     (1980).
57. UMEZAWA, H., TAKEUCHI, T., IINUMA, H., KUNIMOTO, S., TAKEUCHI,
     M., HAMADA, M., IKEDA, Y., IWASAWA, H., NAGANAWA, H. &
     KONDO, S. *J. Antibiotics*, in press.
58. ISHIZUKA, M., TAKEUCHI, T., MASUDA, T., FUKASAWA, S. &
     UMEZAWA, H. *J. Antibiotics 34:* 331 (1981).
59. IMADA, A., KITANO, K., KINTAKA, K., MUROI, M. & ASAI, M.
     Nature 289: 590 (1981).
60. SYKES, R. B., CIMARUSTI, C. M., BONNER, D. P., BUSH, K.,
     FLOYD, D. M., GEORGOPAPADOKOU, N. H., KOSTER, W. H., LIN,
     W. C., PARKER, W. L., PRINCIPE, P. A., RATHNUM, M. L.,
     SLUSARCHYK, W. A., TREJO, W. H. & WELLS, J. S. *Nature 291:*
     489 (1981).
61. OHNO, M., KOBAYASHI, S., IIMORI, T., WANG, YI-F. & ISAWA, T.
     J. Am. Chem. Soc. 103: 2405 (1981).
62. KOBAYASHI, S., IIMORI, T., ISAWA, T. & OHNO, M. *J. Am. Chem.*
     Soc. 103: 2406 (1981).
63. UMEZAWA, H., OKANISHI, M., KONDO, S., HAMANA, K., UTAHARA,
     R., MAEDA, K. & MITSUHASHI, S. *Science 157:* 1559 (1967).
64. UMEZAWA, H. in "Advances in Carbohydrate Chemistry and Bio-
     chemistry," vol. 30 (R. S. Tipson and D. Horton, eds.)
     Academic Press, New York (1974) p. 183.
65. UMEZAWA, H. *Jap. J. Antibiotics 32 (Suppl.):* S1 (1979).
66. MIYASAKA, T., IKEDA, D., KONDO, S. & UMEZAWA, H. *J. Anti-*
     biotics 33: 527 (1980).
67. REINHART, K. L. *Jap. J. Antibiotics 32 (Suppl.):* S32 (1979).

68. OKA, Y., ISHIDA, H., MORIOKA, M., NUMASAKI, Y., YAMAFUJI, T.,
     OSONO, T. & UMEZAWA, H. *J. Antibiotics 34:* 777 (1981).
69. MAZIERES, N., PEYRE, M. & PENASSE, L. *J. Antibiotics 34:*
     544 (1981).
70. OKAMI, Y., HOTTA, K., YOSHIDA, M., IKEDA, D., KONDO, S. &
     UMEZAWA, H. *J. Antibiotics 32:* 964 (1979).
71. HORIUCHI, Y., IKEDA, D., KONDO, S. & UMEZAWA, H. *J. Anti-
     biotics 33:* 1577 (1980).

Session I
LARGE SCALE PRODUCTION AND PURIFICATION OF BIOMOLECULES
Chairmen: H. Weetall and H. Yamada

INTEGRATION OF LARGE SCALE PRODUCTION AND PURIFICATION

OF BIOMOLECULES

K. Venkatasubramanian

H. J. Heinz Company and Rutgers University
Pittsburgh, Pennsylvania, U.S.A.

Despite the many dramatic advances in biotechnology over the past few years, little attempt has been made to analyze large scale manufacturing schemes for biomolecules. In particular, the production and purification operations have not been studied in an integrated fashion. In many instances, the isolation and purification of the biomolecule could be as extensive and expensive as the synthesis itself. In this paper, the question is addressed from a commercial standpoint.

Using a relatively simple case, the production of enriched fructose corn syrup (EFCS), the difficulties in the scale up of purification processes are illustrated. Critical factors in the design of a large scale semi-continuous chromatographic separation system are highlighted. Different design options and their impact on the overall economics of the process are discussed. Next we consider more complex bioprocessing operations and present a brief analysis of the relative costs of production and purification. Based on this analysis, critical challenges in the development of overall reaction-separation systems are outlined.

LARGE SCALE SEPARATION OF GLUCOSE/FRUCTOSE MIXTURES

Among the mixtures of corn-derived sugars produced on a commercial scale, high fructose corn syrup (HFCS) is the most important one. HFCS is produced from dextrose syrup by enzymatic isomerization. It represents the largest commercial application of immobilized enzyme technology to-date. Since the isomerization reaction is reversible, the maximum achievable conversion is only about 50% on pure dextrose feeds. This corresponds to an

equilibrium constant of one at 60°C. In practice, it is quite difficult to prepare a pure dextrose substrate economically from corn.
Therefore it is necessary to operate at a conversion level lower
than the theoretical maximum conversion. Typically, the product
stream (HFCS) contains about 42% fructose, 52% unconverted dex
trose, and small quantities of oligosaccharides.

In order to obtain products with higher levels of fructose,
it is necessary to concentrate fructose selectively. Many common
separation techniques are not applicable for this purpose, since
they do not discriminate readily between two isomers of essentially
the same molecular size. However, fructose preferentially forms
a complex with different cations such as calcium. It has been possible to develop commercial processes by taking advantage of this
property of fructose and combining it with more established separation technologies such as ion exchange.

There are basically two different commercial processes available today for the large scale purification of fructose. In both
instances resins in the preferred cationic form are employed in
packed bed systems. One process employs an inorganic resin leading to selective molecular adsorption of fructose (1). Chromatographic fractionation using organic resins is the basis for the
second commercial separation process (2). When an aqueous solution
of dextrose and fructose is fed to a fractionating column, fructose
is selectively held by the resin to a greater degree than glucose.
Deionized water is used as the eluant. Typically, the separation
is achieved in a column packed with a bed of a low crossed-linked
fine mesh sulfonic type cation exchange resin using calcium as the
preferred salt form. The enriched fraction contains 90% or more
of fructose, referred to as very enriched fructose syrup (VEFCS).
This can be blended back with the feed material to obtain products
having fructose content between 42 and 90%; these are referred to
as enriched fructose syrups (EFCS).

DESIGN AND SCALE-UP CONSIDERATIONS

A number of apriori considerations must be borne in mind in
the scale-up and commercialization of such a fractionation process.
Since the separation is typically a batch operation, one has to
be concerned about its compatibility with the front end of the
operation, which typically is continuous. This problem is circumvented by operating the separation column in a simulated continuous fashion employing a moving bed concept. In a further attempt
to develop a truly continuous system for enrichment of fructose,
we have developed a new process which employs the principles of
electrodialysis and electroosmosis (3). This process has the additional advantage of saving significant quantities of diluent water
and hence evaporation costs. However, the real cost of the system

is governed by the flux rates through these membranes and the life of the membrane under actual use. Further work is progressing in our laboratory to address both these questions in order to assess the economic viability of this process.

Another concern relates to the process synthesis of isomerization and fractionation. The treatment of the raffinate stream which is rich in glucose is an important consideration since this would require additional isomerization for its further conversion to fructose. In addition to dextrose recycle, the presence of oligosaccharides in the feedstream often complicates the overall problem. If the level of oligosaccharides is allowed to build up beyond a certain value, it could inhibit the activity of the isomerization reaction (4); and they would impart undesirable flavors to the finished product. Since the isomerase enzyme is also inhibited by calcium, its elution from the resin bed could be detrimental to enzyme activity. Therefore some form of cation removal system must be considered as part of the overlal design.

Since water is used as the elution medium, it has a great impact on the overall evaporation load on the system. Very low solids concentration can also lead to possible microbiological problems. The most important design parameter dictating the economics of the overall process is the maximization of solids yield at an acceptable purity while reducing the dilution effect of the eluant rinse to a minimum, i.e., maximizing the efficiency of usage of feed and water. The yield is important for all the obvious reasons as well as the cost of reisomerization of each kilogram of solids.

Several procedures are available for achieving the above goals; these include recycling techniques, higher equalization of the resin phase with proper distribution in a packed column, and addition of multiple entry/exit points in the column. These approaches generally increase the purity and yield and are sometimes referred to as feed enrichment. A small apparent increase in the purity of the feed to the column results in a much larger gain in production through increased yield at a given product purity. In practice this translates into a) maximization of the ratio of sugar volume fed per volume of resin per cycle, b) minimization of the ratio of water volume required per volume of resin per cycle, and c) provision for careful fluid distribution to the column.

PROCESS ECONOMICS

Table 1 shows the relative cost of EFCS production. It can be readily noted that evaporation cost dominates the picture followed by the cost of enzymatic reisomerization of dextrose.

K. VENKATASUBRAMANIAN

TABLE 1

RELATIVE COST OF EFCS PRODUCTION

Category	Cost (%)
Water	0.60
Chemicals	2.10
Equipment Amortization	18.04
Resin Amortization	15.35
Evaporation	31.12
Maintenance	0.90
Labor	5.33
Dextrose Reisomerization	26.30
Water Disposal	0.16
Total	99.90

In Table 2 the added cost of going to EFCS from HFCS is shown.
Expressed in this manner it can be seen that the added enzyme cost
can be up to one-fourth of the total cost in enriching the fructose product. These costs are followed by the amortization cost
reflecting the relatively high capital investment required to
operate this process.

TABLE 2

ADDED COST FOR HFCS - TO - EFCS

Category	Cost (%)
Fractionation Cost	18 - 27
Added Enzyme Cost	4 - 5
Other Added Costs	2 - 3
Total	24 - 35

*Does not include corn cost.

As the extent of fructose in the feedstock to the fractionator increases, the purification load on the column decreases. However, at the same time the cost of isomerization increases. Therefore, one must balance these two opposing costs to arrive at the optimum effluent concentration from the isomerization reactor. Similarly, it is necessary to consider such balances in the case of dextrose recycling leading to optimal usage of the enzyme. A number of flow schemes are possible, but the economics of each one has to be carefully evaluated before the optimal flow configuration is chosen. It is clear that any rational design and economic analysis cannot look at the separation problem without considering its integration into the front-end of the process especially with respect to isomerization. Because of the large number of recycled streams involved both internally within the fractionation column as well as the linkage of this process to the rest of the unit operations, it is often not possible to understand the interactive effects of opposing elements in the process. We have employed a detailed computer model to simulate the entire process and to assess the impact of instituting a change at one point on the entire system.

PURIFICATION OF OTHER ACTIVE BIOMOLECULES

Whereas the fructose purification process is characterized by a large volume/low value product stream, many other biotechnological processes are characterized by relatively low volume/high value products. Antibiotics production and purification falls in the later category. The principles of design and process integration are essentially the same; but the impact of the relative roles of conversion and purification costs vary depending upon the system. Considering for example the case of penicillin, which is a well established fermentation, the overall process cost is dictated by the fermentation cost rather than by the purification cost. The ratio of these two costs being about 80:20 (5). This is due to the fact that penicillin can be produced at very high concentrations of product in the fermentation broth and that the purification scheme is greatly simplified by the selective solvent extraction of penicillin from the broth. Another reason the fermentation cost is high is due to the high cost of the raw material, especially the precursor to penicillin, i.e., phenyl acetic acid.

Many newer antibiotics on the other hand have relatively low product concentration at the end of the fermentaiton cycle and hence the cost of recovery becomes disproportionately large in comparison to separation and purification. In Table 3 we give the estimated costs of fermentation and purification for a new antibiotic which is of the cephalosporin type. It is worth noting here that the recovery cost is over three-quarters of the total cost. In this instance the cost of cell removal and disposal is substantial. However, by applying newer technologies, such as mem-

TABLE 3

NEWER ANTIBIOTIC PRODUCTION COSTS

Category	Cost (%)
Fermentation	24.4
Cell Separation	24.1
Further Purification	51.5
Total	100.0

brane technology, it is possible to reduce the cost of cell separation significantly in comparison to the more conventional vacuum
drum filtration. It is clear from these examples that an apriori
analysis of the relative cost of the different unit operations
could point the direction towards more meaningful process development and scale-up.

BIOCATALYSIS BY IMMOBILIZED CELLS

Of late there is considerable effort being expended on the
development of immobilized living cell systems for carrying out
a variety of complex catalytic processes. Much emphasis has been
placed on the immobilization methodology and reactor development.
As these processes become more sophisticated it is imperative that
the downstream processing operations be equally compatible with
the reactor design in order to achieve meaningful economies in the
overall processing system. It is obvious that the product concentration out of the reactor will play an important role in determin-
intg the complexity of the purification process. Most of the works
published in the literature to-date are excessively concerned
about volumetric productivity without adequate attention paid to
the product concentration. Work done in our laboratory shows that
it is not always possible to obtain an equivalent product concentration with immobilized cell systems in comparison to the free
cell fermentations.

One of the special advantages of an immobilized cell system
might be that the amount of cell mass that is produced is considerably lower in comparison to free fermentation. Hence, considerable savings could be accrued in the removal and disposal
of spent biomass. In developing cost comparisons between free
cell and immobilized cell systems this aspect must be considered.

Most conventional fermentations are carried out with complex media. This in turn results in relatively more difficult purification procedures. With the development of an integrated bioreactor system using immobilized cells, one could afford to use better defined, cleaner media including synthetic media. Another aspect which is often ignored is the need for complete consumption of the nutrients, which would otherwise contribute to additional difficulties in the purification process.

In the past most purification operations have not been developed to a sophisticated state simply because the driving force for doing so was minimal. In other words, since the biomolecule was often produced in relatively slow batch fermentation processes, there was little incentive to develop dedicated continuous purification operations. This could be changed with the advent of carefully designed biocatalytic reactors. In its limiting version this could result in continuous reactor-separator devices in which only the desirable product is selectively removed from the reactor. It must be emphasized that process economic analysis performed without examining all these aspects and combinations could be misleading.

ACKNOWLEDGMENTS

It is a pleasure to acknowledge the assistance of H. Keller and J. Laraway of Techni-Chem, Inc. and A. Shah for their assistance in the preparation of this paper. The impeccable typing assistance of S. Kowalski is gratefully acknowledged.

REFERENCES

1. JENSEN, R. H. *Abst. Am. Inst. Chem. Eng. 85th Nat. Mtg.*, Philadelphia (1978).
2. SERBIA, G. R., U.S. Patent 3,004,906 (1962).
3. VENKATASUBRAMANIAN, K., JAIN, S. & GIFFRIDA, U.S. Patent (claims allowed) (1981).
4. VENKATASUBRAMANIAN, K. & HARROW, L. S. *Ann. N.Y. Acad. Sci. 326:* 141 (1979).
5. SWARTZ, R. *Ann. Rep. Ferm. Processes 3:* 75 (1979).

ROLE OF ACTIVE TRANSPORT OF INDUCER IN ENZYME BIOSYNTHESIS

AND PROCESS SCALE-UP

W. R. Vieth, K. Kaushik and K. Venkatasubramanian

Department of Chemical and Biochemical Engineering
Rutgers University
New Brunswick, New Jersey, USA

This paper describes a study of the role of active transport of a key species in microbial enzyme biosynthesis. The aim is to provide a quantitative understanding by integrating accepted mechanistic descriptions into complete, testable engineering models that represent phenomenologically accurate simulations for enzyme biosynthesis.

The candidate system chosen for detailed analysis and controlled fermentation studies was lactose induction of β-galactosidase in *E. coli*. Three major elements involved in regulation were emphasized: translocation of inducer across the plasma membrane; the effect of cyclic AMP in catabolite repression; and regulation of intracellular cAMP concentration by external glucose levels (Fig. 1).

The role of cAMP was isolated; and a catabolite repression index interrelated with intracellular ATP concentration was formulated. Active transport of lactose was considered to be coupled to co-transport of protons, in accord with the Mitchell hypothesis. The model was verified using our own fermentation data and also corroborated by comparison with data on lactose translocation from other authors. It was demonstrated that, for adequate modeling of enzyme biosynthesis, a transport model must be included in the system equations for fermentaiton.

NEW ELEMENTS OF BIOSYNTHESIS MODEL: LACTOSE TRANSPORT

The kinetics of lactose translocation are determined by the rates of binding and translocation involved in the mechanism for

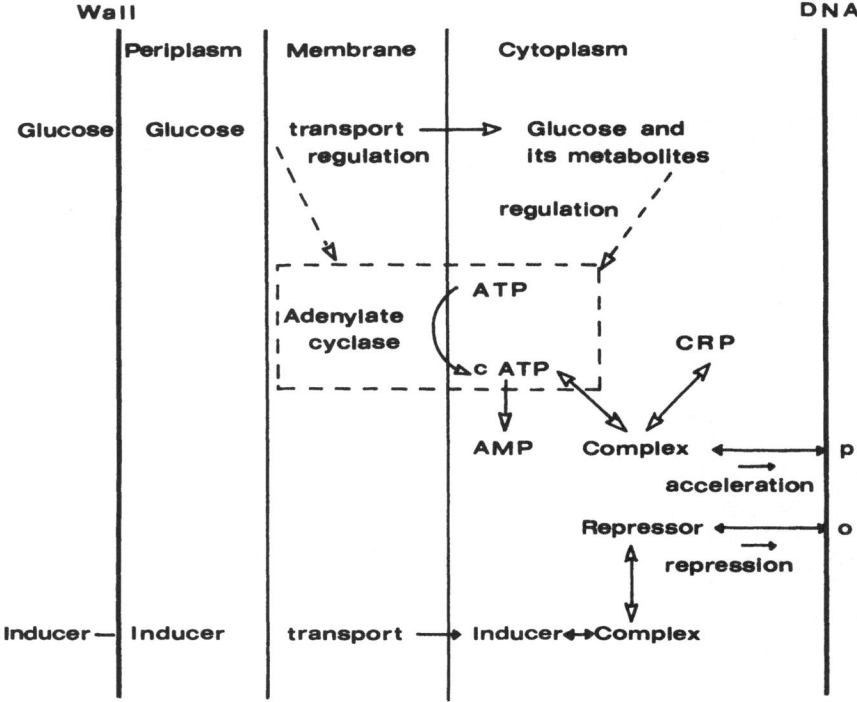

Fig. 1. Control mechanisms involved in regulation of inducible
        enzyme biosynthesis. The box (--) represents the
        regulation of cAMP by glucose.

transport. If the rates of the elementary steps are disproportionately different, the kinetics can be approximated by a rate determining step. Fig. 2 shows schematically a generalized sequence
of kinetic steps that appear to correspond most closely to available experimental evidence on lactose-proton symport (1 - 3).
Transport across the plasma membrane is assumed to involve a carrier, probably an unidentified permease. In Fig. 2 the carrier,
denoted as C, carries a net negative charge in the free state and
is uncharged when coupled with a proton. The carrier resides only
within the membrane and can change position so as to face either
the outside medium (States 1, 3, 5) or the intracellular medium
(States 2, 4, 6). If N is the total number of sites available for
binding and translocation on the lac carrier, then the number of
sites at any time in states 1 through 6 is denoted as $N_1, N_2 \ldots N_6$.
Physical movement of the carrier is not implied. The sequences
1-3 and 2-4 represent binding of H^+ to the carrier outside and inside the membrane, respectively. Similarly, 3-5 and 4-6 represent
lactose binding to the carrier. For efflux of lactose, the net
sequence is 4-6-5-3-1-2; and for influx it is 3-5-6-4-2-1. Step

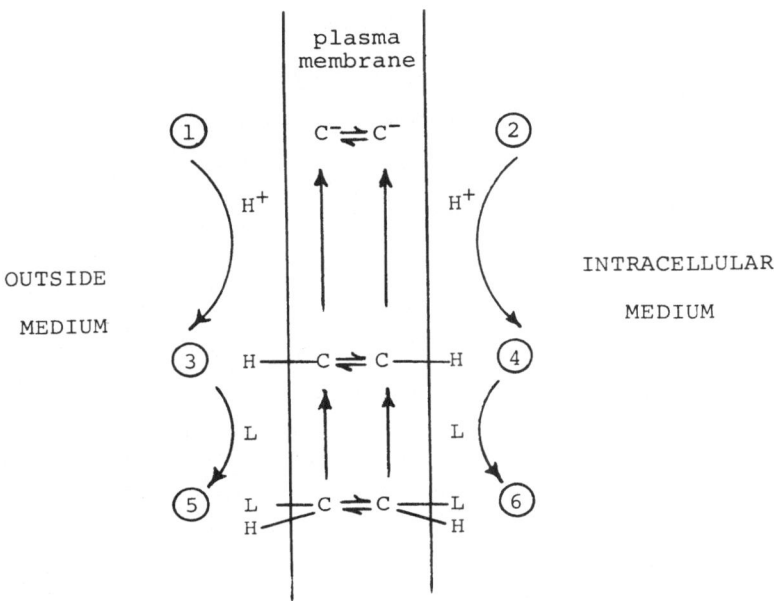

Fig. 2.   Schematic showing steps involved in lactose (L) and
          proton (H^+) transport by means of an intramembrane
          carrier (C). The carrier is charged in the free state
          (C^-) and can move so as to face either the outside
          medium (State 1, 3, 5) or the intracellular medium
          (States 2, 4, 6). Arrows represent changes from one
          state to another occurring either in the medium or in
          the membrane.

3-4 dissipates the proton gradient without lactose transport and
is probably quite a small effect under normal conditions.

     Vesicles, intact cell membranes with the entire membrane ma-
chinery but without cytoplasmic constituents, were used as models
for whole cells. We employed literature experimental data (5, 6)
on both whole cells and vesicles in developing the model. During
exchange, cells preloaded with ^{14}C lactose were diluted into equimolar substrate with unlabeled lactose, and the loss of labeled
lactose monitored. Note that for exchange to occur, 4-6-5-3 and
3-5-6-4 can also occur; whereas for influx or efflux the whole
cycle must be completed. It was found experimentally that exchange
was very fast, about ten times the rate of influx or efflux. This
indicated that the lactose section of the loop (4-6-5-3) was fast
compared to the H^+ part, and the rate-determining step for efflux
influx was in steps 1-3, 1-2, or 2-4.

TABLE 1

SYSTEM EQUATIONS

$$\frac{d}{dt}(S_{2B}) = \frac{-C}{Y_2}\ \mu_2 \tag{Eq. 1}$$

$$\frac{d}{dt}(S_{1B} = \frac{-C}{Y_1}\ \mu_1 \tag{Eq. 2}$$

$$\frac{dC}{dt} = \mu C \qquad\qquad , \text{ where } \mu = \mu_1 + \mu_2 \tag{Eq. 3}$$

$$\frac{d}{dt}(C\ M) = k_{+m}\ F\ Q\ (\mu + b)\ C - k_{-m}\ C\ M \tag{Eq. 4}$$

$$\frac{d}{dt}(C\ E) = k_{+E}\ C\ M \tag{Eq. 5}$$

$$\frac{d}{dt}(L_{in}) = \frac{A(t)\ S_{1B}}{B(t) + S_{1B}} \tag{Eq. 6}$$

$$Q = \frac{1 + b_1\ L_{in}}{1 + b_2\ L_{in} + b_3} \tag{Eq. 7}$$

$$F = \left[\frac{K_5 + K_7\ \alpha}{1 + K_5\ K_7\ \alpha}\right]\left[\frac{1 + K_5 + K_7}{K_5 + K_7}\right] \tag{Eq. 8}$$

$$\alpha = \frac{k'\ K_1'\ (E_o)\ (S)/k_2'\ (cAMP_m)}{1 + K_1'\ (S) + S_{2B}\left[K_2' + K_1'\ K_3'\ (S)\right]} \tag{Eq. 9}$$

Assuming that the lactose part of the loop was at equilibrium, rate expressions were derived based on 1-3, 1-2, or 2-4 as the rate-determining step. It was also known that influx and efflux was affected by both components of the electrochemical potential, Δ pH and $\Delta\ \Psi$ (the membrane potential). Therefore, any rate expression not involving H^+ concentrations or the equilibrium constant for 1-2 (K_{12}) was eliminated. Logically, we then arrived at the rate expression shown in simplified lumped form in Table 1 (Eq. 6). Because this equation was based on a balance for saturable sites, it showed Monod-like characteristic behavior (which was reassuring). Nonetheless, the ultimate value of this expression depends on its validation by experimental data. (Note that A(t) and B(t) in Eq. 6 are time-dependent in *batch* fermentation. For most cases, this dependency is weak and the use of A and B as constant parameters will be a good approximation.) In Table 1 it should be noted that:

$$\mu_1 = \frac{\mu_{m1} \; S_{1B}}{K_{S1} + S_{1B}} \quad \text{and} \quad \mu_2 = \frac{\mu_{m2} \; S_{2B}}{K_{S2} + S_{2B}}$$

As an example, consider the normal case of D-lactate oxidation (5), where $\Delta\Psi$ = -60 mV and Δ pH = -120 mV, expressed in equivalent electrochemical units. For this base case and with the outside pH buffered, the cell interior will be alkaline; thus, H_{in}^+ will be low. Also, since the carrier is negatively charged, a negative membrane potential will keep sequence 1-2 toward 1; or, in other words, H_{in}^+ will be low. A(t) is 115 nmol/mg protein/min. Now consider facilitated diffusion where $\Delta\Psi$ = 0 and Δ pH = 0. The interior is less alkaline compared to our reference case, i.e., H_{in}^+ increases. Similarly, A(t) should decrease. This is precisely what is observed experimentally, A(t) being 53 nmol/mg protein/min for this case. Similarly, for all the cases studied, the equations show the right trends. Fig. 3 illustrates one of the simulations we performed. Agreement of the model with the data of Kaczorowski *et al.* (5, 6) is quite good.

Eq. 1 through 5, and 8 of Table 1 are closely similar to previous published results from our laboratory (7, 8). The focus of this paper is on the incorporation of Eq. 6, Eq. 7 (inducer transport), and Eq. 9 (cAMP regulation) into the analysis.

NEW ELEMENTS OF BIOSYNTHESIS MODEL: GLUCOSE REGULATION OF INTRACELLULAR cAMP

cAMP is synthetized from adenosine triphosphate (ATP) by the enzyme-catalyzed reaction:

$$ATP \xrightarrow{\text{adenylate cyclase}} cAMP + PP_i$$

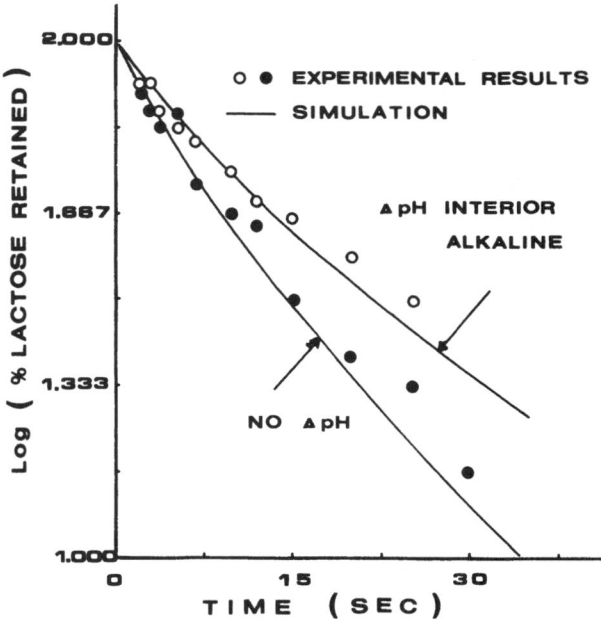

Fig. 3. Efflux of lactose under a pH gradient (interior of cell
         alkaline). 0 and ● are experimental points obtained
         from Kaczorowski, *et al.* (5.6) for lactose efflux from
         E. coli vesicles. Solid lines are numerical solutions
         of the mathematical model for transport. Initial
         parameter values: No Δ pH : A = 180.0 nmoles/mg
         protein/min, B = 8.0 nM; Interior Alkaline : A = 120.0,
         B = 3.0.

where PP_i is pyrophosphate. The enzyme, adenylate cyclase, is
bound to the plasma membrane of the cell. Depletion of cAMP from
the cell occurs by action of a phosphodiesterase, which catalyzes
the conversion of cAMP to 5'AMP (9), and by excretion of cAMP into
the external medium. Abou-Sabé and Mento (10) concluded from other
related experiments on glucose transport that glucose interacts
during transport with the adenylate cyclase system, forming an
allosteric effector and thereby inhibiting its activity. The relative proximity of the glucose transport system (the phospho-enol-
pyruvate-sugar-phospho-transferase or PTS) to the adenylate cyclase
molecules lends credibility to this hypothesis (11). Rather than
formulating a catalytic model restricted to competitive or noncom-
petitive inhibition, we developed a general model that was applicable to both cases.

The general mechanistic sequence of reactions is as follows:

$$E + S \underset{}{\overset{K_1'}{\rightleftharpoons}} E.S. \underset{}{\overset{k'}{\rightleftharpoons}} E + P$$

$$E + G \underset{}{\overset{K_2'}{\rightleftharpoons}} E.G.$$

$$E.S. + G \underset{}{\overset{K_3'}{\rightleftharpoons}} E.S.G.$$

$$E.G. + S \underset{}{\overset{K_4'}{\rightleftharpoons}} E.S.G.$$

Let (S) = ATP concentration, (E) = free enzyme concentration, (E_o) = total enzyme concentration, and (G) = glucose concentration. If it is assumed that the rate controlling step is formation of cAMP, then the other binding steps will be at equilibrium with constants K_1', K_2', K_3', and K_4'. Now, $(E_o) = (E) + (ES) + (EG) + (EGS)$; and the rate of synthesis of cAMP, dPs/dt, is given by Eq. 10.

$$\frac{dPs}{dt} = k'(ES) = \frac{k' K_1' (E_o)(S)}{1 + K_1'(S) + K_2'(G) + K_1'K_3'(S)(G)} \qquad \text{(Eq. 10)}$$

If degradation of cAMP and excretion are considered to be first order processes, then the rate of depletion of cAMP, dPd/dt, is

$$\frac{dPd}{dt} = - K_e'(P) - k_d'(P) = - K_2'(P) \qquad \text{(Eq. 11)}$$

where k_e' and k_d' are excretion and degradation rate constants and $k_2' = k_e' + k_d'$. The concentration of cAMP in the cell at time t is given by the integration of Eq. 12.

$$\frac{dP}{dt} = \frac{k' K_1' (E_o)(S)}{1 + K_1'(S) + K_2'(G) + K_1'K_3'(S)(G)} - k_2'(P) \qquad \text{(Eq. 12)}$$

At steady-state, $dP/dt = 0$, and the cAMP concentration, P_{ss}, is

$$P_{ss} = \frac{k' K_1' (E_o) (S) / k_2'}{1 + K_1' (S) + (G) (K_2' + K_1' K_3' (S))} \qquad \text{(Eq. 13)}$$

when (G) = 0, then P_{ss} becomes

$$P_{ss} = \frac{k' K_1' (E_o) (S) / k_2'}{1 + K_1' (S)} \qquad \text{(Eq. 14)}$$

The level of ATP in the cell (S) is maintained at a constant level by other cellular mechanisms and is dependent on the growth conditions of the culture. For consideration of the cAMP effects, (S) is therefore a constant. This result is incorporated into Eq. 9 of Table 1. The rate process described above is relatively fast compared to the slowly varying concentration of glucose. The effect of slow changes in glucose level would be reflected in variations in the steady-state level of cAMP. Note that the cAMP concentration at steady-state shows a hyperbolic decrease, from a constant value at zero glucose level to a low value as the glucose concentration increases. The same variation has also been observed experimentally (12).

MODEL VERIFICATION AND DISCUSSION

Quantitative verification of the model is essential in establishing its validity. Our verification involves comparison of model predictions with experimental data on two levels. First, as just discussed, experimental data from the literature were used (4) to validate the transport model in an idealized system where transport was studied in isolation (5, 6). Then, data from our experiments on induction of β-galactosidase in *E. coli* were compared with model results for a range of cases which would occur in typical fermentations. Details of the experimental procedures, assay methods, statistical analysis, and derivation of the model equations are described elsewhere (13).

Briefly, the procedure for validation involved integration of the system equations to obtain profiles of the variables which were measured in the experimental fermentation runs. A non-linear parameter estimation program was used to estimate values for the parameters and perform statistical tests to compare model predictions with experimental data. The tests indicated that the model results fit experimental data quite well; the correlation coefficients were generally between 0.93 to 0.99. We have reported (7, 8) that a transport equation was essential to simulate the biosynthesis of glucose isomerase. This requirement was expected since the effective concentration of inducer responsible for regulation was the intracellular concentration. Hence, it was not sufficient to use the bulk concentration as the effective inducer concentration. As we have seen, the internal and external concentrations were related by a non-linear expression (Eq. 6).

An interesting experimental variation is suggested by some induction experiments on pure lactose. It is evident from the transport model that if the transported substrate is not consumed in the cell, it will accumulate and induce a high rate of enzyme biosynthesis. IPTG is a nonmetabolizable (gratuitous) inducer; and introduction of IPTG into a batch culture of glucose-grown cells gives us important insights into the behavior of the model. It is possible to determine the threshold level at which IPTG causes induction of β-galactosidase activity. Table 2 shows low levels of IPTG introduced into a fermentation carried out with glucose as the carbon source. At this threshold level, the activity was zero for all cases, i.e., no induction was observed. Table 3 shows induction in a similar glucose-grown culture subjected to a feed of IPTG at about six times the threshold level. Note that the threshold level (6×10^{-2} mM) was about six hundred times below the normal inducing level of lactose used in the fermentation (approximately 35 mM). It is known (14) that the accumulation ratio of IPTG is ten to a hundred times the bulk concentration, and that the *in vitro* potency of induction for IPTG and lactose

TABLE 2

RESPONSE TO IPTG (THRESHOLD LEVEL)

Number	Fermentation Time* (hr)	Cell Concentration (g/L)
1	16.0	3.30
2	19.0	3.23
3	20.0	--
4	22.5	--
5	24.5	--
6	26.0	--
7	40.0 .	3.23

*Initial glucose concentration 15 g/L; IPTG concentration 0.01443 g/L or 6.057×10^{-5} M; IPTG introduced at 18.5 hr.

TABLE 3

RESPONSE TO IPTG (FED BATCH)

Number	Fermentation* Time (hr)	Cell Concentration (g/L)	Enzyme Activity (U)	Glucose Concentration (g/L)
1	2.25	0.281	0	15.0
2	3.25	0.281	0	16.6
3	5.25	0.265	0	17.0
4	7.25	0.265	0	--
5	10.25	0.273	0	16.6
6	24.25	2.63	-	--
7	25.75	3.27	622	--
8	27.0	3.74	1,117	5.0
9	28.5	4.06	1,531	0.0
10	30.0	4.14	2,433	--
11	31.75	4.14	3,566	--
12	33.75	4.10	3,601	--

*Initial glucose concentration 15 g/L; IPTG concentration 0.0919 g/L or 38.565×10^{-5} M; IPTG introduced at 24.25 hr.

are not very different. Now, if the model equations are solved using a hypothetical lactose concentration, and the same constants for transport as for the lactose case, the equivalent lactose threshold concentration which causes no induction can be calculated. In addition, the equivalent lactose concentration which generates the same profile as is observed in Table 3 can also be estimated.

If the transport constants are correctly obtained and if the model is valid, the equivalent threshold level of lactose should be about 1-5 mM. This is exactly the result predicted by the model. In addition, the equivalent lactose concentration for the induction profile shown in Table 3 is about 15 mM. This corresponds to an accumulation ratio of 40, corroborating the value (10 to 100) quoted earlier (14) and adding credibility to our transport model.

After obtaining verification for growth and substrate utilization models from separate experiments with growth on glucose and lactose, and having confirmed the importance of the transport model,

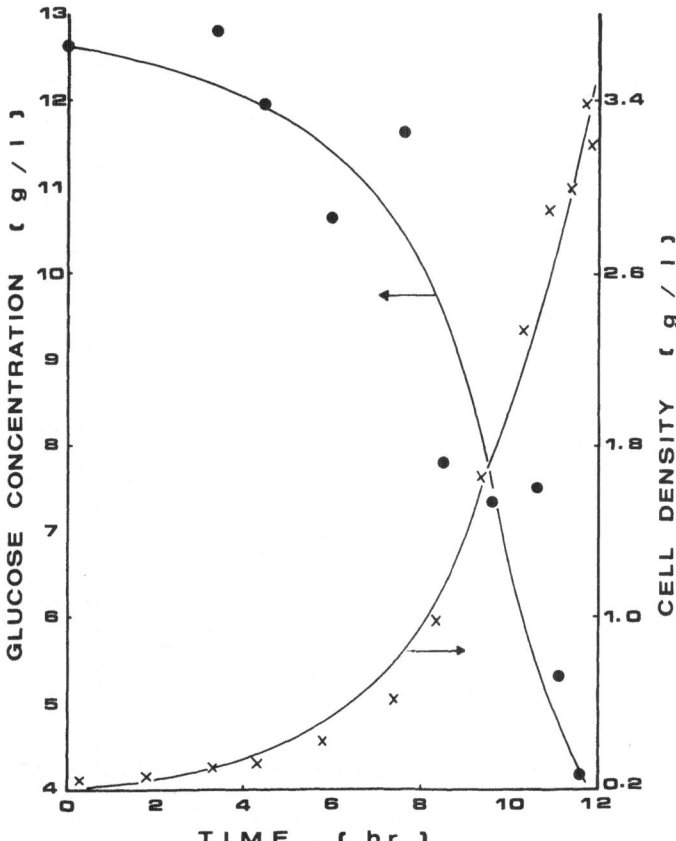

Fig. 4. Cell growth (*E. coli* PP01) and substrate depletion on
         mixed substrate (glucose + lactose). Solid lines are
         simulation curves which represent solution of system
         equations.

the entire set of model equations was tested by comparing model
predictions with experimental results on growth, depletion of substrates, and enzyme induction on mixed substrates. Monod-type
induction-repression, catabolite repression, and transport of inducer are operative, and hence, the model must accurately co-
ordinate all these control mechanisms and predict the time course
of the substrate, cell, and enzyme concentrations in the fermentation. Fig. 4 and 5 show the simulation curves and the experimental
data. The model results compare very closely with the experimental
data (correlation coefficient 0.98), attesting to the validity of
the model. Note that these simulations represent a successful
scale-up of the microliter-scale transport experiments of Kaczorowski *et al.* (5, 6) to a 5 L fermentor scale of operation. This

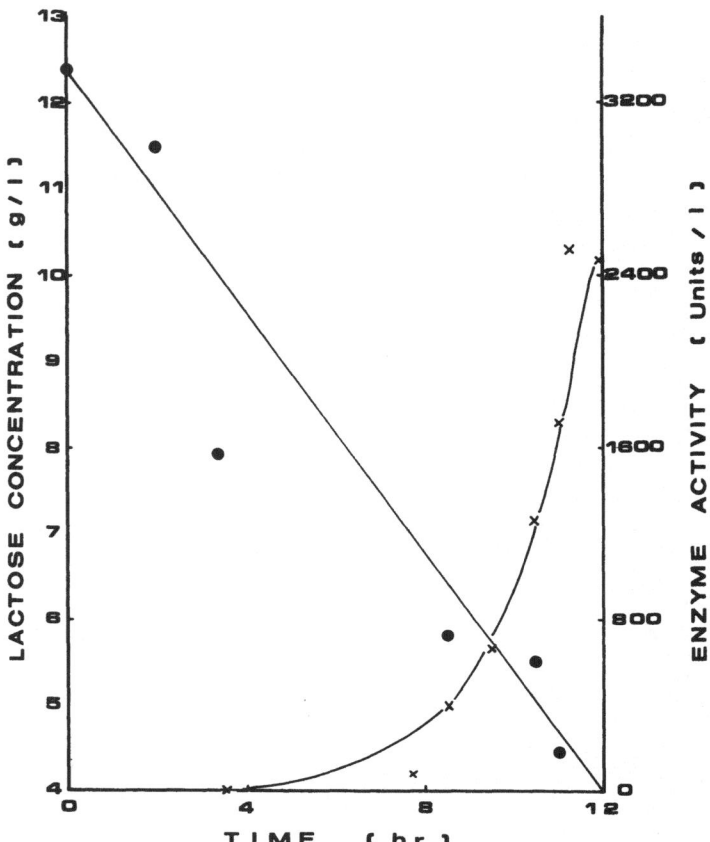

Fig. 5. Enzyme activity profile and lactose depletion during
fermentation on glucose/lactose mixture. The solid
lines are simulation curves representing solution of
system equations.

result cannot be achieved unless the inducer transport is included
in the model, as some elements of the set of parameters otherwise
take on negative values and cause the simulation to fail (Table
4).

    We conclude that the observed agreement of our unified model
with several types of experimental data obtained on several scales
of investigation, together with its predictive power for mixed substrates, constitutes strong evidence for its validity for a large
range of fermentation conditions. It follows, therefore, that for
adequate modelling of enzyme biosynthesis, a transport model for
inducer is an essential element.

TABLE 4

PARAMETER VALUES

(a)   Growth on glucose (subscript 2) or lactose (subscript 1)

μ_{m2} = 0.42 hr^{-1} Correlation Coefficient = 0.99

Y_2 = 0.156;

μ_{m1} = 0.496 hr^{-1} Correlation Coefficient = 0.99

Y_1 = 0.174;

(b)   Mixed substrate fermentation

μ_{m1} = 0.1582 hr^{-1}

K_{s1} = 14.09 g/L K_{s2} = 10.45 g/L

μ_{m2} = 0.357 hr^{-1} Y_2 = 0.292

(c)   Induction on lactose   Correlation Coefficient = 0.932

μ_{m1} = 0.389 hr^{-1} k_{-M} = 8.0   b = 0.7

Y_1 = 0.1739 b_3 = 50.0   A = 4.867

b_1 = 5.0 k_E = 1000.0   B = 25.09

b_2 = 5.0 k_M = 200.0

(d)   Induction on mixed substrate   Correlation Coefficient = 0.979

μ_{m1} = 0.1976 μ_{m2} = 0.199 K_5 = 0.984

K_{s1} = 10.22 K_{s2} = 10.73 K_7 = 1.00

Y_1 = 0.22 Y_2 = 0.22   B = 25.00

k_E = 1000.0 k_M = 200.0 b_3 = 50.0

b = 0.67 k_{-M} = 8.0

A = 2.00 b_1 = 4.99

ACKNOWLEDGMENTS

The authors acknowledge with gratitude the support of the National Science Foundation, Grant-in-Aid CPE 80-06978 and the assistance of S. Hirose of our laboratory.

NOMENCLATURE

A	transport modulus (moles/L/hr)
B	transport constant (moles/L)
C	cell concentration (g/L)
E	enzyme concentration (U/L)
E_o	adenylate cyclase concentration (total (U/L)
F	catabolite repression index (dimensionless)
K_1', K_2', K_3'	equilibrium constants
k'	rate constant (hr^{-1})
L_{in}, L_o	lactose concentration, intracellular and bulk levels (g/L)
M	mRNA concentration (mg RNA/g dry cells)
Q	fraction of repressor-free operators, induction index (dimensionless)
S	ATP concentration (g/L)
S_{1B}	inducer concentration in bulk (lactose) (g/L)
S_{2B}	bulk glucose concentration (g/L)
Y_1, Y_2	yield coefficients on substrates 1 and 2 (dimensionless)
b	empirical constant (hr^{-1})
b_1, b_2	constants in Eq. 7 (g/L)
b_3	empirical constant (dimensionless)
cAMP	cyclic AMP concentration (g/L)
k_e', k_d', k_2'	rate constants for excretion, degradation, and total depletion of cAMP (sec^{-1})
k_{+E}	rate constant for enzyme production (U/mg mRNA/hr
k_{+M}	rate constant for mRNA production (hr^{-1})
k_{-M}	rate constant for deactivation of mRNA (hr^{-1})
t	time (hr)
α	dimensionless cAMP concentration
$\mu_1, \mu_2, \mu, \mu_{m1} \mu_{m2}$	growth rates on substrates 1, 2, the total growth rate, and the maximum growth rates on substrates 1 and 2 (hr^{-1})

REFERENCES

1. MITCHELL, P. *Nature 191:* 144 (1961).
2. MITCHELL, P. *Biol. Rev. Cambridge Phil. Soc. 41:* 445 (1966).
3. MITCHELL, P. *J. Bioenerg. 4:* 63 (1973).
4. KAUSHIK, K. R., VIETH, W. R., & VENKATASUBRAMANIAN, K. *Abst. Div. Micro. Biochem. Technol. 178th Nat. Mt. Am. Chem. Soc., Washington* (1979).
5. KACZOROWSKI, G. J., ROBERTSON, D. E., GARCIA, M. L., PADAN, E., PATEL, L., LEBLANC, G., & KABACK, H. R. *Ann. N.Y. Acad. Sci. 358:* 307 (1980).
6. KACZOROWSKI, G. J., ROBERTSON, D. E., & KABACK, H. R. *Biochemistry 18:* 3697 (1979).
7. GONDO, S., VENKATASUBRAMANIAN, K., VIETH, W. R., & CONSTANTI-NIDES, A. *Biotechnol. Bioeng. 20:* 1797 (1978).
8. KAUSHIK, K. R., GONDO, S., & VENKATASUBRAMANIAN, K. *Ann. N.Y. Acad. Sci. 326:* 57 (1979).
9. PERLMAN, R. & PASTAN, I. *J. Biol. Chem. 243:* 5420 (1968).
10. ABOU-SABE, M. & MENTO, S. *Biochim. Biophys. Acta 385:* 294 (1975).
11. WRIGHT, L. F. & KNOWLES, C. J. *FEMS Microbiol. Lett. 1:* 259 (1977).
12. SCHLANDERER, G. *Proc. 2nd Intl Conf. Microbial Growth* 64 (1977).
13. KAUSHIK, K. R., thesis, Rutgers University, (1981).
14. KOTYK, A. & JANACEK, K., "Cell Membrane Transport," Plenum Press, New York.

MULTI ENZYME SYSTEMS IN MEMBRANE REACTORS

C. Wandrey, R. Wichmann and A.-S. Jandel

Institute of Biotechnology, Nuclear Research Center
Juelich, Federal Republic of Germany

Continuous operation is possible not only with carrier fixed enzymes but also with native (soluble) enzymes retained in reactors by means of an ultrafiltration membrane. This concept is especially useful for coenzyme depending multi enzyme systems, when the coenzyme also can be retained in the reactor. This is achieved by covalent coupling, e.g. NAD, to water soluble polyethylene glocol (PEG) of MW 10,000 to 40,000. Thus, L-leucine was produced from the corresponding α-keto acid at a productivity of 243 g/L/d for 24 days.

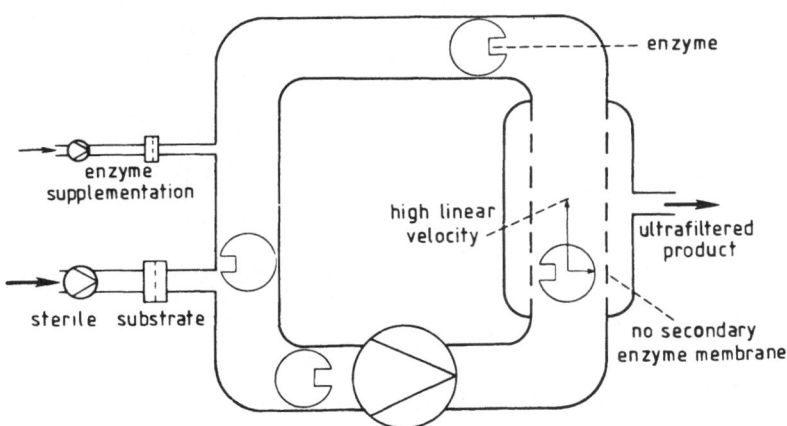

Fig. 1. Enzyme membrane reactor.

REACTOR CONCEPT

In a membrane reactor (1) continuous homogeneous catalysis
becomes possible as illustrated in Fig. 1. This is due to the
difference in size between the biocatalyst and the product. By
means of a recirculating pump the reaction mixture is conveyed
along a hollow fiber at high linear velocity. Secondary enzyme
membranes, due to concentration polarization, can be avoided in
spite of product flux across the membrane. Substrate is added at
constant flux across a sterile filter by means of a dosing pump.
Due to some catalyst inactivation and incomplete enzyme retention
respectively, some enzyme supplementation is necessary in order
to operate the system at constant productivity. The enzyme supplementation can be controlled via product analysis for constant
conversion. Such a system has a number of advantages: homogeneous
catalysis with no activity loss during enzyme fixation, no transport limitations, no contaminating chemicals, low investment per
unit activity, high activity per volume, constant productivity,
and ultrafiltered product. Enzyme membrane reactors are easy to
clean, sterilize, and control. Membrane cost is normally no limitation, but the stability of the enzyme(s) in solution has to be
reasonable. Up till now, as with carrier fixed enzymes, it has
been mainly hydrolases that have been used in membrane reactors.
The racemic resolution of amino acids by means of acylase was com-
mercialized in an enzyme membrane reactor in 1981 (2).

The real potential of membrane reactors becomes evident with
coenzyme depending systems. Since coenzymes, such as NAD (transport metabolites), are only effective if they can move between two
enzymes, continuous homogeneous catalysis in a membrane reactor
is a promising system if the coenzyme can be retained in the reactor together with the enzymes. This is achieved by covalent
coupling of the cofactor to a water soluble polymer, like PEG with
a MW of 10,000 to 40,000. Details for the preparation of the NAD-
derivatives are given by Bueckmann (3).

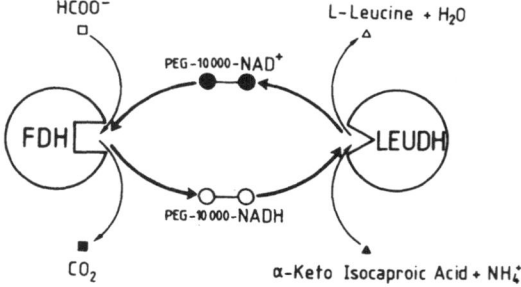

Fig. 2. L-amino acid bioreductive amination from the correspond-
Ing α-keto acid.

REACTION SYSTEM

The reductive amination of an α-keto acid to produce the corresponding L-amino acid was chosen as an example for a coenzyme depending system. The reaction is formulated for L-leucine in Fig. 2. α-Ketoisocaproic acid and other α-keto acids can be produced classically by the hydrolysis of the corresponding nitrils (4). L-leucine dehydrogenase (LEUDH) was purified from *Bacillus sphaericus*. Formate dehydrogenase (FDH) was isolated from *Candida boidinii* (5). The thermodynamic properties of the reactants favor the quantitative consumption of substrate, while the cheap co-substrate (formate) gives a non- disturbing by-product (CO_2). The kinetic properties of all substrates and products were studied in detail (6). The main results are that the coenzyme derivatives, obtained in yields up to 80% (7), show similar activity (and increased stability) in comparison to native NAD. A typical example is given in Fig. 3. The coenzyme derivatives show only very small decrease in activity and increase in K_m in comparison with the native coenzyme. The saturation concentration with respect to the coenzyme can easily be achieved.

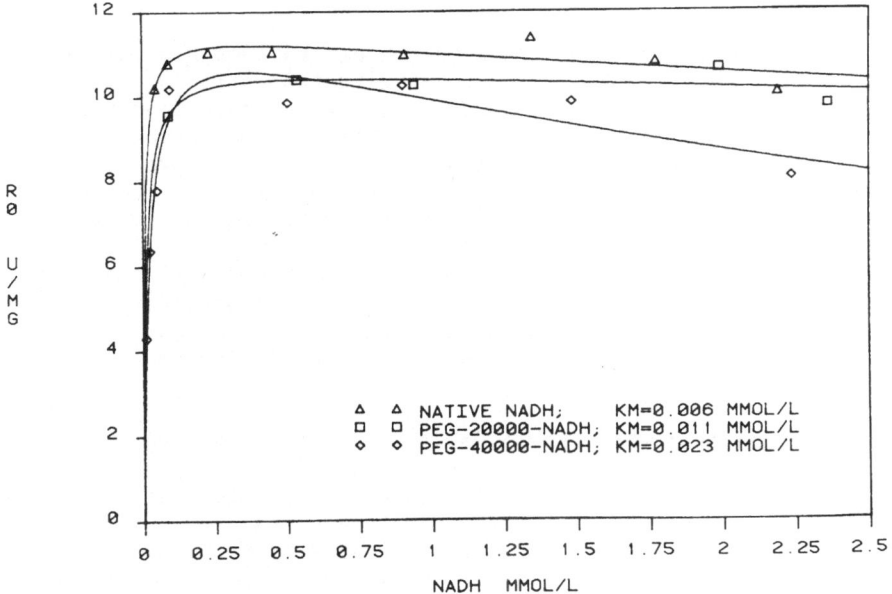

Fig. 3. L-leucine dehydrogenase activity, shown as initial rate (RO) as a function of coenzyme concentration at pH 8 and 25°; α-ketoisocaproic acid 100 mmol/L; NH_4 400 mmol/L.

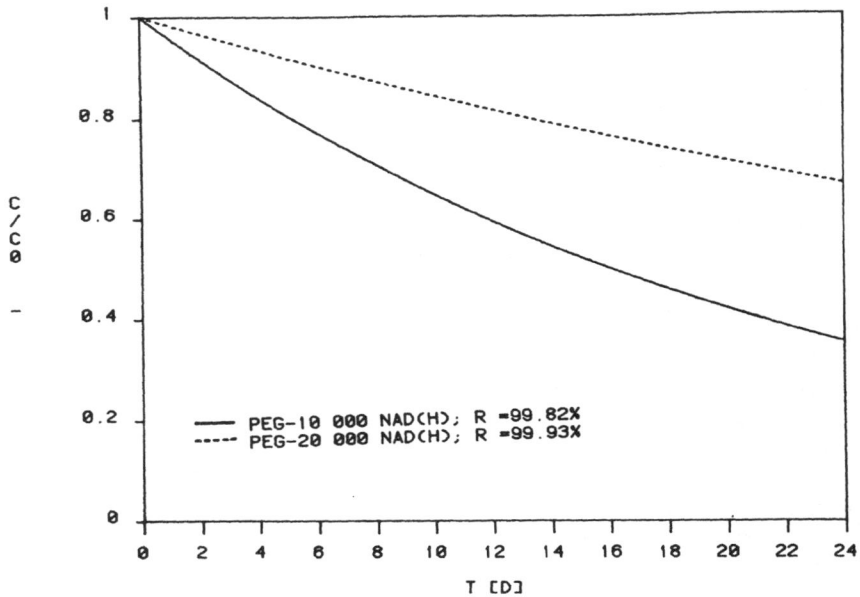

Fig. 4.   Coenzyme retention in a membrane reactor; Amicon YM5 mem-
          brane, residence time 1 hr. R is % retention for one
          residence time without enzyme present.

          After the activity of the coenzyme derivatives was demon-
strated, the retention by means of an ultrafiltration membrane was
checked. Incomplete retention of the coenzyme would have the same
effect on reactor performance as would chemical inactivation. The
retention was measured in a continuously operated membrane reactor
(without enzymes) at a residence time of 1 hr. As can be seen
from Fig. 4, a coenzyme derivative of PEG with a MW of 10,000
shows excellent retention. The loss per residence time is only
0.18%. This value can be decreased further to 0.07% with PEG of
20,000 MW. Over a period of 24 days the loss accumulates to quite
a significant value as can be seen from Fig. 4. In contradiction
to normal ultrafiltration processes the retention in membrane reactors has to be extremely high.

          To prove that the concept was applicable for a long period
with enzyme present, L-leucine was produced for 24 days in an
enzyme membrane reactor of 10 ml volume and an Amicon YM5 membrane.
The loss of enzyme and coenzyme activity was measured (8). After
9 and 19 days appropriate amounts of fresh enzyme and coenzyme
were added, so that the conversion could be increased again. Over
a period of 24 days the average conversion was 77%. The average
productivity was 243 g/L/d (Fig. 5). The enzyme and coenzyme stability are summarized in Table 1. The total inactivation was

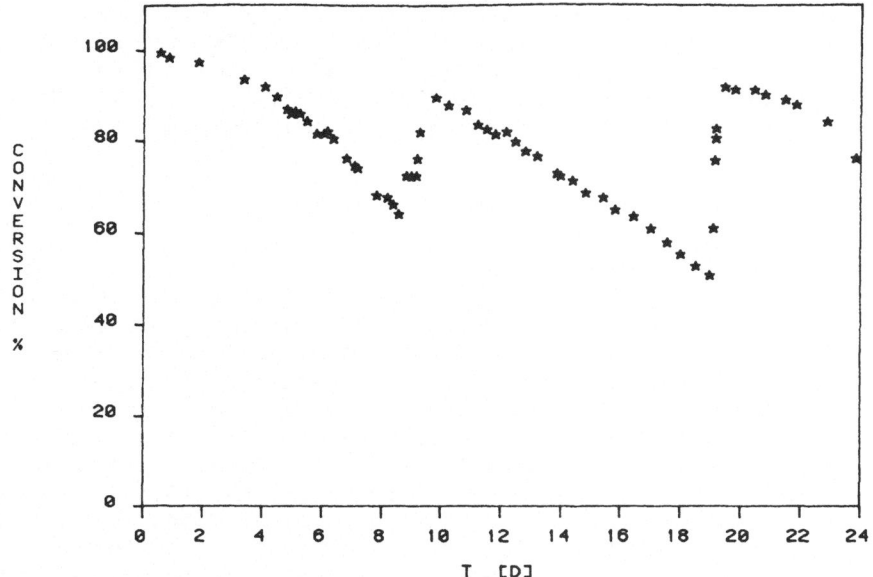

Fig. 5.   L-Leucine production in a membrane reactor with catalyst
          supplementation after 9 and 19 days. Conditions: resi-
          dence time 1 hr; pH 8; 25°; α-ketoisocaproic acid 100
          mmol/L; NH_4 – formate 400 mmol/L; LEUDH 5,000 U/L; FDH
          5,000 U/L; PEG-10,000-NAD(H) 0.55 mmol/L.

TABLE 1

STABILITY OF ENZYMES AND COENZYMES FOR L-LEUCINE FROM α-KETO
ISOCAPROATE*

Inactivation	LEUDH	FDH	Coenzyme
%/day	2.39	6.18	10.9**
U/kg product	1070	2080	278***

  *Continuous operation for 24 days.
 **4.3% loss due to incomplete retention and 6.6% to deactivation.
***278 mg NAD/kg product or 18,200 mols product/mol NAD used.

2.39 %/d for L-leucine dehydrogenase and 6.18 %/d for formate dehydrogenase. This gave an average conversion of 77% with an activity of about 1,000 U LEUDH/kg L-leucine and 2,000 U FDH/kg L-leucine. The coenzyme demand was 278 mg NAD/kg L-leucine. This could also be expressed as 18,200 moles product per mole NAD consumed.

CONCLUSION

The cost of coenzyme per unit weight of product is no longer the limiting economic factor. Nevertheless, the retention of coenzyme can be increased still further if larger MW coenzyme derivatives (PEG-40,000) are used (Fig. 3). These larger derivatives show approximately the same activity as smaller ones. The system was successfully scaled-up and tested in a 1,750 ml reactor at a production rate of about 0.4 kg L-leucine/day (9). A condition for successful operation is the availability of formate dehydrogenase and L-leucine dehydrogenase at competitive prices (10, 11). The reductive amination of α-keto acids can be carried out at very high conversions due to the favorable very small K_m values of the substrates. Furthermore, the entire substrate can be used without the need for intermediate racemisation, as in the racemic resolution of amino acids. The concept is not limited to the production of L-amino acids; α-hydroxy acids (12) can be produced from the corresponding α-keto acids. Many NADH depending reactions are applicable if the corresponding enzyme has sufficient stability in solution.

ACKNOWLEDGMENT

Cooperation with the Gesellschaft fuer Biotechnologische Forschung, Braunschweig-Stoeckheim, is gratefully acknowledged. A. F. Bueckmann synthesized the coenzyme derivatives. W. Hummel and H. Hustedt and co-workers isolated and purified the enzymes. We thank the Bundesministerium fuer Forschung und Technologie of the Federal Republic of Germany for generous financial support.

REFERENCES

1. MICHAELS, A. S. in "Progress in Separation and Purification," vol. 2, Wiley, New York (1968) p. 297.
2. DEGUSSA, AG, personal communication.
3. BUECKMANN, A. F. GermanPatent 28 41 414 (1979).
4. US Patent 410 8875 (1978).
5. KULA, M. R., *et al*., this volume.
6. WICHMANN, R., thesis Technische Universitaet Clausthal (1981).
7. BUECKMANN, A. F., KULA, M.-R., WICHMANN, R. & WANDREY, C. *J. Appl. Biochem*., in press.

8. WICHMANN, R., *et al.*, this volume.

9. LEUCHTENBERGER, W., BUECKMANN, A. F., KULA, M.-R., WICHMANN,
    R. & WANDREY, C. *Abst. 2nd Euro. Cong. Biotechnol.*,
    Eastbourne, England, (1981).

10. HUMMEL, W., SCHUETTE, H. & KULA, M.-R. *Eur. J. Appl.
    Microbiol. Biotechnol 12:* 22 (1981).

11. KRONER, K. H., SCHUETTE, H., STACH, W. & KULA, M.-R *J. Chem.
    Technol. Biotechnol.*, in press.

12. WANDREY, C., WICHMANN, R., LEUCHTENBERGER, W., KULA, M.-R.
    & BUECKMANN, A. F. German Patent 29 30 087 (1979).

SCALE-UP OF PROTEIN PURIFICATION BY LIQUID-LIQUID EXTRACTION

M.-R. Kula, K. H. Kroner, H. Hustedt and H. Schutte

Gesellschaft fur Biotechnologische Forschung mbH.
Braunschweig-Stockheim, Federal Republic of Germany

New demands and renewed interests in large scale protein isolation processes have arisen from recent advances in enzyme technology, e.g. enzymic transformation with cofactor regeneration (1)
or new developments in the production of biologically active proteins, such as interferon or proteo-hormons by recombinant-DNA
technology (2).

The following operations are involved in the isolation and
purification of intracellular proteins:

    a. cell desintegration
    b. clarification of the extract
    c. removal of interfering substances
    d. extensive purification if needed
    e. concentration and preservation

In this paper attention is focused on the "clarification of
crude extracts" and "removal of interfering substances." On a laboratory scale biochemical separation techniques are highly advanced.
However, scale-up of most techniques has not yet been extensively
studied and appears in many cases, e.g. chromatography or centrifugation, to be severely limited by the mechanical properties of
resins or materials. Therefore, we are very much interested in
alternative technology with the potential for continuous processing and capable of relatively simple scale-up to production size.
In this respect we investigated the extraction and purification of
proteins by partition in aqueous two-phase systems (3, 4). Partition in aqueous two-phase systems is a gentle technique. Proteins
remain in solution at all times; denaturation at the liquid inter-

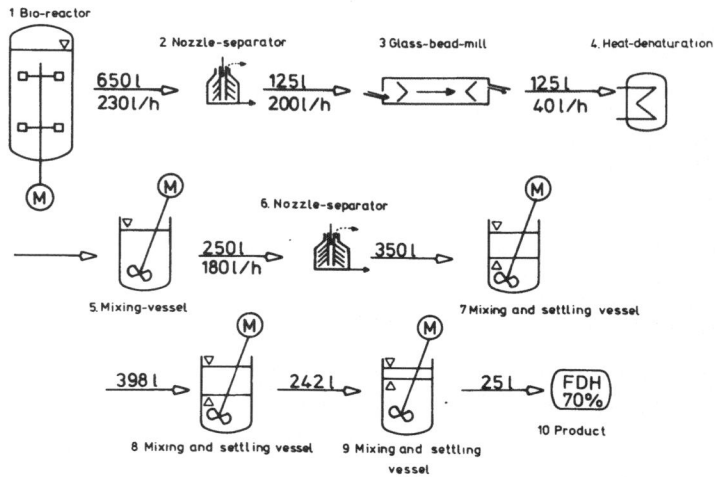

Fig. 1.   Flow sheet for the isolation and purification of formate
          dehydrogenase from *Candida boidinii* using liquid-liquid
          extraction.   Explanation of numbers: 1)   cultivation of
          Candida boidinii 2) harvesting of cells; 3) disintegra-
          tion of cells; 4) heat denaturation; 5) mixing of the 1st
          phase system including cell debris; 6) separation pro-
          cedure; 7) mixing and settling of the 2nd phase system
          (removal of nucleic acids, polysaccharides and some pro-
          teins); 8) mixing and settling of the 3rd phase system
          (enrichment of formate dehydrogenase in the bottom phase,
          reduction of the amount of contaminating proteins); 9)
          mixing and settling of the 4th phase system (concentra-
          tion of the enzyme in the top phase); 10) enzyme product,
          technical grade ($\geq$70% purity, <70% recovery of the ini-
          tial amount), stable for several months at 4°C.

phase is negligible due to the very low interfacial tension of
such systems.

     After the initial experiments, which demonstrated that the
basic concept of extracting proteins from cell debris in aqueous
phase systems could be verified (3), we studied scale-up and phase
separation using commercially available equipment (4, 6-9). The
general strategy for process development is outlined below, using
formate dehydrogenase isolation as an example (see Fig. 1).

     As described by Wandrey et al. (10) formate dehydrogenase is
an important enzyme for cofactor regenerating processes. For pilot
plant experiments 100 g quantities of the enzyme were needed; it
could not be bought or prepared at reasonable cost using conven-

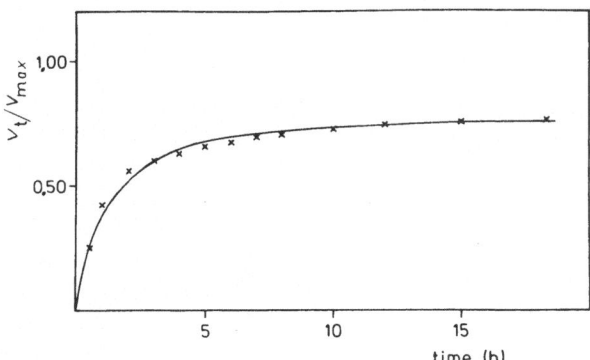

Fig. 2. Separation of formate dehydrogenase from *Candida boidinii*
         cell homogenate in a polyethylene glycol (PEG)/salt sys-
         tem under gravity. The relative volume of the enzyme
         containing top phase is plotted versus time. V_{max} =
         maximum volume of the top phase measured by centrifuga-
         tion; V_t = settled volume at time (t). System conditions
         are the same as in Fig. 3; settling tank is a 150 L
         glass vessel of 2.5 hight to diameter.

tional procedures. Formate dehydrogenase was produced by cultivation of the yeast *Candida boidinii* on methanol (11) as the carbon
and energy source, and represented a major portion of the soluble
protein. Cells were harvested in a nozzle separator and immediate-
ly disintegrated by treatment in a glass bead mill. The extraction
process is outlined in Fig. 1 and discussed in detail in Ref. 12.
In the first extraction step removal of cell debris is accomplished
in a polyethylene glycol/salt system. Here cell debris remains suspended in the bottom phase. Separation of the phases can be carried out either by gravity, as demonstrated in Fig. 2, or by centrifugation. As shown in Fig. 3 a separation efficiency of 0.98
can be obtained using a nozzle separator with feed rates in the
range of 3 L/min. This corresponds to a residence time of approximately 16 sec. The final yields of enzyme extracted in the top
phase are very high and come close to the yield calculated from
the partition coefficient of the enzyme and the volume ratio.
The separation efficiency is defined as (11):

$$S_e = P_T - \frac{1-P_B}{1+V'_f}$$

where P_T and P_B are the purity factors of over- and under-flow
respectively (%/100) and V'_f is the ratio of over- and under-flow.

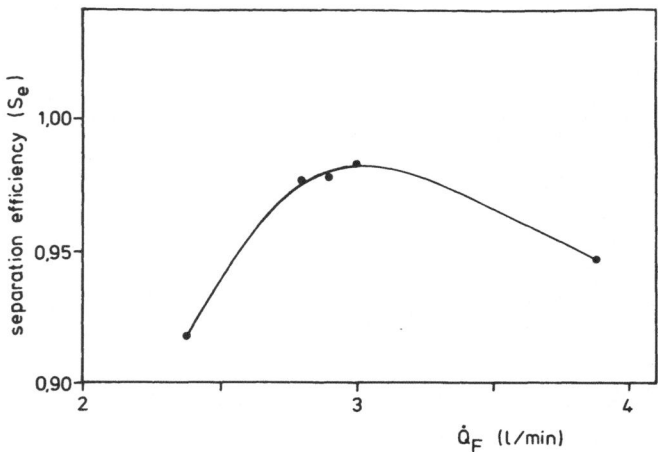

Fig. 3.   Performance of a nozzle separator for the separation of
          formate dehydrogenase from *Candida boidinii* cell homo-
          genate using PEG/salt systems. The separation efficiency
          is plotted versus throughput of the dispersion in a α-
          Laval YEB 1334 nozzle separator. System conditions: 18%
          PEG 400; 7% PEG 1550; 8% potassium phosphate, pH 7.8; 20%
          cell homogenate. Operational conditions: 3 nozzles of
          0.5 mm bore diameter.

          In contrast separation under gravity with cell debris con-
taining systems is a relatively slow process and takes about twenty
hours as seen from Fig. 2. The enzyme yield in the top phase is
lower than calculated which results from the incomplete separation
under gravity during the operation period. Apparently, some small
droplets of the top phase get entrapped in the viscous bottom
phase and are only separated efficiently by using additional g-
forces. That the enzyme is not degraded or inactivated by longer
process times in the phase system is evidenced by the fact that
very good recovery is obtained by centrifugation after 17 hours
storage. Once the cell debris is removed the viscosity of the
phase system, especially of the bottom phase, becomes much lower.
In this situation separation under gravity occurs quite fast and
is very efficient. It can be accomplished within 60-90 min. in
a settling tank (9, 12). Out of the four extraction steps during
isolation and purification of formate dehydrogenase three steps
are operated under gravity; this reduces the energy consumption
of the separation process considerably. All extractions are performed at room temperature. The overall yield of 70% is quite
high; and all steps could be calculated and predicted with high
precision from laboratory data.

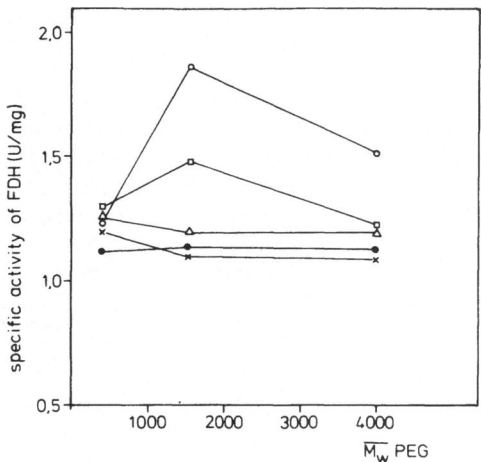

Fig. 4. Dependence of the selectivity of extraction of formate
         dehydrogenase in PEG/salt systems on the average molecu-
         lar weight of PEG and the concentration of potassium
         chloride. The specific activity of the formate dehydro-
         genase extracted is plotted versus the molecular weight
         of PEG, for different potassium chloride concentrations.
         System conditions: 15% potassium phosphate, pH 7.8; the
         polyethylene glycol concentration varied between 10 and
         25% and was selected to yield the same length of the tie-
         line in each system. Potassium chloride concentration:
         (●) none, (x) 0.5 M, (△) 1.0 M, (□) 2.0 M, (o) 2.5 M.
         The partition coefficient of formate dehydrogenase in
         PEG 400 containing systems is >1; in PEG 1550 and PEG
         4000 systems it is <1.

     Currently 50 kg batches of *Candida boidinii*, containing
approximately 150 g of enzyme protein, are extracted in single experiments. The recovered enzyme has a specific activity of about
2 U/mg and a purity of approximately 70%. This rather high purity
is needed because the reaction velocity of the pure enzyme is relatively slow and because too high a protein concentration should
be avoided when the enzyme is used in a membrane reactor. The
selectivity of extraction of formate dehydrogenase in PEG/salt
systems as a function of the average molecular weight of the polyethylene glycol and the kind and concentration of ions employed
is shown in Fig. 4. The modulation of the partition coefficients
in PEG/phosphate systems by potassium chloride is exploited in the
second and third extraction step of the process when large portions
of contaminating proteins as well as nucleic acids and polysaccharides are removed from the enzyme. The discrimination between en-

zyme protein and contaminating proteins is best for polyethylene glycol with an average molecular weight of 1500 daltons.

It should be noted that the polyethylene glycol added to the cell homogenate for the first extraction is reused in the following steps. A first approximation of the cost of the process indicates that the enzyme can be prepared considerably cheaper than using conventional techniques (12). The reduction in cost results from savings in manpower and energy. Further scale-up of the process appears possible. The potential for continuous processing is clearly evident. It can be expected that intracellular proteins can be produced on an industrial scale utilizing liquid-liquid separation.

REFERENCES

1. WICHMANN, R., WANDREY, C., BUCKMANN, A. F. & KULA, M.-R. *Biotechnol. Bioeng.*, in press.
2. WEISSMANN, C. *et al.*, this volume.
3. KULA, M.-R., KRONER, K. H., HUSTEDT, H., GRANDA, S. & STACH, W., German Patent 26 39 129, US Patent 4,144,130.
4. KULA, M.-R., KRONER, K. H. & HUSTEDT, H., in "Advances in Biochemical Engineering" (A. Fiechter, ed.), Springer Verlag, in press.
5. ALBERTSSON, P. A. "Partition of Cell Particles and Macromolecules," 2nd Ed., Wiley, New York (1971).
6. KRONER, K. H., HUSTEDT, H., GRANDA, S. & KULA, M.-R. *Biotechnol. Bioeng. 20:* 1967 (1978).
7. HUSTEDT, H., KRONER, K. H., STACH, W. & KULA, M.-R *Biotechnol. Bioeng. 20:* 1988 (1978).
8. HUSTEDT, H., KRONER, K. H., MENGE, U. & KULA, M.-R. in "Enzyme Engineering" vol. 5 (H. Weetall and G. Royer, eds.), Plenum Press, New York (1980) p. 45.
9. KULA, M.-R., KRONER, K. H., HUSTEDT, H. & SCHUTTE, H. *Ann. NY Acad. Sci. 369:* 341 (1981)
10. WANDREY, D., WICHMANN, R. & JANDEL, A. S., this volume.
11. SCHUTTE, H., FLOSSDORF, J., SAHM, H. & KULA, M.-R. *Eur. J. Biochem. 62:* 151 (1976).
12. KRONER, K. H., SCHUTTE, H., STACH, W. & KULA, M.-R. *J. Chem. Technol. Biotechnol.*, in press.

RAPID AND LARGE-SCALE PURIFICATION OF ANGIOTENSIN-I FORMING

ENZYMES AND MILK CLOTTING ENZYMES BY AFFINITY CHROMATOGRAPHY

K. Murakami, S. Hirose and H. Kobayashi

Institute of Applied Biochemistry, The University of
Tsukuba, Ibaraki-ken, Japan

The purpose of this paper is to describe affinity columns for
the simple isolation of valuable acid (carboxyl) proteinases which
have never before been isolated or which have required many complicated purification steps. Various affinity columns, including N-
acylated pentapeptide from *Actinomycetes* (pepstatin A) or its
derivatives (1), were used depending on their affinity for each
of the carboxyl proteinases.

An angiotensin-I forming enzyme (renin), which plays a key
role in blood pressure regulation, was isolated from hog (2) and
human (3) kidney or bovine anterior pituitary with a million fold
purification. Milk-clotting enzymes from calf (4) and microorganism (5) commercial preparations were pruified to homogeneity in
a one step purification on the affinity column. The column of
pepstatin A (isovaleryl pepstatin) or its derivatives was prepared
by the method of Murakami and Inagami (6). The affinity column
was indispensable for the rapid and large-scale isolation of renin
and milk-clotting enzymes. Pure brain renin, 0.45 mg, was isolated from 50 kg of bovine anterior pituitary at a 1.7 million fold
purification after 4 chromatographic steps, including an affinity
column of pepstatin A-aminohexyl-agarose (6). More than 10,000
fold purification was achieved by one pass through the affinity
column. Pepstatin A-aminohexyl-agarose was also used for the purification of a milk-clotting enzyme from calves (chymosin). The
affinity column, with 3 ml of wet gel, made it possible to isolate
90 mg of pure chymosin from 20 g of rennet tablet from Chr. Hansen
Laboratories (4). However, another microbial milk-clotting enzyme
from Mucor miehei was strongly adsorbed by the pepstatin A-
aminohexyl-agarose column and was not eluted in a stable preparation. This may have been due to a very high affinity of the micro-

bial enzyme for pepstatin A.

To conquer this difficulty two affinity columns, including N-acetylpepstatin (*streptomyces* pepsin inhibitor, SPI) or N-isobutyrylpepstatin (pepsinostreptin), were prepared and found to be effective for the rapid purification of microbial milk-clotting enzymes (rennets) from *Mucor miehei* and *Endothia parasitica* (5). These two enzymes were purified to a homogeneous state from commercial preparations by only one pass through the affinity columns. *M. miehei* rennet was purified about 12.5 fold over the crude enzyme with a yield of more than 90% by use of N-acetylpepstatin affinity gel. One ml of this gel made it possible to isolate 20 mg of pure enzyme from 3 ml of crude rennet solution. *E. parasitica* rennet was purified about 11.0 fold over the crude enzyme with a yield of almost 100% by use of *N*-isobutyrylpepstatin affinity gel. One ml of this gel could isolate more than 20 mg of pure enzyme from 1.6 ml of the crude rennet solution.

We are greatly indebted to S. Murao of Osaka Pref. University for the supply of SPI and to A. Kakinuma for the supply of pepsino-streptin.

REFERENCES

1. UMEZAWA, H. & AOYAGI, T. in "Proteinases in Mammalians Cells and Tissues," North-Holland Biomedical Press (1977) p. 638.
2. INAGAMI, T. & MURAKAMI, K. *J. Biol. Chem. 252:* 2978 (1977).
3. YOKOSAWA, H., HOLLADAY, L. A., INAGAMI, T., HAAS, E. & MURAKAMI, K. *J. Biol. Chem. 255:* 3498 (1980).
4. KOBAYASHI, H. & MURAKAMI, K. *Agric. Biol. Chem. 42:* 2227 (1978).
5. KOBAYASHI, H. & MURAKAMI, K. *Anal. Biochem.,* in press.
6. MURAKAMI, K. & INAGAMI, T. *Biochem. Biophys. Res. Commun. 62:* 757 (1975).

SEPARATION AND PURIFICATION OF ENZYMES VIA CONTINUOUS

PH-PARAMETRIC PUMPING

S. Y. Huang, C. K. Lin and L. Y. Juang

Department of Chemical Engineering, National
Taiwan University, Taipei, Taiwan, 107, China

Parametric pumping is a separation process that proceeds by periodic variation of an intensive variable, such as temperature, pH, electric field, etc. A fluid, consisting of a mixture having different levels of a selected variable, is pumped through a packed bed in which the adsorption and desorption occur in a certain period. Reciprocating pumping of the fluid through an adsorbent results in the separation of the component of interest from the others. In 1975, Shaffer et al. (1) employed batchwise pH-parametric pumping for separating an enzyme system using Sepharose-CHOM as an adsorbent for affinity chromatography. In 1979, Chen et al. (2) investigated the continuous fractionation of protein mixtures by pH-parametric pumping, which promoted the column productivity and reduced the eluant quantity. The potential of employing continuous pH-parametric pumping coupled with affinity chromatography for enzyme separation was demonstrated by these workers. In this work the factors affecting the separation efficiency, e.g., flow rate, pH, nature of feed liquids, and enzyme concentration in the feed, were studied.

Trypsin and chymotrypsin were from porcine pancreas. Chitin was prepared from crab shell (3). Native inhibitors, CHOM (chicken ovomucoid) and DKOM (duck ovomucoid), were bound to the chitin with glutaraldehyde to form an affinity adsorbent for trypsin and chymotrypsin, respectively. These ligands were prepared by a slightly modified acetone method (4). ε-Amino-n-caproyl-D-tryptophan methyl ester (ACTME) (5) was bound to chitosan to form an adsorbent for bromelain.

The apparatus for pH-parametric pumping was similar to that of Chen et al. (2), with columns of 1.3 cm I.D. by 16 cm tall and

2.6 cm I.D. by 40 cm tall. The pH of the top and bottom feeds
were 8.0 and 2.5, 8.0 and 2.0, and 8.7 and 3.0 for trypsin, chymotrypsin, and bromelain respectively. The superficial velocity of
the feed ranged from 0.56 to 5.31 cm/sec. The parametric pumping
was carried out as follows: a) a high pH feed was passed through
the column to give a bottom product; during this step the desired
enzyme was adsorbed on the adsorbent; b) a low pH buffer was
charged from the top of the column to release the desired enzyme
from the adsorbent; c) the low pH feed was pushed back up through
the column to give a top product which carried the concentrated
enzymes; enzyme remaining on the adsorbent was released further;
and d) a high pH buffer from the bottom reservoir was pumped into
the column, the enzyme contained in the liquid was adsorbed onto
the adsorbent; and the displaced liquid was received by the top
reservoir. The operation was continued automatically from cycle
to cycle.

The adsorption equilibria between trypsin and chitin-CHOM and
between chymotrypsin and chitin-DKOM were determined at different
pH levels. In both cases the driving forces between high pH and
low pH were fairly large indicating that the chitin-CHOM and
chitin-DKOM were suited for fractionation of the enzymes. For
trypsin pH levels of 8.3 and 2.0 showed the highest separation
factor (4, 5). A step change of flow rate from 4 mL/min to 2.5
mL/min did not increase the enzyme concentration of the top
product because axial dispersion of fluid resulted in back mixing
of the enzyme.

REFERENCES

1. SHAFFEN, A. G. & HAMRIN, C. E. *Am. Inst. Chem. Eng. J. 21:*
    782 (1975).
2. CHEN, H. T., WONG, Y. W. & WU, S. *Am. Inst. Chem. Eng. J.*
    25: 320 (1979).
3. STANLEY, W. L., WATTERS, G. G., KELLY, S. H. & OLSON, A. C.
    Biotechnol. Bioeng. 20: 135 (1978).
4. LINEWEAVER, H. & MURRAY, C. W. *J. Biol. Chem. 171:* 565 (1974).
5. BOBB, D. *Prep. Biochem. 2(4):* 347 (1972).

Session II
APPLICATIONS OF BIOCATALYSTS FOR
NEW REACTIONS AND ORGANIC SYNTHESIS
Chairmen: A. L. Laskin and I. Chibata

ENZYMATIC SYNTHESIS OF PENICILLINS AND CEPHALOSPORINS BY

PENICILLIN ACYLASE

R. Okachi, Y. Hashimoto, M. Kawamori, R. Katsumata,
K. Takayama and T. Nara

Tokyo Research Laboratory, Kyowa Hakko Kogyo Co., Ltd.
Tokyo, Japan

The reverse reaction of penicillin acylase was utilized to synthesize clinically useful penicillins or cephalosporins.

Kluyvera citrophila has been selected as one of the strains for producing potent intracellular penicillin acylase activity with broad substrate specificity. *Pseudomonas melanogenum* produces a different type of penicillin acylase which is specific for α-amino acyl side chain precursors. Investigations of the conditions for cultivation of the bacteria and also of the enzymatic reactions revealed the optimal conditions suitable for the enzymatic synthesis of penicillins and cephalosporins by intact cells of these microorganisms. β-Lactamase, co-produced by the microorganisms under certain conditions, was deleted by alkaline treatment as well as by mutational techniques.

Penicillin acylase, benzylpenicillin amidohydrolase E.C. 3.5.1.11 first reported in *Penicillium chrysogenum* and *Aspergillus oryzae* (1), catalyzes the hydrolytic deacylation of natural penicillins to yield 6-aminopenicillanic acid (6-APA) and acyl side chain compounds. Penicillin acylase also catalyzes the reverse reaction synthesizing natural penicillins (2) and some semi-synthetic penicillins (3,4) from 6-APA and appropriate acyl side chain precursors under certain conditions. However, since the synthetic activity of penicillin acylase is weak, compared to the hydrolytic activity, the practical application of penicillin acylase for the synthesis of penicillins or cephalosporins has not been reported so far.

In light of this background, the selection of microorganisms capable of producing potent enough penicillin acylase to

81

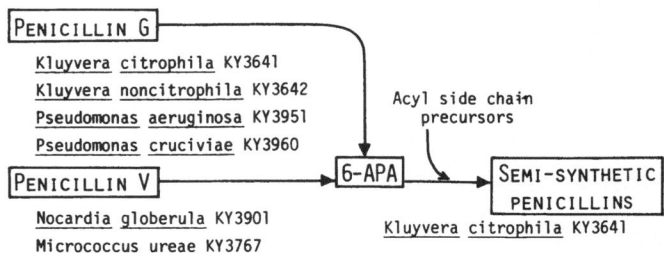

Fig. 1. Screening of microorganisms possessing penicillin acylase activity.

TABLE 1

OPTIMAL CONDITIONS FOR SYNTHESIS OF β-LACTAM ANTIBIOTICS
BY *K. CITROPHILA*

6-APA or 7-ADCA* (mg/ml)	Acyl Side Chain Precursor (mg/ml)	Cell (mg/ml)	Reaction**	Products (mg/ml)
6-APA 10	D-phenylglycine methylester HCl 25	30	35°C, 4hr; pH 6.5	ampicillin 11.2
6-APA 20	D-p-hydroxy-phenylglycine methylester-HCl 20	30	30°C, 2 hr, pH 6.0	amoxicillin 12.0
7-ADCA 15	D-phenylglycine methylester HCl 20	25	35°C, 4 hr, pH 6.5	cephalexin 11.7
6-APA 5	phenylacetic acid 10	20	35°C, 2 hr, pH 6.5	penicillin G 5.5

 *7-ADCA is 7-amino deacetoxy cephalosporanic acid;
**in M/30 phosphate buffer.

synthesize some useful β-lactam antibiotics has been started (5).
Some bacteria and a few actinomycetes were selected from our

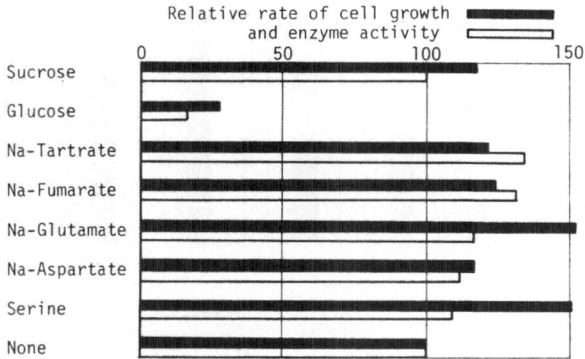

Fig. 2. Effects of additives at 5g/dl on cell growth and enzyme
         activity of *K. citrophila*. Basal medium: polypeptone
         1%, yeast extract 1%, and NaCl 0.25%.

screening (6) for acylase-producing strains which effectively removed acyl side chains from penicillin G or penicillin V (Fig. 1).
With these microorganisms, the ability of ampicillin production
by the reverse reaction of penicillin acylase was investigated.
Kluyvera citrophila was found to be a promising strain for the
synthetic reactions.

The optimal conditions for enzymatic synthesis of penicillins
and cephalosporins with the reverse reaction of penicillin acylase
was further evaluated (7). As shown in Table 1, 6-APA or 7-ADCA
was used as a β-lactam nucleus; and the intact cells of *K.*

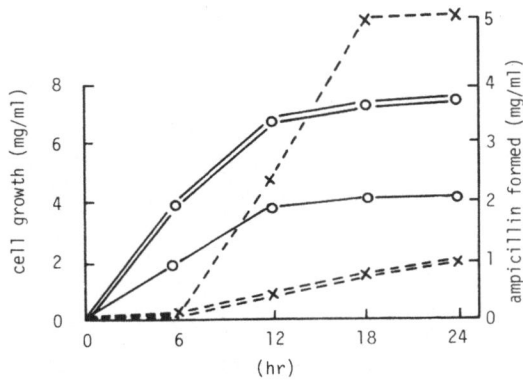

Fig. 3. Effects of pH control on cell growth (o) and ampicillin
         formation (x). Not controlled indicated by single line
         and controlled by double line.

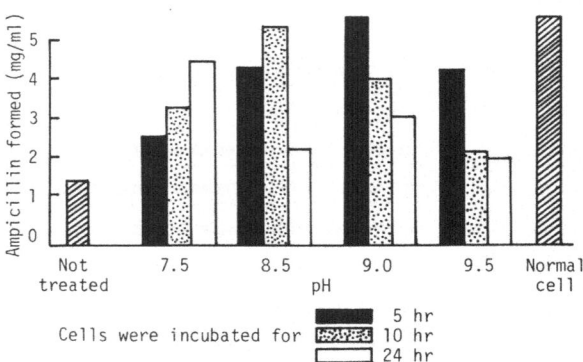

Fig. 4. Inactivation of penicillinase by alkaline treatment at
         40°C.

citrophila could attack the respective acyl side chain compound
in phosphate buffer at weakly acidic pH yielding penicillins or
cephalosporins (5,8). The penicillin acylase of *K. citrophila*
exhibited broad substrate specificity towards acyl side chain precursors in both hydrolysing and synthesizing reactions.

     Some problems involved in the enzymatic synthesis by *K.
citrophila* were investigated further (9). It was revealed that
the cell growth was stimulated more than 50% by addition of sodium
glutamate or serine to the fermentation medium. On the other
hand, the enzyme activity per unit cell was stimulated by addition
of some organic acid salts, such as sodium tartrate or sodium
fumarate (Fig. 2). But under these conditions, the pH of the
culture broth rose to over 9.5 by the later phase of the fermentation and prohibited further growth. Therefore, continuous control
of the pH of the culture broth with organic acid was examined.

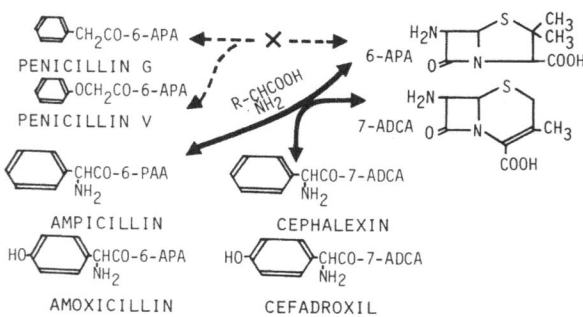

Fig. 5. Penicillin acylase specific to α-amino acyl side chain
         compounds; enzyme found in *Pseudomonas melanogenum*,
         maltophilia, *ovalis*, and *geniculosa* and in *Xantho-
         monas physalidicola* and *cucurbitae*.

TABLE 2

OPTIMAL CONDITIONS FOR SYNTHESIS OF α-AMINO β-LACTAM
ANTIBIOTICS BY *P. MELANOGENUM*

6-APA or 7-ADCA (mg/ml)	Acyl Side Chain Precursor (mg/ml)	Cell (mg/ml)	Reaction*	Products (mg/ml)
6-APA 10	D-phenylglycine methylester HCl 25	10	35°C, 4 hr, pH 5.5	ampicillin 10.5
7-ADCA 10	D-p-hydroxy-phenylglycine methylester HCl 20	20	35°C, 4 hr, pH 6.0	cephadroxil 10.4
6-APA 5	phenylacetic acid 10	10	35°C, 2 hr, pH 5.5	--

*in M/30 phosphate buffer.

When the culture broth was continuously controlled at pH 7.5 with
1 M tartaric acid, the cell growth was stimulated more than 50%
as shown in Fig. 3. But, the cells cultivated with pH control
lost the ability to accumulate ampicillin. Also, irrespective of
the pH control, the young cells which were cultivated less than
6 hr could not accumulate any ampicillin in the reaction mixture.

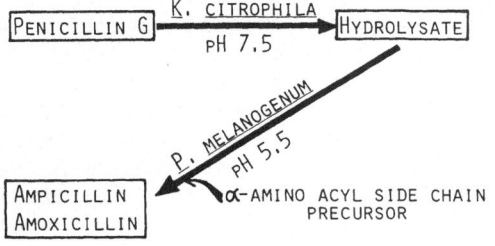

Fig. 6. Two step enzymatic synthesis of α-amino penicillins.

TABLE 3

β-LACTAMASE ACTIVITY OF AMPICILLIN ACYLASE
PRODUCING BACTERIA

Strain	β-Lactamase*
Pseudomonas melanogenum KY3987	3.45
Pseudomonas maltophilia KY4676	2.10
Pseudomonas geniculosa KY4678	1.50
Xanthomonas cucurbitae KY4217	1.85
Xanthomonas physalidicola KY4220	1.10

*Degradation of amoxicillin (mg/ml, 30 min).

The reaction mixtures were chromatographed on TLC. The results showed that penicilloic acid was the main product under such
conditions. Thus, it was concluded that the penicillinase (or β-
lactamase) in the bacterial cells was very active under such conditions and caused the degradation of the synthesized ampicillin.
The inactivation of the intracellular β-lactamase activity was
attempted. The nutrient broth, containing the microbial cells
carrying the penicillinase, was adjusted stepwise to alkaline pH
with aqueous ammonia and then incubated at 40°C for the times
shown in Fig. 4. The cells treated at pH 9.0 for 5 hr accumulated
the same amount of ampicillin as did normal cell; and the treatment seemed to be effective in inactivating specifically the
intracellular β-lactamase. Under these conditions the penicillin
acylase was not affected.

Further screening of microorganisms capable of synthesizing
ampicillin revealed that certain bacteria possessed penicillin
acylase with limited substrate specificity (10). As mentioned
previously, the penicillin acylase of K. citrophila exhibited
broad substrate specificity; but the enzymes found in some strains
of Pseudomonas or Xanthomonas exhibited affinity only on penicillins and cephalosporins which carry α-amino moiety in their acyl
side chains, as shown in Fig. 5. This enzyme is called ampicillin
acylase (11) because the affinity towards ampicillin was first reported by us. In 1972 similar enzymes were reported by T.
Takahashi in Xanthomonas citri and some other bacteria and
designated as α-amino acid ester hydrolase (12).

TABLE 4

CHARACTERISTICS OF β-LACTAMASE DEFICIENT MUTANTS DERIVED FROM *P. MELANOGENUM*

Strain	β-Lactamase* (%)	Minimal Inhibitory Concentration**		
		Amoxicillin	Carbenicillin	Cephaloridin
Parent KY3987	100	200	1.6	400
Mutant KY8539	45	100	25	400
Mutant KY8540	0	25	<0.1	25
Mutant KY8541	0	12.5	<0.1	100
Mutant T-432	42	0.8	0.1	12.5
Mutant T-417	59	0.2	<0.1	3.1

*Amoxicillin degradation activity.
**Assayed by agar dilution assay (γ/ml)

TABLE 5

INDUCTION OF β-LACTAMASE IN *P. MELANOGENUM*

Strain	β-Lactamase (%)*	
	Without Inducer	With Inducer**
Mutant KY8539	38	100
Mutant KY8540	0	0
Mutant KY8541	0	0
Parent KY3987	85	100

*Relative activity of amoxicillin hydrolysis
**Induced by penicillin V

　　Pseudomonas melanogenum was selected for further evaluation.　Results showed that the strain attached the α-aminophenyl-
acetic acid of p-hydroxy-α-aminophenylacetic acid to β-lactam
nucleus and produced rather high levels of ampicillin and cepha-
droxil, respectively, at pH 5.5 to 6.0 at 35°C and within 4 hr
(Table 2).　The reaction required smaller concentrations of cells
in the reaction mixture than did *K. citrophila*.　This microorganism synthesized some other α-amino penicillins and cephalosporins; but the strain did not synthesize penicillin G or V from
6-APA and phenylacetic acid or phenoxyacetic acid, respectively.

　　Utilizing *K. citrophila* and *P. melanogenum*, the two step
enzymatic synthesis of α-amino penicillins from benzyl penicillin
was accomplished, as shown in Fig. 6.　Penicillin G hydrolysate,
which contains 6-APA and phenylacetic acid, can be used as a substrate for the synthetic reaction.　The hydrolysate was centrifuged to remove the *K. citrophila* cells.　Then, the *P. melanogenum* cells were added with a suitable amount of α-amino acyl
side chain precursor.　In this case, it was not necessary to isolate 6-APA from the hydrolysate, because the latter microorganism
did not utilize phenylacetic acid as substrate in the synthetic
reaction.

　　Subsequent work on the ampicillin acylase producing microorganisms revealed that those strains also produced intracellular
β-lactamase under certain conditions (Table 3).　The cells obtained under those conditions hydrolysed penicillins at relatively

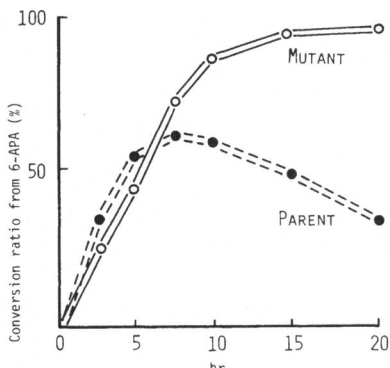

Fig. 7. Amoxicillin synthesis by β-lactamase deficient mutant

higher rates. Mutational techniques were employed to obtain β-
lactamase deficient mutants without affecting the penicillin
acylase activity of the parent strain. In Table 4, some of the
characteristics of β-lactamase deficient mutants derived from *P.
melanogenum* are summarized. Two mutants, KY 8540 and KY 8541,
completely lost the hydrolytic activity against amoxicillin; but
others still retained the β-lactamase activity to a certain degree. β-Lactamase deficient mutants appear to be more susceptible
to β-lactam antibiotics than do their parents with a few
exceptions. The induction of β-lactamase was not observed in two
strains; although in a certain mutant, as well as in the parent,
the β-lactamase was induced by penicillin V (Table 5). These results eliminate the possibility of production of β-lactamase during
the reactions when the former two mutants are employed in the enzymatic synthesis. The penicillinase deficient mutant KY 8540 obtained by the above experiments was tested in the enzymatic synthesis of amoxicillin. As shown in Fig. 7, a remarkable improvement was observed, particularly in the later phase of the synthesis reaction.

REFERENCES

1. SAKAGUCHI, K. & MURAO, S. *J. Agric. Chem. Soc. Japan 23:*
       411 (1950).
2. ROLINSON, G. N., BATCHELOR, F. R., BUTTER-WORTH, D., CAMERON-
       WOOD, J., COLE, M., EUSTACE, G. C., HART, M. V., RICHARDS,
       M., & CHAIN, E. B. *Nature 187:* 236 (1960).
3. KAUFMANN, W., BAUER, K., & OFFE, H. A. *Antimicr. Agents Ann.
       1960:* 1 (1960).
4. COLE, M. *Biochem. J. 115:* 747 (1969).
5. NARA, T., OKACHI, R., & MISAWA, M. *J. Antibiot. 24:* 321
       (1971).

6.   NARA, T., MISAWA, M., OKACHI, R., & YAMAMOTO, M. *Agric. Biol. Chem. 35:* 1676 (1971).
7.   OKACHI, R., MISAWA, M., DEGUCHI, T., & NARA, T. *Agric. Biol. Chem. 36:* 1193 (1972).
8.   SHIMIZU, M., MASUIKE, T., FUJITA, H., KIMURA, K., OKACHI, R., & NARA, T. *Agric. Biol. Chem. 39:* 1225 (1975).
9.   TAKASAWA, S., OKACHI, R., KAWAMOTO, I., YAMAMOTO, M., & NARA, T. *Agric. Biol. Chem. 36:*   1701 (1972).
10. OKACHI, R., KATO, F., MIYAMURA, Y., & NARA, T. *Agric. Biol. Chem. 37:* 1953 (1973).
11. VANDAMME, E., J. & VOETS, J. P. in "Advances in Applied Microbiology", vol. 17 (D . Perlman, ed.) Academic Press, New York (1974) p. 311.
12. TAKAHASHI, T., YAMAZAKI, Y., KATO, K., & ISONO, M. *J. Am. Chem. Soc. 94:* 4035 (1972).

ENZYMATIC ACYL TRANSFER IN PENICILLIN AND CEPHALOSPORIN CHEMISTRY

J. Konecny, A. Schneider and M. Sieber

Pharmaceutical Division, CIBA-GEIGY
BASEL, Switzerland

Enzymatically catalyzed N-acylations of β-lactam compounds, like 6-aminopenicillanic acid (6-APA), by esters and amides have been attracting increasing attention in the last decade (1-3). While chemical introduction of multifunctional side chains like D-α-aminoacids requires the introduction and removal of protecting groups, the enzymatic route is direct. Other attractive features of this approach include the mild conditions which minimize the destruction of the labile β-lactam ring and, in a process like cephalexin production, the advantageous coupling of this reaction with the preceding enzymatic deprotection of the cephalosporin amino group (4). Potential advantages include the use of racemates in place of the expensive D-aminoacids. The mechanism of the reaction has a bearing on various practical issues, including the substrates and methods used for enzyme screening.

Although most of the literature is descriptive and relates to work with crude enzyme preparations or whole cells, there is now considerable evidence that at least four of the enzymes react like chymotrypsin (10), some other proteases (11), and some β-lactamases via an acyl-enzyme intermediate. This includes the esterases from *Acetobacter turbidans* and *Xanthomonas citri* (5,6), and the well-known penicillin acylases from *E. coli* and *B. megaterium* (8,9). General considerations and experience in peptide synthesis and hydrolysis are therefore relevant to this area of chemistry.

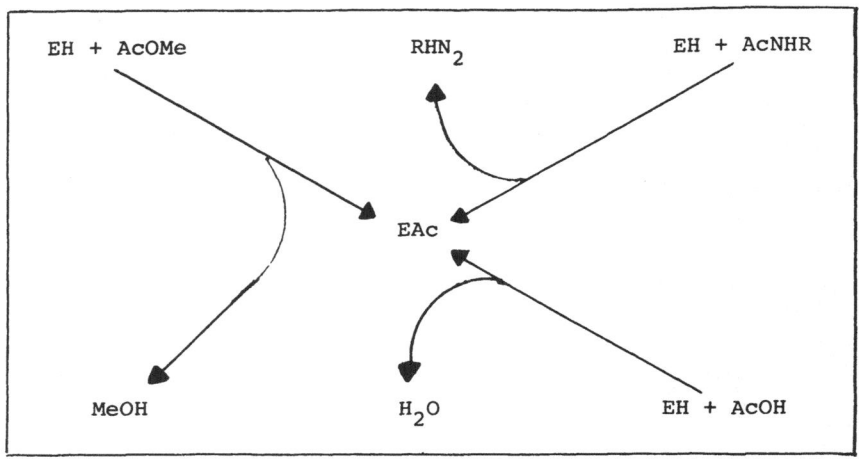

Fig. 1. Schematic representation of the reactions of the enzyme
         EH with an ester, amide, and acid as acyl donors. The
         reactions are reversible; and the common acyl-enzyme
         intermediate reacts with water, amine, or alcohol as
         acyl receptors.

MECHANISM

Contrary to what its name suggests, penicillin amidohydrolase
from *E. coli* interacts with the acyl side chain and not with the
penicillin nucleus. Various amides and esters of phenylacetic
acid are hydrolyzed at comparable rates (12,13). The reactions
which all four of the above enzymes (EH) catalyze may be described
as reversible acyl transfer from an ester (AcOMe), amide (AcNHR),
or acid (AcOH) to suitable acceptors, such as amines (RNH_2), water,
or alcohols (MeOH). The scheme in Fig. 1 shows the linking of
these pathways through the common acyl-enzyme intermediate EAc,
each of the pathways being a reversible two-step process.

$$EH + AcX \underset{x_{-1}}{\overset{x_1}{\rightleftharpoons}} EH:AcX \underset{x_{-2}}{\overset{x_2}{\rightleftharpoons}} EAc + XH \qquad\qquad (Eq.\ 1)$$

These reactions are coupled with pH-dependent equilibria,
namely, the ionization of the acid AcOH to AcO^- and the protona-
tion of RNH_2 or any other amino groups, including those of the
enzyme. In view of the magnitudes of the equilibrium constants
(14), the thermodynamic aspects of the reactions are important.

Since water is the solvent, acyl transfer from ester to amine
is always associated with hydrolysis of the ester and the product
amide. The latter is a thermodynamically unstable intermediate

(7,9,14,15,16) in neutral or alkaline solutions. At sufficiently
low pH the equilibria permit the formation of amide from the amine
and acid (7,14).

KINETIC EVIDENCE

The outlined scheme is a generalization of that proposed by
Kato (6). Tractable kinetic expressions are obtained (8,9) on the
plausible assumption that the bond breaking step x_2 in Eq. 1 is
slow and that the species involved in the x_1/x_{-1} steps are effectively at equilibrium. The rate equations are then obtained by
solving for the steady state concentration of EAc and of the free
enzyme EH for each species X. The distinction, that the acyl
donors AcX compete for the enzyme EH while the acyl acceptors XH
compete for EAc, has important kinetic consequences.

$$\frac{d(AcX)}{dt} = - \frac{d(XH)}{dt} = x_{-2} \, (EAc) \, (XH) - (x_2/K_x)(EH) \, (AcX) \quad (Eq. \ 2)$$

The mathematical solutions account for the hitherto unexplained kinetics of penicillin G hydrolysis (competitive inhibition by phenyl acetic acid, non-competitive inhibition by 6-APA)
catalyzed by the enzymes from *E. coli* (17) and *B. megaterium*
(18). The kinetic evidence also accounts for the inhibition of
ester hydrolysis by amine and alcohol (non-competitive), by amide
and acid (competitive), and by other features of the acyl transfer
reactions involved in the synthesis of cephalexin and deacetyl
cephacetril (6,8,9). Thus, at low concentrations of RNH_2 repression of acid formation is accompanied by a corresponding increase
amide formation. The sum of the two rates is approximately
equal to the rate of ester hydrolysis in the absence of the amine.
Deviations from the predicted behavior are, however, apparent at
higher concentration of the amine (6,9).

The data in Fig. 2 show the course of the acylation of 6-
deacetyl-7-amino cephalosporanic acid by $CNCH_2CO_2CH_3$ at two concentrations of the acyl receptor (9). The produced amide is a
very poor substrate for the enzyme. The data illustrate a) the
predicted change of the ratio of acid and amide formation with
increasing initial concentration of RNH_2 and b) the predicted
acceleration of acid formation with progressive depletion of the
amine in the course of the reaction.

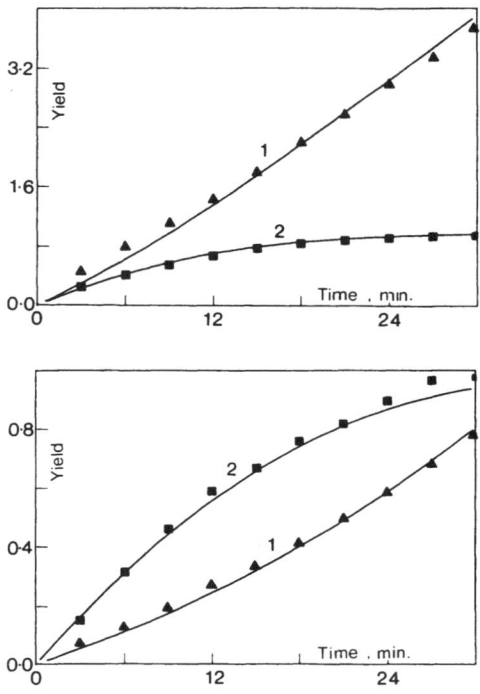

Fig. 2. Acid (▲) and amide (■) formation in the reaction of
         225 mM CNCH$_2$CO$_2$CH$_3$ with 3 mM (top) and 18 mM (bottom)
         deacetyl-7-ACA at pH7 and 25° C. Acylase 0.087 and 0.28
         U/ml, respectively. Points experimental, curves calcu-
         lated. Yields expressed as mole/mole deacetyl-7-ACA
         present initially.

                            OTHER ASPECTS

        The pH versus rate profile of the acylation of chymotrypsin
is bell-shaped, while that of the deacylation is sigmoid-shaped
(10). Depending on the charge of the reacting species, the
effects of salts on the individual reaction steps are also likely
to diverge. Such effects have in fact been observed (9,19) and
are at least partly responsible for the deviations from the theoretical equations. The phenomena deserve further study. In addition to ionic strength, specific interactions and general base
catalysis by anions of weak acids seem to be involved.

        Protonation of the amino groups of substrates like cephalexin
has a profound effect on reaction rates (9,13). Comparison of the
available data shows that the ratios of estereolytic/amidolytic
activities (9,13,20) vary considerably depending on the enzymes

and substrates. The effect of pH on this ratio for a given acyl group is a matter of considerable interest.

Unlike the hydrolysis of penicillin G, the acyl transfer reactions must be terminated in a non-equilibrium state. This imposes various requirements on process design, especially when the catalyst is sensitive to low pH and when the formation of acid is still fast.

To minimize product costs it is desirable to maximize yields and minimize the waste of unreacted amine and of expensive acyl donors. The keys to success are obviously a) a catalyst with low amidolytic activity with respect to the product and b) efficient recovery systems. Methods used in the laboratory to enhance the synthesis of amide bonds include the reduction of water concentration by adding miscible solvents or equilibrium shifts to exploit low solubilities (5) or selective partition in bi-phasic systems (21). Such measures, combined with acidophilic enzymes, may permit the synthesis of amides from acids in place of esters in certain situations. Preliminary results from our laboratory indicate that diffusion and partition of the reagents change the ratio of initial rates of amide and acid formation in reactions catalyzed by the immobilized proteins, as expected.

ACKNOWLEDGMENTS

We are indebted to E. Flaschel of the Swiss Federal Institute of Technology for the simulations.

REFERENCES

1. VANDAMME, E. J. & VOETS, J. P. *Adv. Appl. Microbiol. 17:* 311 (1974).
2. OKACHI, R. *Nippon Nogei Kagaku Kaishi 53:* R 169 (1979).
3. MATSUMOTO, K. *Hakko Kogyo 38:* 216 (1980).
4. FUJII, T., MATSUMOTO, K. & WATANABE, T. *Process Biochem.* *11*(8): 21 (1976).
5. TAKAHASHI, T., KATO, K., YAMAZAKI, Y. & ISONO, M. *J. Antibiotics Suppl. 30:* 230 (1977).
6. KATO, K. *Agric. Biol. Chem. 44:* 1083 (1980).
7. SVEDAS, K., MARGOLIN, A. L., BORISOV, E. L. & BEREZIN, I. V. *Enzyme. Microbial. Technol. 2:* 313 (1980).
8. KONECNY, J. *Biotechnol. Lett. 3:* 107 (1981).
9. KONECNY, J., SIEBER, M. & SCHNEIDER, A. *Biotechnol. Lett. 3:* 507 (1981).
10. WOLD, F. "Macromolecules: Structure and Function," Prentice-Hall, Englewood Cliffs, New Jersey (1971) p. 76.

11. WIDMER, F. & JOHANSEN, J. T. *Carlsberg Res. Commun. 44:* 37 (1979).
12. KUTZBACH, C. & RAUENBUSCH, E. *Z. Physiol. Chem.* 354: 45 (1974).
13. MARGOLIN, A. L., SVEDAS, V. K. & BEREZIN, I. V. *Biochim. Biophys. Acta 616:* 283 (1980).
14. SVEDAS, V. K., MARGOLIN, A. L. & BEREZIN, I. V. *Enzyme Microbial. Technol. 2:* 138 (1980).
15. MARCONI, W., BARTOLI, F., CECERE, F., GALLI, G. & MORISI, F. *Agric. Biol. Chem. 39:* 277 (1975).
16. RHEE, D. K., LEE, S. B., RHEE, J. S. & RYU, D. D. Y. *Biotechnol. Bioeng. 22:* 1237 (1980).
17. BALASINGHAM, K., WARBURTON, D., DUNNILL, P. & LILLY, M. D. *Biochim. Biophys. Acta 276:* 250 (1972).
18. CHIANG, C. & BENNETT, R. E. *J. Bacteriol. 93:* 302 (1967).
19. KATO, K., KAWAHARA, K., TAKAHASHI, T. & IGARAZI, S. *Agric. Biol. Chem. 44:* 821 (1980).
20. KATO, K., KAWAHARA, K., TAKAHASHI, T. & KAKINUMA, A. *Agric. Biol. Chem. 44:* 1075 (1980).
21. SEMENOV, A. N., BEREZIN, I. V. & MARTINEK, K. *Biotechnol. Bioeng. 23:* 355 (1981).

ENZYMATIC PROCESSES FOR THE SYNTHESIS OF OPTICALLY ACTIVE AMINO ACIDS

H. Yamada

Department of Agricultural Chemistry, Kyoto University, Kyoto, Japan

The production of optically active amino acids and their derivatives is required in the world, especially for the food and pharmaceutical industries. There are two processes for the biological production of amino acids established in Japan. They are fermentation and enzymation.

In recently developed enzymatic processes (enzymation), the synthesis of L-lysine from DL-2-amino-ε-caprolactam (1) and the synthesis of L-cysteine from DL-2-aminothiazoline 4-carboxylate (2) have been reported. During the investigation of the biological and enzymatic production of amino acids, we have developed three new processes. These are to synthesize L-tryosine, L-tryptophan, L-cysteine and their related amino acids by multifunctional pyridoxal enzymes, L-serine by serine hydroxymethyltransferase, and D-p-hydroxyphenylglycine and its related D-amino acids by dihydropyrimidinase. This paper outlines the syntheses of these amino acids by microbial enzymes.

SYNTHESIS OF L-TYROSINE, L-TRYPTOPHAN, L-CYSTEINE AND THEIR RELATED AMINO ACIDS BY MULTIFUNCTIONAL PYRIDOXAL ENZYMES

β-Tyrosinase (EC 4.1.99.2), tryptophanase (EC 4.1.99.1) and cysteine desulfhydrase (EC 4.4.1.1) are the enzymes which respectively catalyze the degradation of L-tyrosine, L-tryptophan and L-cysteine, and require pyridoxal 5'-phosphate (PLP) as a cofactor. Crystalline preparations of these enzymes have been prepared in our laboratory from *Eschrichia intermedius*, *Proteus rettgeri* and *Aerobacter aerogenes;* and their properties have been established in some detail (3-5). With the crystalline en-

TABLE 1

COMPARATIVE SUBSTRATE AFFINITIES AND ACTIVITIES OF β-TYROSINASE
IN CATALYSIS OF DIFFERENT REACTIONS*

Compound	Role	Product measured	Km or Ki (mM)	Vmax (μmol/min/mg)
α, β-Elimination Reactions				
L-Tyrosine	Substrate	Pyruvate	0.23	1.9
L-Serine	Substrate	Pyruvate	34	0.35
S-Methyl-L-cysteine	Substrate	Pyruvate	1.8	1.2
β-Chloro-L-alanine	Substrate	Pyruvate	4.5	18.2
L-Alanine	Inhibitor	Pyruvate	6.5	
L-Phenylalanine	Inhibitor	Pyruvate	2.0	
Phenol	Inhibitor	Pyruvate	0.04	
Pyrocatechol	Inhibitor	Pyruvate	0.46	
Resorcinol	Inhibitor	Pyruvate	0.16	
β-Replacement Reactions				
L-Serine	Cosubstrate	L-Tyrosine	35	0.33
S-Methyl-L-cysteine	Cosubstrate	L-Tyrosine	1.8	0.82
β-Chloro-L-alanine	Cosubstrate	L-Tyrosine	4.5	1.4
Phenol	Cosubstrate	L-Tyrosine	1.2	

*For detail see (3) and (4).

TABLE 2

RELATIVE SYNTHETIC RATES OF TYROSINE-RELATED AMINO ACIDS FROM
AMMONIA, PYRUVATE AND PHENOL DERIVATIVES BY β-TYROSINASE*

Substrate	Product	Relative velocity of synthesis
Phenol	L-Tyrosine	100
Pyrocatechol	3-Hydroxy-L-tyrosine (L-dopa)	60.0
Resorcinol	2-Hydroxy-L-tyrosine	58.2
o-Fluorophenol	3-Fluoro-L-tyrosine	66.3
m-Fluorophenol	2-Fluoro-L-tyrosine	23.2
o-Chlorophenol	3-Chloro-L-tyrosine	15.3
m-Chlorophenol	2-Chloro-L-tyrosine	33.4
o-Bromophenol	3-Bromo-L-tyrosine	2.0
m-Bromophenol	2-Bromo-L-tyrosine	4.2
o-Iodophenol	3-Iodo-L-tyrosine	0
m-Iodophenol	2-Iodo-L-tyrosine	1.5
o-Methylphenol	3-Methyl-L-tyrosine	2.0
m-Methylphenol	2-Methyl-Ltyrosine	9.8
o-Ethylphenol	3-Ethyl-L-tyrosine	0
m-Ethylphenol	2-Ethyl-L-tyrosine	2.0
o-Methoxyphenol	3-Methoxy-L-tyrosine	1.1
m-Methoxyphenol	2-Methoxy-L-tyrosine	30.6

*For detail see (5).

zymes we found that they catalyze a variety of α,β-elimination
(Eq. 1), β-replacement (Eq. 2) and the reverse of the α,β-elimina-
tion (Eq. 3) reactions as follows:

$$L\text{-}RCH_2CHNH_2COOH + H_2O \longrightarrow RH + CH_3COCOOH + NH_3 \qquad \text{(Eq. 1)}$$
$$L\text{-}RCH_2CHNH_2COOH + R'H \longrightarrow L\text{-}R'CH_2CHNH_2COOH + RH \qquad \text{(Eq. 2)}$$
$$R'H + CH_3COCOOH + NH_3 \longrightarrow L\text{-}R'CH_2CHNH_2COOH + H_2O \qquad \text{(Eq. 3)}$$

where for β-tyrosinase: R = phenolyl, -OH, -SH, -Cl; R' =
phenolyl; for tryptophanase: R = indolyl, - OH, -SH, - Cl; R' =
indolyl; and for cysteine desulfhydrase: R = - SH, -OH, -Cl; R'
= mercaptan radicals.

The catalytic properties of β -tyrosinase, as an example, will
be described in some detail. Table 1 shows comparative substrate
affinities of β-tyrosinase in the catalysis of α β-elimination and
β-replacement reactions. In these reactions, L-tyrosine, L-
serine, S-methyl-L-cysteine, and β-chloro-L-alanine are substrates:
L-alanine and L-phenylalanine are competitive inhibitors. In the

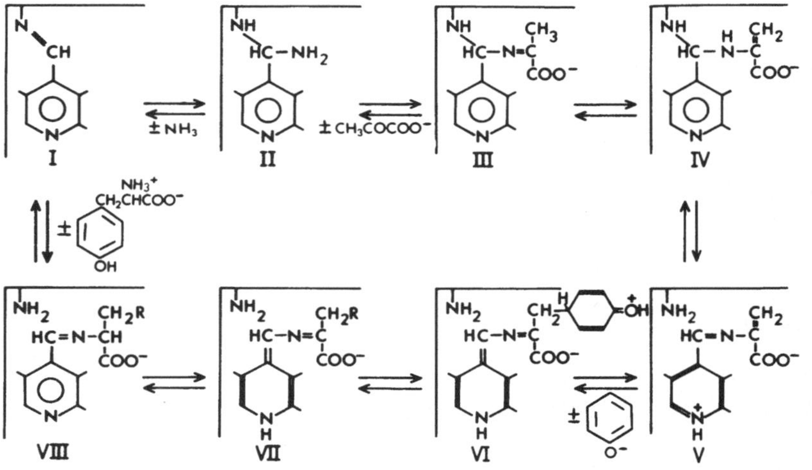

Fig. 1. Schematic representation of the mechanism for the re-
         actions catalyzed by β-tyrosinase.

β-replacement reaction, phenol is the second substrate to synthesize L-tyrosine. When pyrocatechol, resorcinol, pyrogallol, and
hydroxyhydroquinone were added to the reaction mixture, in place
of phenol, 3-hydroxy-L-tyrosine (L-dopa), 2-hydroxy-L-tyrosine,
2,3-dihydroxy-L-tyrosine, and 2,5-dihydroxy-L-tyrosine, respectively were synthesized (3,4). Table 2 shows the synthesis of L-
tyrosine and its related amino acids by the reverse of the α,β-
elimination reaction (3-5).

     The mechanism of the β-tyrosinase catalyzed reactions may be
explained by the general mechanism for PLP-dependent reactions as
shown in Fig 1, in which the enzyme-α-aminoacrylate complex is a
key intermediate for all three reactions (5). The same mechanism
may be adopted to explain the reactions catalyzed by tryptophanase
and by cysteine desulfhydrase.

     Based on results obtained with crystalline β-tyrosinase,
tryptophanase and cysteine desulfhydrase, we developed enzymatic
processes to produce L-tyrosine, L-tryptophan, L-cysteine and
their related amino acids. As a practical application, bacterial
cells containing high enzyme activities were prepared and directly
used as the catalysts. L-Tyrosine, L-dopa, L-tryptophan, 5-
hydroxy-L-tryptophan, and L-cysteine were synthesized in extremely good yields (Table 3). Like the β-tyrosinase reaction,
syntheses of a variety of derivatized amino acids of L-tryptophan
and L-cysteine, such as 5-amino-L-tryptophan, S-alkyl-L-
cysteines and so on, were possible.

TABLE 3

SYNTHESIS OF L-TYROSINE, L-DOPA, L-TRYPTOPHAN, 5-HYDROXY-L-TRYPTOPHAN AND L-CYSTEINE BY MULTIFUNCTIONAL PYRIDOXAL ENZYMES***

Product	Yield		Substrates	Reaction Enzyme Microorganism
	(g/L)	Molar yield (%)		
L-Tyrosine	58	88*[a]	Sodium pyruvate Ammonium acetete Phenol	Reverse of α,β-elimination β-Tyrosinase *Erwinia herbicola*
L-Tyrosine	53.5	78*[b]	DL-Serine Ammonium acetate Phenol	β-Replacement β-Tyrosinase *Erwinia herbicola*
L-Dopa	58.5	-**	Sodium pyruvate Ammonium acetate Pyrocatechol	Reverse of α,β-elimination β-Tyrosinase *Erwinia herbicola*
L-Dopa	53	71*[b]	DL-Serine Ammonium acetate Pyrocatechol	β-Replacement β-Tyrosinase *Erwinia herbicola*
L-Tryptophan	100	100[d]	Sodium pyruvate Ammonium acetate Indole	Reverse of α,β-elimination Tryptophanase *Proteus rettgeri*

TABLE 3 (Continued)

Product	Yield		Substrates	Reaction Enzyme Microorganism
	(g/L)	Molar yield (%)		
5-Hydroxy-L-tryptophan	23.3	57*[e]	Sodium pyruvate Ammonium acetate 5-Hydroxyindole	Reverse of α, β-elimination Tryptophanase *Proteus rettgeri*
L-Cysteine	50	88*[f]	β-Chloro-L-alanine Sodium sulfide	β-Replacement Cysteine desulfhydrase *Enterobacter cloacae*

*Values were calculated on the basis of sodium pyruvate (a), DL-serine (b), indole (d), 5-hydroxyindole (e), or β-chloro-L-alanine (f).

**Sodium pyruvate was fed at 2 hr intervals to keep 5 g/L for 48 hr.

***For details see (3, 4, 6-8).

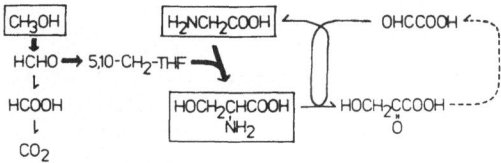

Fig. 2. Outline of the serine pathway for the synthesis of L-
         serine.

SYNTHESIS OF L-SERINE BY SERINE HYDROXYMETHYLTRANSFERASE OF A METHYLOTROPHIC BACTERIUM

Methylotrophs are known to grow on methanol and to assimilate methanol through the serine pathway (Fig. 2) or the ribulose mono-phosphate pathway. In the principal part of the serine pathway, formaldehyde liberated from methanol reacts with glycine to form L-serine under the catalysis of serine hydroxymethyltransferase (EC 2.1.2.1). L-Serine production by methylotrophs having a serine pathway has been reported (9-11), but the yield of L-serine is still not sufficient.

To increase the yield of L-serine using an enzymatic method, we screened a variety of methylotrophic bacteria, and an isolated strain, *Hyphomicrobium* sp. KM 146, was selected as a promising enzyme source. The bacterium grew well on methanol (74 g cells/L) with high enzyme activity, when methanol was fed to keep about 2% in the culture medium. Synthesis of L-serine was carried out from glycine and methanol as starting substrates in Tris-HCl, pH 8.0, using the cells or cultured broth directly as the enzyme. Under the optimum conditions, approximately 24 g of L-serine was synthe-sized from 100 g of glycine, 24 g of methanol and 30 g of the bacterial cells in 1 liter of reaction medium (12).

SYNTHESIS OF D-*p*-HYDROXYPHENYLGLYCINE AND ITS RELATED D-AMINO ACIDS BY DIHYDROPYRIMIDINASE

Dihydropyrimidinase (EC 5.3.2.2) is involved in the metabo-lism of pyrimidines and catalyzes the hydrolytic ring-opening re-action of dihydropyrimidines to N-carbamyl-β-amino acids. The enzyme from animal tissues has been partially purified and shown to be identical with hydantoinase, which catalyzes the hydrolytic ring-opening reaction of hydantoins to N-carbamyl amino acids (13). Dudley *et al.* (14) reported that when 5-substituted hydantoins were used as substrates, only D-forms of hydantoic acids were formed under catalysis by this enzyme.

The distribution of this enzyme activity in microorganisms has recently been investigated in our laboratory. We found the activity widely distributed in bacteria, particularly in *Aerobacter*, *Corynebacterium*, *Agrobacterium*, *Bacillus*, *Pseudomonas*, and *Streptomyces*. In this screening, we used DL-5-(2-methylthioethyl) hydantoin as the substrate. The reaction products with several strains were confirmed to be N-carbamylmethionine of D-form (15).

The bacterial enzyme was isolated as crystals from the cells of *Pseudomonas putida*; and its physicochemical and catalytic properties were investigated in detail (16). Like the animal enzyme, the bacterial enzyme showed the highest activity and affinity toward dihydropyrimidines. The substrate specificity of the enzyme is shown in Fig. 3. A variety of aliphatic and aromatic 5-monosubstituted hydantoins were substrates for the enzyme. The hydrolysis was specific for D-forms of these substituted hydantoins. The steric hinderance by the bulky substituted group introduced on the hydantoin molecule may give the stereospecificity for the D-form.

Based on the above mentioned findings with crystalline enzyme, we examined the enzymatic process for the production of D-forms of amino acids (17-20). The process successfully developed involves two chemical steps and one enzymatic step. The substrate hydantoins are generally synthesized through the method of Bucherer and Bergs with various aldehydes as the starting materials. Then, the D-hydantoins are hydrolyzed enzymatically to N-carbamyl-D-amino acids. Practically, bacterial cells with high enzyme activity can be used directly as the catalyst. Preparation of these cells is carried out by culturing the bacterial cells in

Substrate		Relative rate of hydrolysis	Substrate		Relative rate of hydrolysis
5- H-	hydantoin	13	5- 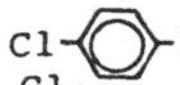 Cl-◯-hydantoin		19
" CH$_3$-	"	45	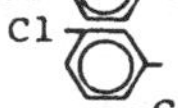 " Cl-◯(Cl)- "		10
" (CH$_3$)$_2$CH$_2$-	"	15	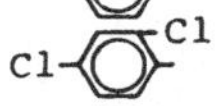 " Cl-◯-Cl "		7
" (CH$_3$)$_2$CHCH$_2$-	"	48	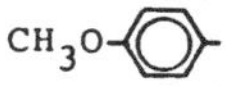 " CH$_3$O-◯- "		4
" CH$_3$SCH$_2$CH$_2$-	"	48	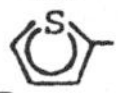 " ◯(S)- "		48
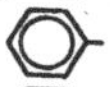 " ◯-	"	25	5,5- 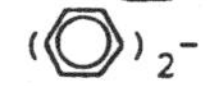 (◯)$_2$- "		0
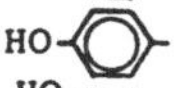 " HO-◯-	"	16	" (CH$_3$)$_2$- "		0
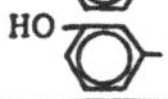 " HO(HO)◯-	"	5	Dihydrouracil		100

Fig. 3. Substrate specificity of dihydropyrimidinase from
         Pseudomonas putida. See (16) for details.

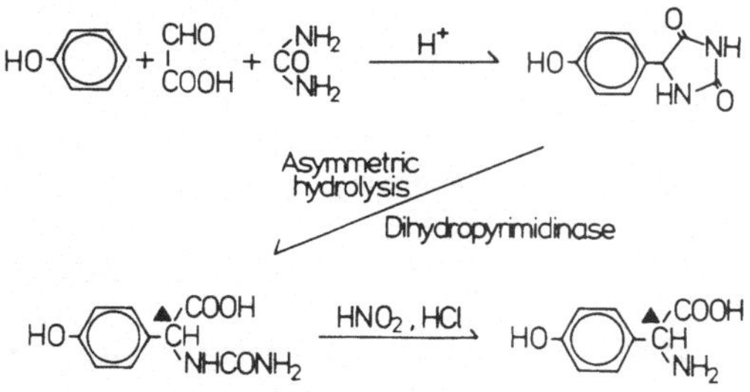

Fig. 4. Process for the synthesis of D-p-hydroxyphenylglycine.

a medium supplemented with hydantoins or their related compounds
as the inducer. Under the conditions used for the enzymatic reaction of pH 8-10, the L-forms of the remaining hydantoins are
racemized rapidly and spontaneously through the mechanism of basic
catalysis (This is a well-known property of hydantoin derivatives).
Therefore, 100% of the hydantoins can be theoretically transformed
to N-carbamyl-D-amino acids by this enzymatic step. Decarbamyla-
tion to D-amino acids is carried out by treating the N-carbamyl-
D-amino acids with equimolar nitrite under acidic conditions.
Complete retention of the configuration is achieved through this
step (Fig. 4).

As a result of these fundamental examinations of the production of D-amino acids, we developed a new process for the production of D-p-hydroxyphenylglycine. D-p-Hydroxyphenylglycine is an
important component of the semisynthetic penicillin, Amoxicillin,
or the cephalosporins, Cefadroxyl and Cefatrizine. This amino
acid so far had been synthesized chemically as a racemic mixture,
which was then optically resolved through a rather complicated
process. First, we examined a simple and economic method for the
chemical synthesis of the substrate, DL-5-(p-hydroxyphenyl)-
hydantoin, and established a method which involves a new amido-
alkylation reaction of phenol with glyoxylic acid and urea under
acidic conditions (20). Figure 4 shows the overall process for
the production of D-p-hydroxyphenyl-glycine. This process may be
one of the most economical processes for large scale production
of D-p-hydroxyphenylglycine.

REFERENCES

1. FUKUMURA, T. *Agric. Biol. Chem. 41:* 1327 (1977).
2. SANO, K. & MITSUGI, K. *Agric. Biol. Chem. 42:* 2315 (1978).
3. YAMADA, H. & KUMAGAI, H. *Adv. Appl. Microbiol. 19:* 249 (1975).
4. YAMADA, H. & KUMAGAI, H. *Pur Appl. Chem. 50:* 1117 (1978).
5. NAGASAWA, T., UTAGAWA, T., GOTO, J., KIM, C-J., TANI, Y.,
        KUMAGAI, H. & YAMADA, H. *Eur. J. Biochem. 117:* 33 (1981).

6. ENEI, H., NAKAZAWA, H., MATSUI, H., OKUMURA, S. & YAMADA, H.
    FEBS Lett. 21: 39 (1972).
7. ENEI, H., MATSUI, H., NAKAZAWA, H., OKUMURA, S. & YAMADA, H.
    Agric. Biol. Chem. 37: 493 (1973).
8. NAKAZAWA, H., ENEI, H., OKUMURA, S., YOSHIDA, H. & YAMADA,
    H. *FEBS Lett. 25:* 43 (1972).
9. KEUNE, H., SAHM, H. & WAGNER, F. *Eur. J. Appl. Microbiol.
    2:* 175 (1976).
10. TANI, Y., KANAGAWA, T., HANPONGKITTIKUN, A., OGATA, K. &
    YAMADA, H. *Agric. Biol. Chem. 42:* 2275 (1978).
11. MORINAGA, Y., YAMANAKA, S. & TAKINAMI, K. *Agric. Biol. Chem.
    45:* 1425 (1981).
12. IZUMI, Y., TAKIZAWA, H., TANI, Y. & YAMADA, H. *Abstr. 319th
    Reglar Mtg. Agric. Chem. Soc. (Kansai Division) Japan* (1981).
13. WALLACH, D. P. & GRISOLIA, J. *J. Biol. Chem. 226:* 277
    (1957).
14. DUDLEY, K. H., BUTLER, T. C. & BIUS, D. L. *Drug Metab. Dispos.
    2:* 103 (1975).
15. YAMADA, H., TAKAHASHI, S., KII, Y. & KUMAGAI, H. *J. Ferment.
    Technol. 56:* 484 (1978).
16. TAKAHASHI, S., KII, Y., KUMAGAI, H. & YAMADA, H. *J. Ferment.
    Technol. 56:* 492 (1978).
17. TAKAHASHI, S., OHASHI, T., KII, Y., KUMAGAI, H. & YAMADA, H.
    J. Ferment. Technol. 57: 328 (1979).
18. SHIMIZU, S., SHIMADA, H., TAKAHASHI, S., OHASHI, T., TANI,
    Y. & YAMADA, H. *Agric. Biol. Chem. 44:* 2233 (1980).
19. YAMADA, H., SHIMIZU, S., SHIMADA, H., TANI, Y., TAKAHASHI,
    S. & OHASHI, T. *Biochimie 62:* 395 (1980).
20. OHASHI, T., TAKAHASHI, S., NAGAMACHI, T., YONEDA, K. & YAMADA,
    H. *Agric. Biol. Chem. 345:* 831 (1981).

HORSE LIVER ALCOHOL DEHYDROGENASE: AN ILLUSTRATIVE EXAMPLE

OF THE POTENTIAL OF ENZYMES IN ORGANIC SYNTHESIS

J. B. Jones

Department of Chemistry
University of Toronto
Toronto, Ontario, Canada

The benefits of using enzyme-dependent procedures for over-coming difficult organic synthetic problems has been recognized for many years. This is illustrated by the extensive use of micro-biological fermentation methods (1) and the increasing applications of immobilized enzymes and cells (2). In spite of this substantial documentation, enzymic methods have not yet been widely accepted by organic chemists as routine catalysts. However, this situation is changing rapidly; and the most recent literature shows clearly that the aware organic chemist now recognizes the unique synthetic advantages that enzymes offer.

The attractions of enzymes as organic synthesis catalysts are multiple. There is an enzyme-catalyzed equivalent for almost every type of organic chemical reaction. Furthermore, the re-actions are catalyzed under very mild conditions, often at $20^{\circ}C$ and pH 7. This enables problems, such as isomerization, epimeriza-tion, racemization and rearrangement, encountered with many organic reactions on sensitive molecules, to be avoided. The generally high catalytic efficiencies of enzymes is another synthetic attrac-tion. However, the most valuable synthetic properties of enzymes are those deriving from their specificity. The selectivities of reactions achievable, and above all the stereospecificity with which they can be effected, are the most important features of en-zymic catalysts from the organic chemist's viewpoint.

Not all types of enzyme-catalyzed reactions are of equal synthetic value. The six main IUB classification groups (3) listed below permit the most useful enzyme types to be readily identified.

ENZYME REACTION TYPES

1. **Oxidoreductases.** This group includes enzymes catalyzing $C-H \rightarrow C-OH$, $CH(OH) \rightleftharpoons C=O$, and $CH-CH \rightleftharpoons C=C$ conversions. Such reactions are among the keystones of synthetic organic chamistry.

2. **Transferases.** While the aldehyde, ketone, acyl, sugar, and phosphoryl group transfer reactions mediated by this group of enzymes are of great biochemical significance, they are of limited organic chemical applicability.

3. **Hydrolases.** These enzymes catalyze the hydrolysis of a very broad range of functions and structures. They are of considerable general synthetic value for selective hydrolyses and in resolution, particularly for esters of racemic acids.

4. **Lyases.** These enzymes catalyze additions to, or formation of, double bonds such as $C=C$, $C=O$ and $C=N$. Such reactions are very important·in organic synthesis.

5. **Isomerases.** Various isomerization reactions, including racemization, are catalyzed by this group. Generally, organic chemists can achieve such transformations very readily by chemical methods. Use of enzymes thus offers little advantage in this area.

6. **Ligases.** The enzymes of this group, also termed synthetases, catalyze the formation of $C-O$, $C-S$, and $C-N$ bonds. However, while such reactions are of synthetic value, the enzymes are generally too specific to be of widespread organic chemical applicability.

COENZYME REGENERATION

Many enzymes require coenzymes in order to be catalytically active. Since coenzymes are in fact co-substrates that undergo chemical transformation during the reaction, they must be constantly available in their active form for catalysis to cohtinue. Coenzymes are expensive chemicals (4); and it is neither economically possible, nor chemically desirable, to provide them in stoichiometric or greater amounts in the reaction mixture. Generally, only catalytic amounts of coenzymes are used in conjunction with an inexpensive regenerating system (4,5). The problem of coenzyme regeneration is a major consideration in organic synthetic applications of enzymes. Some coenzymes present no problem since they regenerate automatically during normal aerobic aqueous reaction conditions. These include biotin, pyridoxal phosphate, thiamine pyrophosphate, lipoic acid, and oxidized flavin coenzymes. Other coenzymes *viz.* coenzyme A, coenzyme B_{12}, folic acid,

glutathione, reduced flavins, S-adenosyl methionine, nicotinamide
coenzymes, and nucleotide triphosphates, require auxiliary regenerating methods. At present, recycling methods are available only
for the latter two coenzyme types. The problem of ATP regeneration
is basically solved (6); but recycling of the nicotinamide (NAD(P))
coenzymes continues to present major problems. Many excellent
NAD(P)/H regeneration procedures have been developed (4,5,7). However, the oxidized and reduced nicotinamide rings are unstable in
basic (8) and acidic (9) media, respectively. This inherent chemical reactivity is now the major barrier to long-term use of nicotinamide coenzymes. The presently available recycling methods are
already more than adequate; and future work must address the nicotinamide ring instability problem as its first priority.

REQUIREMENTS FOR ROUTINE USE OF ENZYMES IN SYNTHESIS

The following seven requirements are cited: a) The enzymes
must be readily available, preferably commercially, so that chemists can buy them like any other reagents. b) The reactions
catalyzed must be those of fundamental organic chemical, rather
than biochemical, importance; in practice, the enzymes will usually
be those of IUB groups 1, 3 and 4. c) The enzyme should accept
a broad range of substrate structures; this is necessary if the
enzyme is to be generally useful since synthetic chemists deal with
different structures in each new synthesis. d) Modern synthetic
chemistry demands that the stereochemistry of a reaction be controlled. This goal, termed asymmetric synthesis, can be achieved
if any enzyme maintains its ability to discriminate clearly between
different stereoisomers. This requirement of maintaining high
stereospecificity towards a broad structural range of substrates
is a very demanding one since high stereoisomer discrimination by
an enzyme is usually accompanied by narrow structural specificity.
e) A reliable model or method for predicting the specificity of
catalysis with respect to a previously unevaluated substrate is
needed in order for the enzyme to be widely used. f) Chemists
are accustomed to knowing the mechanisms of chemical reactions.
They will want the same information on enzymes. This does not
pose a problem since the mechanisms of the readily available enzymes are well documented. g) The experimental procedure should
use chemical equipment and not require sophisticated biochemical
instruments for the reaction.

While the above requirements are collectively far from triv-
ial, a number of enzyme groups satisfy them (5). Several alcohol
dehydrogenases are among the enzymes of organic synthetic utility.
For the remainder of this paper, I will use horse liver alcohol
dehydrogenase (HLADH) to illustrate how enzymes can provide facile,
and often unique, answers to the types of problems that organic

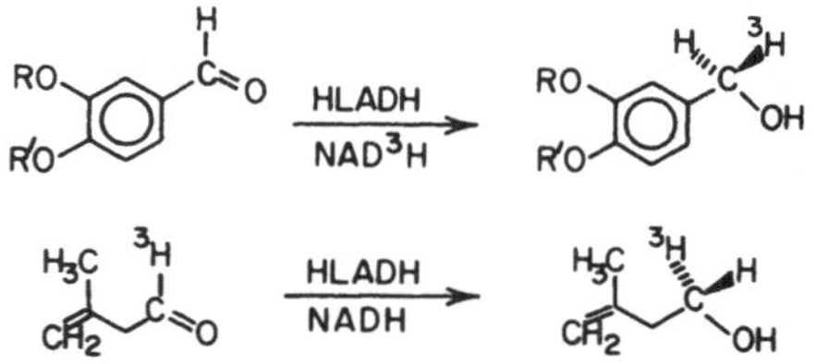

Fig. 1.   HLADH-catalyzed reductions of aldehydes. Since the
           hydride is always delivered to the same face of an alde-
           hyde carbonyl group, either *R*- or *S*-configurations of
           isotopically labelled alcohols can be obtained at will
           by appropriate choice of the labelled reagent.

chemists need to solve, particularly in the area of asymmetric
synthesis.

HLADH IN ORGANIC SYNTHESIS

     HLADH catalyzes oxidoreductions of the type H^+ + C=O + NADH
$\rightleftharpoons$ CH(OH) + NAD$^+$. The high costs of the NAD/H coenzymes of
$1500/mole can be readily overcome in research-scale (up to 10g
of substrate) reactions by several of the recycling methods already
developed (4,5,7), so that the absolute cost per hydride equivalent is reduced to levels of $\leq$ $10.00 per mole. This corresponds
to the per hydride price of sodium cyanoborohydride, a compound
now regarded as a relatively inexpensive reducing agent for research purposes. We use the ethanol coupled-substrate method for
reduction (10), and flavin mononucleotide (FMN) recycling for
oxidation (11) in very simple reaction procedures (5).

     The broad structural specificity of HLADH is well suited for
organic synthesis. It accepts substrates varying from acyclic,
through mono-and bicyclic, to tetracyclic (steroidal) structures
and operates on most of them with high enantiomeric specificity.
It also exhibits prochiral stereospecificity (5). Prochiral
stereospecificity of enzymes is extremely important since it permits asymmetric synthesis from symmetrical starting materials.
For reviews of this topic, see (12, 13.) Furthermore, HLADH
stereospecificity is predictable for both acyclic and cyclic substrates (14,15).

     For aldehyde reduction, the Prelog rule (14) predicts that
the hydride equivalent will always be delivered to only the *Re*-face
(12,13) of the carbonyl group. This permits the controlled syn-
rhesis of isotopically labelled *R*- or *S*-alcohols for biosynthetic
purposes (16,17), as illustrated in Fig. 1. Preparations of the

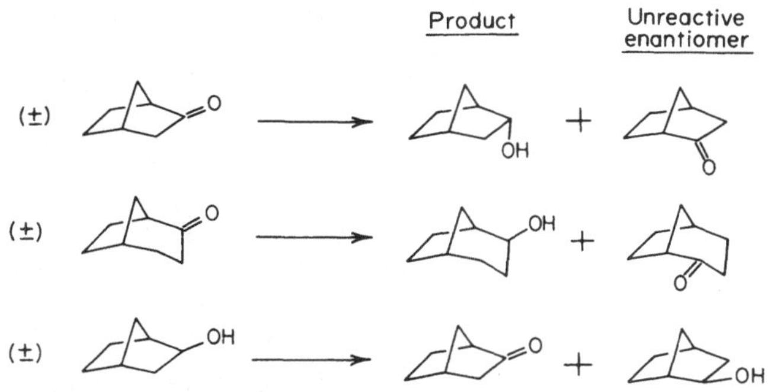

Product   Unreactive
enantiomer

(±)

Fig. 2.   Stereoselective HLADH-catalyzed oxidoreductions of
           racemic bridged bicyclic substrates.

stereoisomers of bridged bicyclic compounds are difficult chemically, but are achieved easily (18) via HLADH-catalyzed transformations of racemic ketones and alcohols (Fig. 2).

A big advantage of enzymic catalysis is the ability to combine several different kinds of specificity and thereby achieve
in one step an overall reaction requiring several separate chemical reactions. This is demonstrated in Fig. 3, where chemospe-
cific oxidation of only the secondary alcohol function is effected
together with complete enantiomeric selectivity (18).

Heterocyclic compounds are useful synthetic intermediates and
HLADH tolerates S and O heteroatoms well. The situation depicted
in Fig. 4 is of particular interest because both enantiomers are
substrates. Normally such nonselective reactions with racemic
substrates are discarded because it is considered that resolution
into enantiomers is impossible. However, each enantiomer is itself
reduced stereospecifically, one to the *trans*-alcohol, the other
to the *cis*-alcohol. These are diastereomers and easily separated. There individual chemical oxidation then gives the optically pure ketone enantiomers (19).

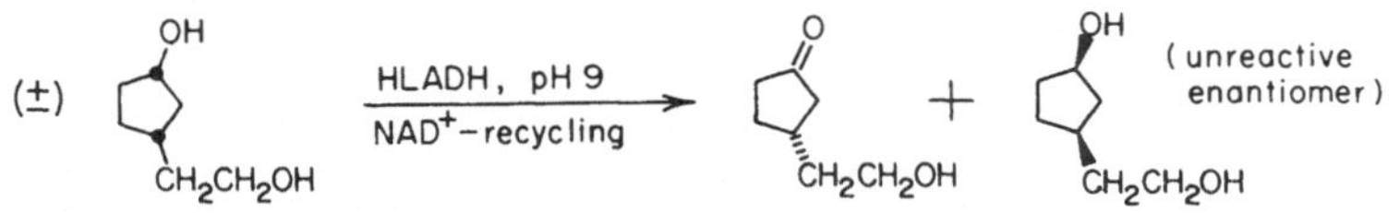

Fig. 3.   Combining different specificities in a one-step reaction.
           Chemospecific oxidation of the secondary alcohol function
           is accompanied by enantiomeric specificity.

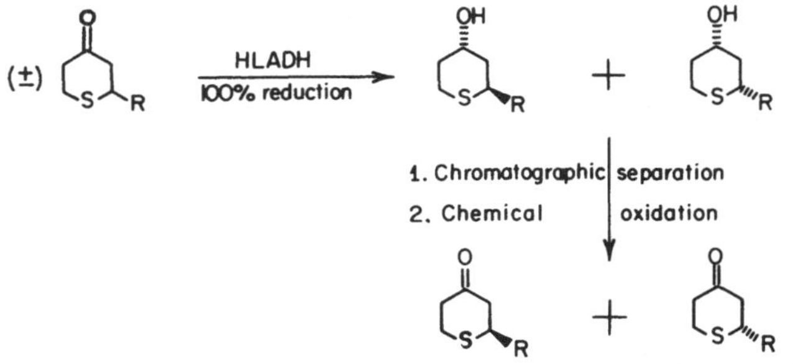

Fig. 4.   HLADH-catalyzed reduction of (±)-2-thiapyranones.

The prochiral stereospecificity of HLADH for enantiotopic (12,13) hydroxyl groups is depicted in Fig. 5. The symmetrical 3-substituted pentane-1,5-diols are oxidized with *pro-S* (12, 13) enantiotopic selectivity, first to a hydroxyaldehyde which is then further oxidized to the 3*S*-lactone products (18). Such stereo-chemical control cannot be matched chemically.

Meso-compounds also have a plane of symmetry, but their asymmetric transformation is possible using enzymes. HLADH-cata-lyzed oxidation of a very broad range of meso-diols give the cor-responding lactone products (20) of 100% enantiomeric purity (Fig. 6). Several of these lactones are excellent intermediates for im-portant products, such as grandisol (the boll weevil sex pheromone), pyrethroid insecticides, macrolide and polyether antibiotics, and prostaglandins, as indicated in Fig. 7.

Stereospecific reduction of highly symmetrical ketones is also possible. The controlled stereospecific reduction of only one carbonyl group of *cis*-2,7-decalindione shown in Fig. 8 can-not be achieved chemically (21).

The above examples are but a few of the broad synthetically useful applications already documented for HLADH. However, no matter how versatile a single enzyme is, it will obviously not meet all the synthetic demands made of it; and other enzymes with complementary specificities are required. Fortunately, it is not always necessary to identify a new enzyme for each new substrate.

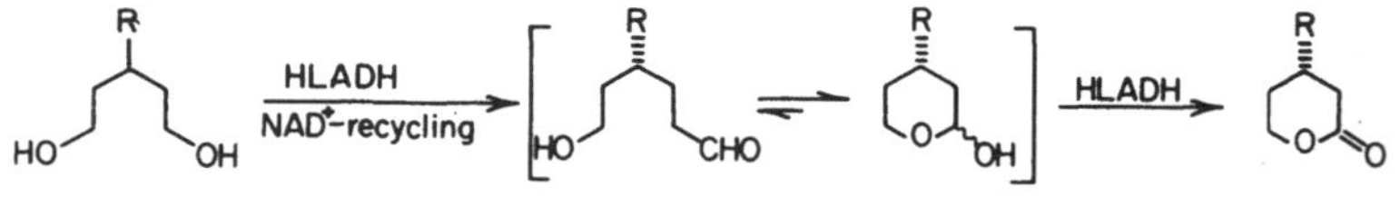

Fig. 5.   Enantiotopically selective oxidations of diols with a
           prochiral center.

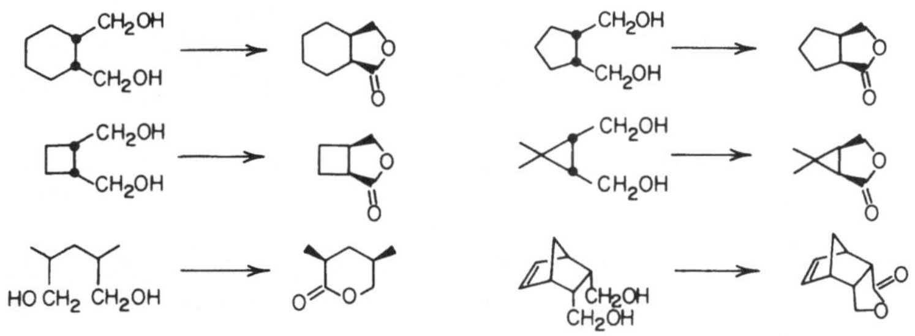

Fig. 6.   Stereospecific HLADH-catalyzed oxidations of *meso*-diols
to enantiomerically pure lactones is a general reaction.
Hydroxyaldehyde intermediates are formed initially in
each reaction (as shown in Fig. 5) but undergo direct *in
situ* enzyme-mediated oxidation to the lactone products
shown.

A few enzymes of overlapping specificities are better suited to
synthetic needs. One such situation is depicted in Fig. 9, where
only three alcohol dehydrogenases can collectively accommodate an
enormous range of substrate structures (5). Furthermore, very
subtle control of the stereochemistry of a product is possible
using different enzymes (5). This is illustrated in Fig. 10.

Enzymes will never replace the chemical methods of synthesis,
nor has this ever been advocated as an ultimate goal. Instead,
enzymes should be regarded as an unique and powerful addition to
the already extensive arsenal of methods available to the organic
chemist. What is recommended, and is now increasingly being
adopted by the leading researchers in organic synthesis, is that
both chemical and enzymic methods be considered as one complete

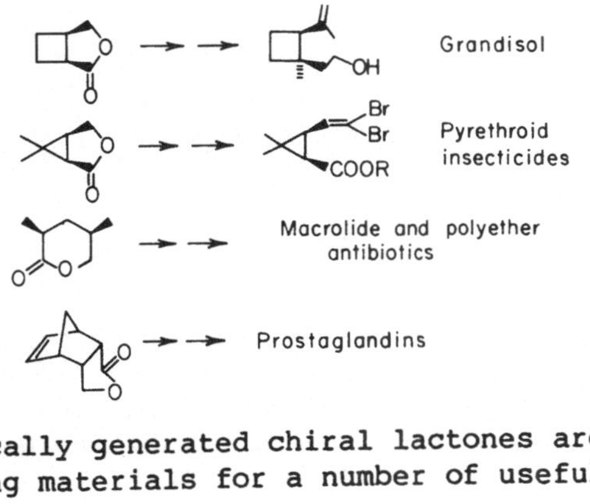

Fig. 7.   Enzymically generated chiral lactones are excellent
starting materials for a number of useful products.

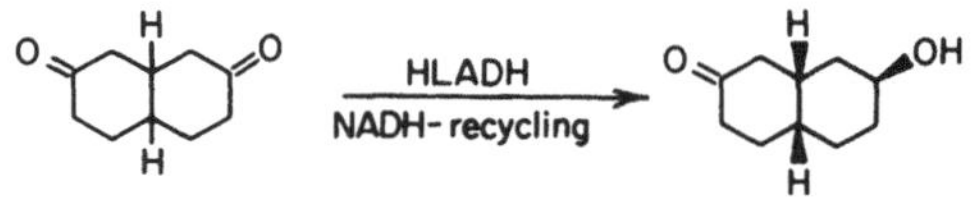

Fig. 8.   Regiospecific and stereospecific reduction of *cis*-2-decalone.

spectrum when syntheses are planned. For many reactions chemical methods will be best; but for some steps the unique advantages of a particular enzyme will dictate its use. The merit of each method must be the only consideration. Industrial processes have always been subject to this criterion; but in academe organic chemists have traditionally been biased against biochemical methods. Happily, these prejudices are now breaking down rapidly and increasingly, the enzymic *vs* chemical reagent choice is being made in the scientific and logical manner that should soon become normal practice.

ACKNOWLEDGMENTS

I am pleased to acknowledge my debt to my coworkers, whose enthusiasm and original contributions uncovered the HLADH data I have reported. I am also grateful to the National Science and Engineering Research Council of Canada and to Hoffmann–La Roche, New Jersey, for generous financial support.

REFERENCES

1. KIESLICH, K. "Microbial Transformations of Non-Steroid Cyclic Compounds", Thieme, Stuttgart (1976).

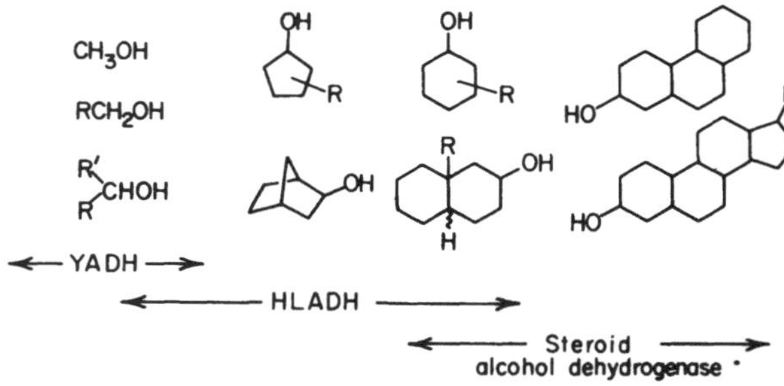

Fig. 9.   By using enzymes of suitably overlapping specificities, only three alcohol dehydrogenases are needed to access a broad range of substrate structures (reproduced by permission of Ellis Horwood Publishers, Chichester, U.K.)

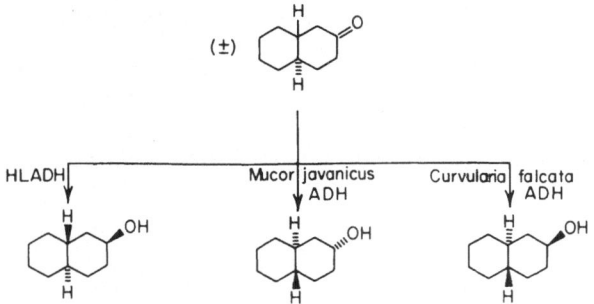

Fig. 10. Three different 2-hydroxydecalin stereoisomers can be obtained from the same (±)-*trans*-2-decalone substrate using enzymes of complementary stereospecificity.

2. MOSBACH, K. *Meth. Enzymol. 44:* (1976).

3. "Enzyme Nomenclature", Elsevier, New York (1973).

4. BARICOS, W. H., CHAMBERS, R. P., & COHEN, W. *Anal. Lett. 9:* 257 (1976).

5. JONES, J. B. & BECK, J. F. *Tech. Chem. (N.Y.) 10:* (1976); JONES, J. B. in "Enzymic and Non Enzymic Catalysis" (P. Dunnill, A. Wiseman, and N. Blakeborough, eds.) E. Horwood/J. Wiley, Chichester/New York (1980) p. 54.

6. BAUGHN, R. L., ADALSTEINSSON, O., & WHITESIDES, G. M. *J. Amer. Chem. Soc. 100:* 304 (1978).

7. WONG, C-H. & WHITESIDES, G. M. *J. Amer. Chem. Soc. 103:* 4890 (1981).

8. JOHNSON, S. C. & TAUZON, P. T. *Biochemistry 16:* (1977).

9. OPPENHEIMER, N. J. & KAPLAN, N. O. *Biochemistry 13:* (1974).

10. ZAGALAK, B., FREY, P. A., KARABATSOS, G. L., & ABELES, R. H. *J. Biol. Chem. 241:* 3028 (1966).

11. JONES, J. B., & TAYLOR, K. E. *Can. J. Chem. 55:* (1976).

12. ELIEL, E. L. *J. Chem. Ed. 57:* 52 (1980).

13. JONES, J. B. *Tech. Chem. N.Y. 10:* 479 (1976).

14. PRELOG, V. *Pure Appl. Chem. 9:* 119 (1964).

15. JONES, J. B. & JAKOVAC, I. J. *Can. J. Chem.*, in press.

16. BATTERSBY, A. R., SHELDRAKE, P. W., STAUNTON, J., & WILLIAMS, D. C. *J. Chem. Soc. Perkin I* 1056 (1976).

17. CORNFORTH, J. W., ROSS, F. P., & WAKSELMAN, C. *J. Chem. Soc. Perkin I* 429 (1975).

18. IRWIN, A. J. & JONES, J. B. *J. Amer. Chem. Soc. 98:* 8476 (1976); 99: 556, 1625 (1977).

19.   DAVIES, J. & JONES, J. B. *J. Amer. Chem. Soc. 101:* 5405
      (1979).
20.   GOODBRAND, H. B. & JONES, J. B. *Chem. Commun.* 469 (1977);
      JAKOVAC, I. J., NG, G. LOK, K. P., & JONES, J. B. *Chem.
      Commun.* 515 (1980).
21.   JONES, J. B. & DODDS, D., unpublished results.

IMMOBILIZED ENZYMES AND SYNZYMES: APPLICATIONS IN ORGANIC

SYNTHESIS

G. P. Royer

Department of Biochemistry, Ohio State University
Columbus, Ohio, USA

Enzymes as catalysts have three remarkable properties: enormous rate enhancing power (10^6-10^{14} times the rates of acid and base catalyzed reactions), specificity, and sensitivity to control. To exploit the first two properties has been the goal of many researchers. However, the use of enzymes for catalysis in industry has been hampered by three problems: enzyme cost, enzyme instability, and insolubility of reactants and/or products. Enzyme immobilization permits reuse of the enzyme, which lowers cost. Stability, in a limited number of cases, can be improved by immobilization. The solubility problem can be addressed by using immobilized enzymes with organic solvent/water mixtures.

An alternate approach is to prepare enzyme-like catalysts from non-biological materials; but this entails a long-term effort because of the complexity of enzyme active centers. However, with enzyme-induced rate enhancements of 10^6-10^{14} a near miss in duplication attempts could be a fabulous success. Immobilized enzymes have been used in the preparation of a variety of organic compounds. Some examples may be seen in Table 1.

In our laboratory we have used immobilized carboxypeptidase Y (I-CPY) for deblocking in peptide synthesis (5). CPY is an exopeptidase from yeast. The enzyme also has esterase activity at pH 9 where the peptidase activity is neglible. This feature permits the use of I-CPY to cleave peptide alkyl esters at the C-terminus without subsequent degradation of the chain (Tables 2 and 3). Blocked peptides are insoluble in water. The I-CPY is not compatible with organic solvents. This problem has been solved by using a poly-(ethyleneglycol) handle to carry the growing peptide chain in water (Table 2). Coupling is carried out in

TABLE 1

APPLICATIONS OF IMMOBILIZED ENZYMES IN ORGANIC SYNTHESIS

Compound(s)	Immobilized Enzyme(s)	Refs.
Antibiotics		
Penicillin	Penicillin amidase	(1)
Cephalosporin	Acetyl esterase	(2)
Steroids		
Cortisol	$11\text{-}\beta\text{-}$Hydroylase	(3)
Prednisolone	$\Delta^{1,2}$-Dehydrogenase	(3)
Amino acids (optically pure)	Acylase	(4)
Peptide syntheses (deblocking)	Carboxypeptidase Y	(5)
Peptide ester synthesis	Chymotrypsin, subtilisin	(6)
Pyrrole porphobilinogen	δ-Aminolevulinate dehydratase	(7)
Chenodeoxycholate	3α-and $7\text{-}\alpha$-Hydroxy steroid dehydrogenase	(8)
Keto acids	L-Amino acid oxidase	(9)
5'-Mononucleotides	5'-Phosphodiesterase, 5'-AMP deaminase	(10)
sn-3-Glycerol phosphate	Glycerol kinase	(11)
Ribulose 1,5-bisphosphate	Phosphoribulose kinase	(12)
Ribose 5-phosphate	6-Phosphogluconate dehydrogenase	(12)

water (pH 6) with a water-soluble carbodiimide and the amino acid ethyl ester in excess. Deblocking is accomplished using I-CPY at pH 9.0. The finished peptide is then released using CNBr cleavage at methionine on the handle (Table 3). Should the target sequence contain methionine, the sulfoxide derivative would be incorporated and reduced to the thio-ether after CNBr cleavage. A variety of sequences have been made with this technique. Optical and chemical

TABLE 2

PREPARATION OF THE HANDLE

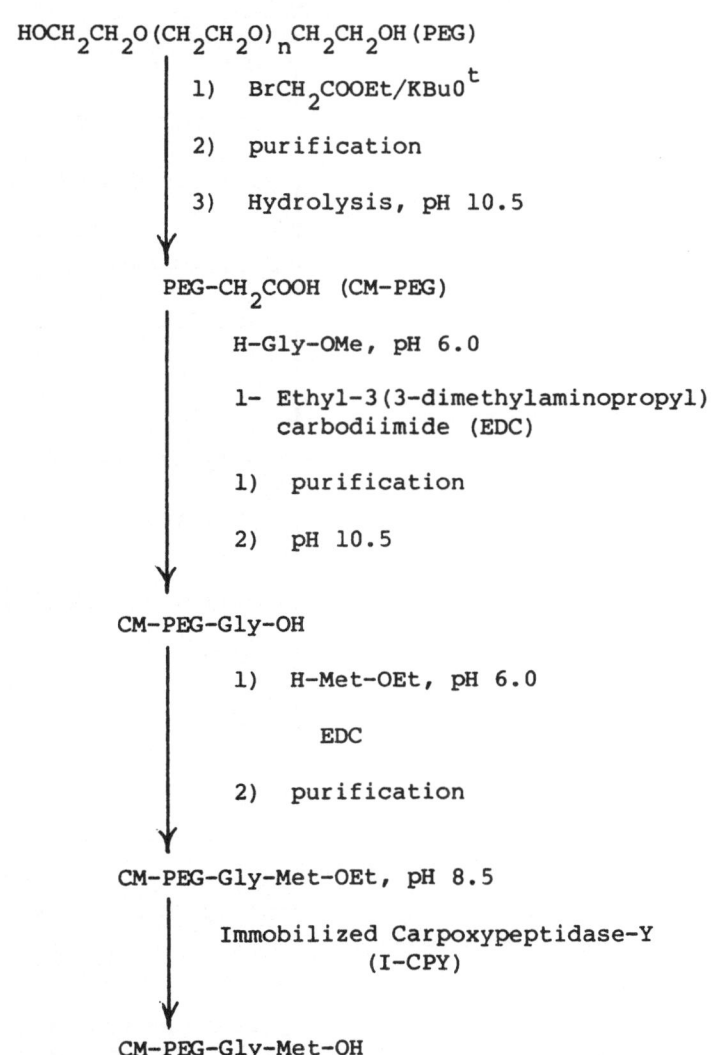

$HOCH_2CH_2O(CH_2CH_2O)_nCH_2CH_2OH$ (PEG)

1) $BrCH_2COOEt/KBuO^t$

2) purification

3) Hydrolysis, pH 10.5

$PEG-CH_2COOH$ (CM-PEG)

H-Gly-OMe, pH 6.0

1- Ethyl-3(3-dimethylaminopropyl)
carbodiimide (EDC)

1) purification

2) pH 10.5

CM-PEG-Gly-OH

1) H-Met-OEt, pH 6.0

EDC

2) purification

CM-PEG-Gly-Met-OEt, pH 8.5

Immobilized Carpoxypeptidase-Y
(I-CPY)

CM-PEG-Gly-Met-OH

purity have been demonstrated in all cases. The I-CPY may also find utility in the cleavage of peptide esters made by conventional methods and for catalysis of peptide bond formation (13). Resolu-

TABLE 3

CHAIN ELONGATION AND RELEASE

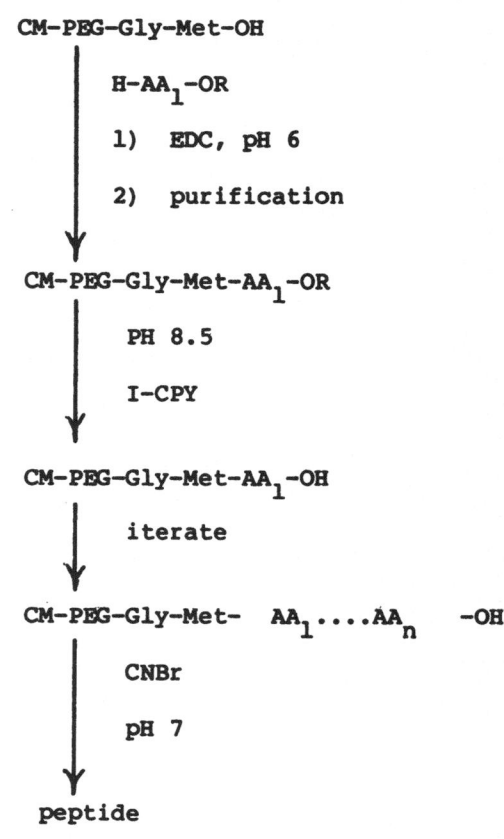

CM-PEG-Gly-Met-OH

    H-AA$_1$-OR

    1) EDC, pH 6

    2) purification

CM-PEG-Gly-Met-AA$_1$-OR

    PH 8.5

    I-CPY

CM-PEG-Gly-Met-AA$_1$-OH

    iterate

CM-PEG-Gly-Met- AA$_1$....AA$_n$ -OH

    CNBr

    pH 7

peptide

tion of racemic mixtures of amino acids and peptide derivatives
would also be possible.

    Like many immobilized enzymes I-CPY has marginal stability;
after eight deblocking runs, one half of the original activity was
lost. The idea of a stable, inexpensive synthetic catalyst with
enzyme-like characteristics has attracted the interest of many
workers (14). Much work has been done on catalysts for esterolysis
reactions. Our interest in peptide chemistry prompted us to in-
vestigate synthetic hydrogenation catalysts in the hope of finding
better ways to remove benzyl type protecting groups. The catalyst
system which we have found to be useful is a palladium/poly

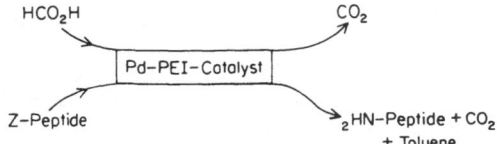

Fig. 1.   Palladium poly (ethylenimine) catalyst approach.

(ethylenimine) (PEI) system which uses formic acid as the H_2 donor
(15) (Fig. 1).   The rationale for the design of the catalyst was
that formate should be attracted to the catalyst matrix and that
the positively charged peptide (as the formate salt) should be
repelled (Fig. 1).   The Pd/PEI catalyst is very effective for the
removal of the benzyloxycarbonyl (Cb)3 group with formic acid as
the hydrogen donor; the effectiveness of the Pd/PEI catalyst is
considerably better than that of palladium black or palladium on
carbon (Fig. 2).   When H_2 gas is used as the hydrogen donor the
rates of the reactions catalyzed by the three catalysts are similar.   This result indicates that formate attraction to the PEI
matrix contributes to the rate enhancement.

REFERENCES

1.   MARCONI, W., CECERE, F., MORISI, F., PENNA, G. O., &
        RAPPEROLE, B. *J. Antibiot. 26:*   228 (1973).
2.   KONECNY, J. & VOSER, W. *Biochim. Biophys. Acta 484:*   367
        (1972).

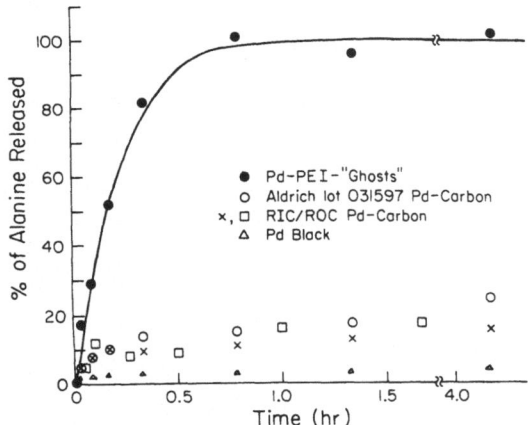

Fig. 2.   Effectiveness of hydrogenation catalysts for removal of
        protecting groups.

3.   MOSBACK, K. & LARSSON, P. O. *Biotechnol. Bioeng. 12:* 19
     (1970).
4.   CHIBATA, I. & TOSA, T. in "Applied Biochemistry and Bio-
     engineering', vol. 1, (L. B. Wingard, E. Katchalski-Katzir,
     and L. Goldstein, eds.) Academic Press, NY (1976) p. 329.
5.   ROYER, G. P. & ANANTHARANIAIAH, G. M. *J. Am. Chem. Soc. 101:*
     3394 (1979).
6.   INGALLS, R. G., SQUIRES, R. G., & BUTLER, L. G., *Biotechnol.
     Bioeng. 42:* 1627 (1975).
7.   GURNE, D. S. & SHEMIN, D. *Science 180:* 1188 (1973).
8.   PYE, E. K. *Proc. US-USSR Conf. Enzymes*, Tallinin (1977).
9.   FINK, D. J., FALB, R. D., & BEAN, M. K. *Am. Inst. Chem. Eng.
     Symp. Ser. No. 172 74:* 18 (1978).
10. NOGUCHI, S., SHIMURA, G., KIMUCA, K., & SAMEJIMA, H. *J. Solid
     Phase Biochem. 1:* 105 (1976).
11. RIOS-MERCADILLO, V. M. & WHITESIDES, G. M. *J. Am. Chem. Soc.
     101:* 5829 (1979).
12. WONG, C.-H., MCCURRY, S. D., & WHITESIDES, G. M. *J. Am. Chem.
     Soc. 102:* 7938 (1980).
13. BREDDAM, K., WIDMER, F., & JOHANSEN, J. T. *Carlsberg Res.
     Commun. 45:* 361 (1980).
14. ROYER, G. P. *Adv. Cat. 29:* 197 (1980).
15. COLEMAN, D. R. & ROYER, G. P. *J. Org. Chem. 45:* 2268 (1980).

IMMOBILIZATION OF LIVING MICROBIAL CELLS AND TRANSFORMATION OF STEROIDS

K. A. Koshcheyenko and G. K. Skryabin

Institute of Biochemistry and Physiology of
Microorganisms, USSR Academy of Sciences
Pushchino, USSR

Living microbial cells were immobilized into polyacrylamide gel (PAAG) and used to perform the following processes of steroid transformation: 1,2-dehydrogenation, 20 β-reduction, 17 β-reduction, 11α- and 11 β-hydroxylation, transformation of sterines, and hydrolysis of steroid ethers (1,2). Cells of *Arthrobacter globiformis* with 3-ketosteroid-Δ^1-dehydrogenase activity and *Sacch. cerevisiae* with 17 β-hydroxysteroid dehydrogenase activity were immobilized by various techniques: entrapment into PAAG, polyvinyl alcohol membranes, and photo-crosslinkable resins, adsorption on carriers, and covalent binding with activated silica gel. In all cases 3-ketosteroid-Δ^1-dehydrogenase activity was practically equal to the activity of the free cells and made up 0.14 μmole/mg cells/min; with 17 β-hydroxysteroid dehydrogenase the activity of the immobilized yeasts was far less than the activity of the free cells.

Variations in the viability and distribution of cells of *A. globiformis* and *Sacch. cerevisiae* during prolonged transformations in buffer solutions were studied in detail. The behavior of cells, which were not incubated in the nutrient medium, after immobilization was compared with that of cells once or twice incubated. After the first incubation the number of viable cells showed a 5.7-fold increase and after the second (in 3 months) incubation a 4.7-fold increase. After two months and 90 transformations the number of viable cells was 35 and 102% and after 6 months it was 18 and 51% for non-incubated and once or twice incubated cells, respectively. Transmission and scanning electron microscopy examinations showed the prevalence of intact cells in gel sublayers in all three cases after 6 months and 200 transformations of hydrocortisone into prednisolone. The enzymic activity was

44, 72 and 100%, respectively, of the initial activity of the granules (95% yield of prednisolone).

Light microscopy of half-thin sections together with transmission electron microscopy were used for the first time to study the pecularities of the propagation of *Sacch. cerevisiae* cells in gels and during stereospecific 17 β-reduction of 3-methoxy- $\Delta^{1,3,5(10),9(11)}$-8,14-seko-extratetraendion-14,17. During semicontinuous transformation in buffer solutions, cell layers in gel granules which were not incubated became differentiated with further decrease of the sublayer. During incubation in the nutrient medium, cells propagated intensively in the sublayer under the gel film. Cell propagation was not observed in the granule center. Anomalous yeast forms occurred in the middle layers of the incubated granules. During 20 days, the morphology of the cells in the sublayers was no different from that of free cells in the exponential growth phase. After incubation in the medium, the 17 β-hydroxysteroid dehydrogenase activity of the granules with cells was 2.5-fold greater and was stabilized by a factor of 10. After 45 days and 45 transformations (20 hr each), the activity was decreased by 15-20%; the product yield was 90%.

A high stability of 17 β-hydroxysteroid dehydrogenase activity of yeast cells and 3-ketosteroid-Δ^1-dehydrogenase activity of *A. Globiformis* 193 was conditioned evidently by their long-term storage in a viable state. Cells propagated in the gel during a single incubation in the nutrient medium as well as in the course of repeat transformations in buffer solution, based on the lysis products of a part of the immobilized cells.

REFERENCES

1. KOSHCHEYENKO, K. A. *Prikl, Biokhim. Mikrobiol.* 17(4):   477 (1981)
2. KOSHCHEYENKO, K. A., SUKHODOLSKAYA, G. V., TYURIN, V. S. & SKRYABIN, G. K. *Euro. J. Appl. Microbiol. Biotechnol.* 12: 161 (1981).

STEROID BIOCONVERSION BY IMMOBILIZED CELLS

C. Glomon, P. Germain, A. Miclo and J. M. Engasser

Laboratoire de Microbiologie Industrielle, ENSAIA, and
Laboratoire des Sciences du Genie Chimique,
CNRS-ENSIC, Institut National Polytechnique de
Lorraine, Nancy, France

Nocardia mutants, unable to grow on steroids as a sole carbon source but producing intermediates of steroid catabolism,
have been isolated and tested in free and immobilized form. A
methyl-perhydroindanone propionic acid (MEPHIP) accumulating mutant catalyses the multistep conversion of androst-4-ene dione
(Δ^4ADO) into MEPHIP with a specific activity of 1 mmole/hr/g cell
and a yield of 90%. Entrapment of cells in 3 mm diameter polyacrylamide gel beads gives the same final yield, but results in
substantial decrease in specific cellular activity due to steroid
diffusion limitations. In the immobilized form, cells exhibit a
very high stability, only a slight reduction in cellular activity
being observed after 9 months of operation (Fig. 1). When exposed
to pure organic solvents, such as hexane and benzene, *Nocardia* re-
tains its steroid converting activity. However, only the single
step dehydrogenation of Δ^4ADO into androst-1,4-diene dione ($\Delta^{1,4}$
ADO) is observed, with an additional reduction in cellular activ-

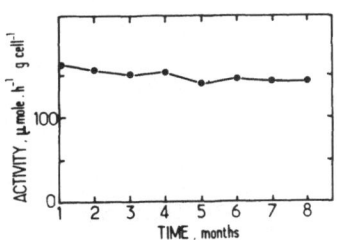

Fig. 1. Kinetics and stability of Δ^4ADO conversion into MEPHIP
by *Nocardia* mutant immobilized in polyacrylamide gel.

ity (25 μmoles/hr/g cell) and stability (2 days half life). The
activity is increased by the addition of phenazine methosulfate,
as an electron acceptor; and the stability is increased by the use
of buffer to protect the cell from the hydrophobic environment.
Polyacrylamide immobilized cells yield a similar 75% dehydrogenation conversion in hexane. Even in the presence of buffer the
cellular activity is limited to 8 days.

Nocardia cells were also immobilized in hollow fiber ultrafiltration modules consisting of bundles of asymmetric membranes.
Cells were first grown inside the macroporous sponge or outside
the membrane on acetate as a carbon source. Then the steroid
solution was perfused through the fiber lumen. In such a reactor
Δ^4ADO was dehydrogenated at a rate of 270 μmoles/hr/g cell, this
activity remaining essentially the same for 5 months (Fig. 2) (1).

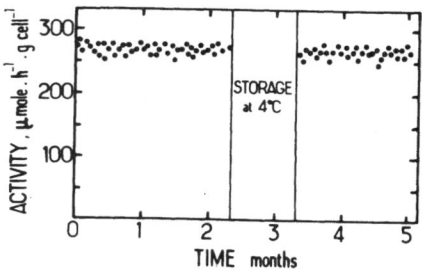

Fig. 2. Kinetics and stability of Δ^{4}ADO conversion into $\Delta^{1,4}$ADO
by Nocardia mutant immobilized in hollow fiber reactor.

REFERENCE

1. GLOMON, C., thesis, Nancy 1981.

STEROID △1-DEHYDROGENASE ISOENZYMES IN *CORYNEBACTERIUM* SPECIES CHOL 73 T 191

W.-R. Mueller, W. Preuss and R. D. Schmid

Departments of Biotechnology and Fat Research
HENKEL KGaA, Duesseldorf, Fed. Rep. Germany

Corynebacterium spec. mutant Chol 73 T 191 degrades cholesterol nearly quantitatively into 20-carboxy-pregna-1,4-diene-3-one (BNC) in the absence of inhibitors (1). From the presence of small amounts of 20-carboxy-pregn-4-ene-3-one (Δ4-BNC), androsta-1,4-diene-3,17-dione (AD) and androst-4-ene-3,17-dione (ADD), the presence of a steroid Δ1-dehydrogenase (E.C. 1.3.99.4) can be inferred. This enzyme has recently been isolated and purified 160 fold from *Nocardia opaca* (2). In *Mycobacterium* in direct evidence for the presence of an isoenzyme system has been presented (3). We now report direct proof for steroid Δ1-dehydrogenase isoenzymes in *Corynebacterium* spec. Chol 73 T 191, based on the R_B position of the specifically stained steroid Δ1-dehydrogenase band after PAA-electrophoresis with bromophenol blue having an R_B value of 1.00. With AD and cholesterol as inducers, the R_B values were 0.09 and 0.71, respectively.

With an isoenzyme preparation purified 3 fold by ammonium sulfate fractionation we have tested the activity towards a variety of steroids. The preparation was active against many steroids containing a 3-oxo-4-ene function. Steroids with a saturated A-ring or a large hydrophobic side chain were not dehydrogenated (4). All pregna-4-ene-3-one derivatives, mostly novel steroid compounds, were oxidized with significant differences in rates (Table 1).

127

TABLE 1

ACTIVITY OF STEROID Δ1-DEHYDROGENASE ISOENZYMES FROM STRAIN
CHOL 73 T 191 TOWARDS STEROID SUBSTRATES

Substrate	Transformation* (AD = 100) (%)
androst-4-ene-3,17-dione (AD)	100
cholest-4-ene-3-one	0
lithocholic acid	0
lithocholic acid, methyl ester	0
chol-4-ene-3-one-24-oic acid, methyl ester	61
3-acetoxy-20-carboxy-pregn-5-ene	0
3-hydroxy-20-carboxy-pregn-5-ene	0
20-carboxy-pregn-4-ene-3-one (Δ4 BNC)	83
20-carboxy-pregna-4,17(20)-diene-3-one	78
20-carboxy-pregn-4-ene-3-one, methyl ester	0
20-carboxamido-pregn-4-ene-3-one	17
20-nitrilo-pregn-4-ene-3-one	50
20-carbaldehydo-pregn-4-ene-3-one	61
pregn-4-ene-3,20-dione (progesterone)	71
20-amino-pregn-4-ene-3-one, HCl-salt	133
pregna-4,20-diene-3-one	61

*Substrates were assayed for initial velocity with crude
 isoenzyme mixture and phenazine metosulfate/cytochrome C at
 pH 8.0 and 20°C. Activity towards AD was 244 U/mg protein
 under standard assay conditions.

We have succeeded in the partial purification of the isoenzyme mixture by PAA electrophoresis. Differences in the activity of these enzymes towards steroid substrates will be reported
later.

REFERENCES

1. HILL, F., PREUSS, W., SCHINDLER, J. & SCHMID, R. D., Eur.
    Patent Appl. 4913, 1979.
2. LESTROVAJA, N. N., DANHARDT, S., & HORHOLD, C. Z. Allg.
    Mikrobiol. 18: 189 (1978).
3. WOVCHA, M. G., BROOKS, K. E. & KOMINEK, L. A. Biochim.
    Biophys. Acta 574: 471 (1979).
4. PENASSE, L. & NOMINE, G. Eur. J. Biochem. 47: 555 (1974).

USE OF IMMOBILIZED ENZYME REQUIRING COFACTOR REGENERATION AND

OF IMMOBILIZED MYCELIUM FOR STEROID MODIFICATION

M. D. Legoy, F. Ergan, P. Dhulster, M. N. Kim and
G. Gellf

Laboratoire de Technologie Enzymatique, ERA
du CNRS, Technical University, Compiegne, France

Two biological approaches have been investigated for specific modifications of steroids: one with a purified enzyme and the other with microbial cells. In the two cases use of organic solvents were needed for the solubilization of substrates and products.

The use of an immobilized enzyme that requires cofactor regeneration is exemplified by 3α-hydroxysteroid dehydrogenase for the specific dehydrogenation of androsterone to androstane dione. The enzyme was immobilized, using a previously described method (1); and the enzyme activity was tested with different organic solvents. The best results were obtained with water-methanol. The functional stability of the immobilized enzyme was continuously monitored at 30°C during 25 days. In comparison the storage stability of free enzyme was only 16 days.

The enzyme needs NAD as cofactor; and from an economical point of view the regeneration of cofactor is a necessity for a continuous process. The cofactor regeneration has been performed by enzymatic (2), electrochemical (3), and chemical (4) methods.

In this system the enzyme and cofactor are co-immobilized in the same support. The NAD regeneration is performed with oxygen through a methoxy derivative of phenazine methosulfate as the electron carrier. With batch processing the O_2 consumption is nearly total after 30 min. For a continuously stirred tank reactor under the same conditions the steady state is obtained after 3 hr; and total inactivation occurs after 90 hrs.

The use of immobilized mycelium is exemplified by the bio-conversion of progesterone by *Aspergillus phoenicis* to produce 11α-hydroxy progesterone, the precursor of cortisone. These experiments were performed with whole cells, to avoid the purification steps and allow reactions requiring cofactors. In this case NADPH was required; and its regeneration could occur as part of the microorganism metabolism. *Aspergillus phoenicis* transforms progesterone to 11α-hydroxy progesterone, dihydro progesterone, 15β-hydroxy progesterone, 6β-hydroxy progesterone, and 16β-hydroxy progesterone. The yield of progesterone transformation to 11α-hydroxy progesterone reaches a maximum after 10 hr incubation. The percentage of dihydroxy progesterone increases gradually when 11α-hydroxy progesterone is degraded to other products. The optimal pH for this activity is about 3; and at this pH the number of products from progesterone increases.

In using this process for continuous production it is necessary to immobilize the fungi. Several methods of immobilization have been tested: kappa-carrageenan, alginate, polyurethan, albumin foam, and gelatin foam. The calcium alginate is the best one for the progesterone transformation; but as far as the selectivity of 11α-hydroxy progesterone is concerned, the kappa-carrageenan and polyurethan supports give better results. The progesterone transformation activity can be increased by the addition of glucose to the reaction medium and by the optimization of the progesterone and solvent concentrations.

REFERENCES

1. LEGOY, M. D., LARRETA-GARDE, V., ERGAN, F. & THOMAS, D.
    J. Solid Phase Biochem. 4: 143 (1979).
2. KLIBANOV, A. M. & PUGLISI, A. V. *Biotechnol. Lett. 2:* 445
    (1980).
3. COUGHLIN, R. W., AIZAWA, M., ALEXANDER, B. F. & CHARLES, M.
    Biotechnol. Bioeng. 515 (1975).
4. LEGOY, M. D., LARRETTA-GARDE, V., LE MOULLEC, J. M., ERGAN,
    F., & THOMAS, D. *Biochimie 62:* 341 (1980).

HYDROXYLATION OF STEROIDS BY IMMOBILIZED MICROBIAL CELLS

A. Tanaka, K. Sonomoto, M.M. Hoq, N. Usui, K. Nomura and S. Fukui

Department of Industrial Chemistry, Faculty of
Engineering, Kyoto University, Kyoto, Japan

It is well known that many fungi hydroxylate steroids and other compounds at specific position(s). Immobilized fungi will be useful as biocatalysts for such hydroxylation processes. However, the mycelial cells of fungi are fragile; and the complex hydroxylation systems may lose activity during immobilization. These facts make the immobilization of active and vegetative mycelia difficult. Recently, Ohlson et al. (1) described steroid hydroxylation using mycelial cells of *Curvularia lunata*, which were derived from the spores entrapped in calcium alginate gels; however, the hydroxylation activity of the immobilized cells was somewhat unstable. We have developed novel and convenient methods to immobilize enzymes, microbial cells, and organelles using photo-crosslinkable resin prepolymers and water-miscible urethane prepolymers (2, 3). These methods can be applied to the entrapment of living fungal spores because of the mildness of the procedures. Spores of *C. lunata* were entrapped in various gels, and incubated for 60 hr to form mycelia in potato-dextrose broth containing Reichstein's Substance S (RSS) as inducer. The immobi-

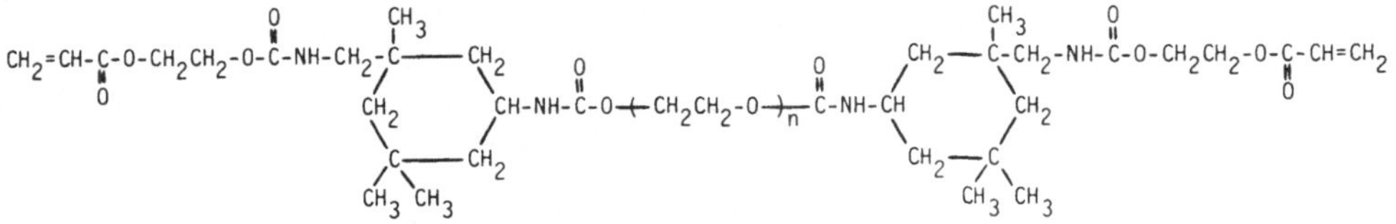

Fig. 1. Structure of hydrophilic photo-crosslinkable resin prepolymers (ENT). MW of poly(ethylene glycol): 1000, 2000, 4000 or 6000.

lized fungal mycelia thus obtained were used to hydroxylate RSS
at the 11ß-position to yield hydrocortisone in a reaction medium
containing 2.5% dimethyl sulfoxide (4).

Among the methods examined, entrapment with hydrophilic
photo-crosslinkable resin prepolymers (ENT) (Fig. 1), especially
with the prepolymer having a main-chain of poly(ethylene glycol)-
4000 (ENT-4000), was found to give the most active preparation;
the activity of the entrapped mycelia was comparable to that of
the free mycelia. The chain-length of the prepolymers affected
remarkably the development and activity of the entrapped mycelia.
Entrapment with longer chain prepolymers (ENT-4000 and ENT-6000)
permitted good development of the mycelia; and the activity of the
preparations was high. The mycelia entrapped with shorter chains
(ENT-1000 and ENT-2000) showed only poor development.

The hydroxylation system of the ENT-4000-entrapped mycelia
was far more stable than that of the free counterpart and could
be re-activated by incubating the entrapped mycelia in potato-
dextrose broth in the presence of RSS (Fig. 2). At least 50 batch

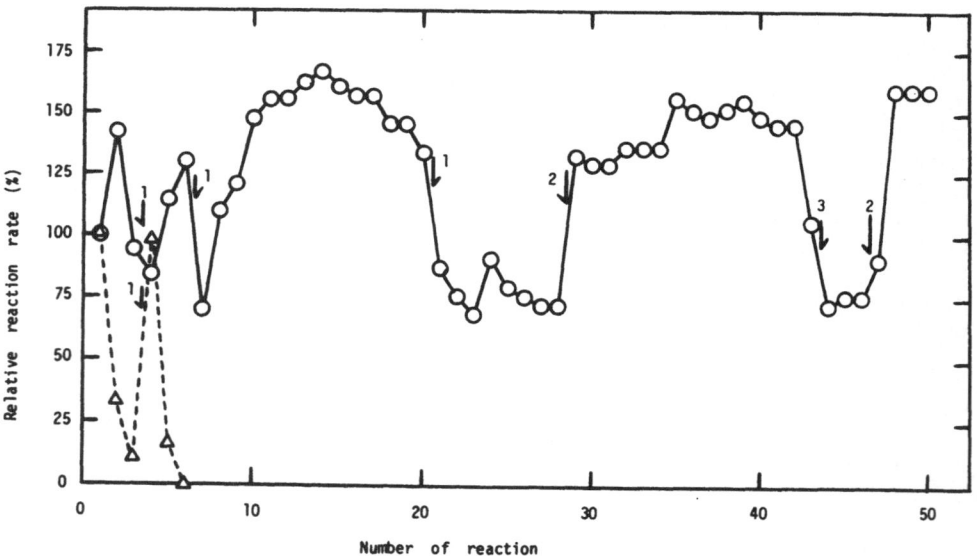

Fig. 2.   Repeated use of *C. lunata* mycelia in hydroxylation of
          RSS. Each reaction was carried out for 48 hr. Arrows:
          1) reactivate for 24 hr with 0.5 mg/ml RSS; 2) reacti-
          vate for 48 hr with 1.0 mg/ml RSS; 3) reactivate for 48
          hr with 1.5 mg/ml RSS. o ENT-4000 entrapped mycelia; Δ
          free mycelia.

reactions could be carried out over 100 days at a conversion ratio of about 60%.

Mycelia of *Rhizopus stolonifer* entrapped with ENT and grown in a similar manner as that of *C. lunata* also hydroxylated progesterone with a good yield to give 11α-hydroxyprogesterone.

REFERENCES

1. OHLSON, S., FLYGARE, S., LARSSON, P. O. & MOSBACH, K. *Eur. J. Appl. Microbiol. Biotechnol. 10:* 1 (1980).
2. FUKUI, S., SONOMOTO, K., ITOH, N. & TANAKA, A. *Biochimie 62:* 381 (1980).
3. FUKUI, S., OMATA, T., YAMANE, T. & TANAKA, A., in "Enzyme Engineering," vol. 5 (H. Weetall and G. Royer, eds.) Plenum, New York (1980) p. 347.
4. SONOMOTO, K., HOQ, M. M., TANAKA, A. & FUKUI, S. *J. Ferment. Technol. 59:* 465, (1981).

ENZYMATIC SYNTHESIS OF SULFUR AND SELENIUM AMINO ACIDS

K. Soda, N. Esaki and H. Tanaka*

Institute for Chemical Research, Kyoto University
and Kyoto College of Pharmacy,* Kyoto 607, Japan.

A variety of sulfur and selenium amino acids occur in nature; and most play important physiological roles. However, the mechanisms of their biosynthesis, in particular from the enzymological standpoint, have remained almost unsolved. L-Methionine γ-lyase (E.C.4.4.1.11) catalyzes the conversion of L-methionine into α-ketobutyrate, methanethiol, and ammonia. The enzyme was purified to homogeneity from *Pseudomonas putida* (=*Ps. ovalis*)(IFO 3738) and crystallized. It has a MW of about 173,000 and contains 4 mol of pyridoxal 5'-phosphate per mol of enzyme as a coenzyme (1). The present work was undertaken to elucidate the enzymatic β- and γ-replacement reactions and to synthesize various sulfur and selenium amino acids enzymatically.

The enzyme catalyzes α,β- and α,γ-elimination reactions of L-Methionine, several derivatives of L-methionine, and L-cysteine, e.g., L-ethionine, L-homocysteine, DL-methionine sulfone, and S-methyl-L-cysteine as follows:

$$RSCH_2CH_2CH(NH_2)COOH + H_2O \rightarrow RSH + CH_3CH_2COCOOH + NH_3$$

$$RSCH_2CH(NH_2)COOH + H_2O \rightarrow RSH + CH_3COCOOH + NH_3$$

When methionine and ethanethiol were incubated with L-methionine γ-lyase, L-ethionine and methanethiol were produced; the γ-replacement reaction between the thiomethyl group of methionine and ethanethiol occured. Various other alkanethiols (C_3-C_7) and arylthioalcohols, e.g., benzenethiol and β-naphthalenethiol, also serve as the substrates; and the corresponding sulfur amino acids and methanethiol are produced. The enzyme catalyzes also the β-replacement reaction between S-methyl-L-cysteine and various thiols,

135

e.g., ethanethiol to yield S-substituted-L-cysteines and methane-thiol, as follows:

$$RSCH_2CH_2CH(NH_2)COOH + R'SH \rightarrow R'SCH_2CH_2CH(NH_2)COOH + RSH$$

$$RSCH_2CH(NH_2)COOH + R'SH \rightarrow R'SCH_2CH(NH_2)COOH + RSH$$

L-Selenomethionine undergoes enzymatic α,γ-elimination to yield α-ketobutyrate, methaneselenol, and ammonia. When a reaction mixture containing L-methionine, benzeneselenol, and enzyme was incubated, a new product was synthesized and identified as Se-phenylselenohomocysteine by a GC–MS method. The enzyme catalyzes also β-replacement reactions between S-methyl-L-cysteine and ben-zeneselenol to yield Se-phenylselenocysteine as follows:

$$RSCH_2CH_2CH(NH_2)COOH + R'SeH \rightarrow R'SeCH_2CH_2CH(NH_2)COOH + RSH$$

$$RSCH_2CH(NH_2)COOH + R'SeH \rightarrow R'SeCH_2CH(NH_2)COOH + RSH$$

REFERENCE

1.   TANAKA et al. Biochemistry 16:   100 (1977).

PRODUCTION OF ALKALOIDS WITH IMMOBILIZED CELLS OF *CATHARANTHUS ROSEUS*

C. A. Lambe, A. Reading, S. Roe, A. Rosevear, and
A. R. Thomson,

Biochemistry Group, AERE, Harwell, Didcot, Oxon, UK

The use of immobilized cells in reactor systems could simplify and increase the efficiency of producing secondary metabolites such as steroids, alkaloids, flavors and growth control compounds from plant cells and complex proteins from animal cells. Possible advantages include minimal biomass production, continuous operation, compact reactors, simplified downstream processing, high levels of containment, and control of cellular microenvironment.

We have studied cells of *Catharanthus roseus (Vinca)*, as a model system. These were routinely maintained on Gamborg B5 medium with 2,4D as the sole hormone. A number of other culture mediums were tested for their effect on alkaloid production; and B5 medium containing indolyl acetic acid and benzyl adenine with 5% sucrose (IBS) was selected for further studies. No extracellular alkaloid was detected with free cells. In IBS, ajmalicine and serpentine production begins during early log phase; and the accumulation reached a peak during the stationary phase, after which turnover occured. The cells were immobilized in a modified polyacrylamide gel sheet, reinforced with cotton mesh. The cell cloths were transferred easily between shake flasks and were used in column reactors. A cell concentration of 54 g wet weight of cells/100g of composite was achieved with up to 90% retention of cell viability.

With immobilized cells in a medium which was regularly changed, the alkaloid was produced continuously and accumulated in the medium. The rate of alkaloid production was dependent on the duration of free cell growth in the IBS medium prior to immobilization. *Vinca* cells immobilized after 14 days growth in the

IBS medium produced ajmalicine and serpentine for 160 days. The total yield of alkaloid was three times the maximum available from free suspension cultured cells.

Thus, we have used immobilized plant cells to produce cell free secondary metabolites over prolonged periods in the absence of cellulolytic agents.

ACKNOWLEDGMENT

This work is supported by the U.K. Department of Industry.

PROPYLENEOXIDE PRODUCTION BY IMMOBILIZED *NOCARDIA CORALLINA*

B-276

K. Furuhashi, S. Uchida, I. Karube,* and S. Suzuki*

Bio Research Center Co., Saitama, and Research
Laboratory of Resources Utilization*,
Tokyo Institute of Technology, Yokohama, Japan

Production of propyleneoxide (PO) from propylene by *Nocardia corallina* B-276 has been reported by the authors (1). *N. corallina* B-276 grew on propylene and produced PO in the stationary growth phase; but the production of PO stopped in a few days. This paper deals with the studies on reaction conditions of PO production by immobilized whole cells of *N. corallina* B-276.

Living whole cells of *N. corallina* B-276, grown on glucose and entrapped in polyacrylamide gel, were used for PO production by batch reaction in flasks. Four ml of gel, which contained 76 mg cells as dry weight, and 20 ml reaction medium were placed in a 500 ml rubber stopped flask and incubated under oscillation at 30°C for 23 hr in the presence of 20% propylene in the gas phase.

The PO-producing activity of the immobilized cells increased gradually as the reaction proceeded and reached its maximum (7 mmol/L/day) on the 3rd-6th day of reaction. The maximum activity of the free suspended cells appeared on the first day. The appearance of the maximum activity of immobilized cells was retarded markedly when the cell concentration in the gel was high. It could be explained that this phenomenon was caused by the limited supply of oxygen into the gels, because a higher partial pressure of oxygen and/or smaller particle size gels increased the activity on the first day of reaction.

In the presence of an appropriate concentration of urea in the medium, higher maximum activity was obtained. At elevated urea concentrations, the PO-producing activity was low; and detachment of cells from the gel was observed. This suggested the

growth of cell mass. When the partial pressure of oxygen was
raised in the presence of urea in the medium, the maximum activity
increased to 15 mmol/L/day; but the activity decreased rapidly
after it reached its maximum.

Continuous production of PO by immobilized cells in a bubble
tower reactor (28 mm x 400 mm) was then studied. In a bubble tower
the product PO was continuously removed by aeration from the reaction mixture, with the concentration of PO being maintained at
a low value. The activity of the immobilized cells increased to
20 mmol/L/day in the medium supplemented with an appropriate concentration of urea without any sudden decrease in the activity
(which was observed in the flask reactions). The rapid decrease
in activity observed in flask reactions was considered to have
been caused by the inhibitory effect of the PO that accumulated
in the closed system. The PO-producing activity of the immobilized cells was maintained almost constant for a week in continuous
operation using the bubble tower reactor.

In conclusion, immobilization and continuous operation using
immobilized cells stabilized the PO-producing activity of *N.
corallina* B-276.

REFERENCE

1. FURUHASHI, K., TAOKA, A., UCHIDA, S., KARUBE, I. & SUZUKI, S.
    European J. Appl. Microbiol. Biotechnol. 12: 39 (1981.)

AIR-OXIDATION OF LINOLEIC ACID BY LIPOXYGENASE-CONTAINING PARTICLES SUSPENDED IN WATER-INSOLUBLE ORGANIC SOLVENT

T. Yamane

Department of Chemical Engineering, Faculty of
Engineering, Kansai University, Osaka, Japan

Some enzymic reactions can be advantageously performed in a biphasic water system (water-insoluble organic solvent) when a) the substrates have low solubilities in water, or b) they are amide- or ester-bond formation reactions in which the thermodynamic equilibrium in water is considerably shifted to the substrates. And if necessary, the enzyme can be immobilized in the aqueous phase in the hope of increasing its stability. The immobilization protects the enzyme against denaturation that might be caused by coming into direct contact with an unfavorable organic solvent at the interface. Biphasic systems were reviewed by Matinek and Berezin (1) and by Klibanov (2); and Carrea *et al.* have transformed steroids by immobilized hydroxy-steroid dehydrogenases in water-organic solvent systems (3). In their experiments the volume ratios among the enzyme-containing gels, the aqueous solution, and the organic solvent presumed permitted only the reaction of the suspended gel particles and the organic solvent emulsified in the continuous aqueous phase. In order to facilitate transfer rates of solutes between different phases, it is more desirable to perform such a reaction with the immobilized enzyme particles suspended in the water-insoluble organic solvent and with water contained only within the particles. As a model system the air-oxidation of linoleic acid by particles containing lipoxygenase-1 (EC 1.13.11.12) has been studied in liquid n-alkane in which linoleic acid was dissolved (1-100%). The product, 13-L-peroxy linoleic acid, was readily reduced by hydroxy conjugated octadienoic acid; the latter has the potential of being a versatile chemical intermediate (4). A reaction catalyzed by immobilized microbial cells in water-insoluble organic solvent also has been exemplified by a steroid bioconversion (5, 6).

The dispersiveness of several kinds of particles which were candidates for enzyme-immobilization materials (porous glass, porous silica, Sephadex G-50, Sepharose 4B, and octyl- and phenyl-Sepharose CL-4B) were tested by stirring them in their fully wetted states in n-heptane in a 50 ml beaker. The reaction conditions for the enzyme adsorbed within octyl Sepharose CL-4B through hydrophobic interactions were investigated batchwise. Each reaction was followed manometrically, the total volume of the agitated reaction mixture was 7.0 ml . Initial reaction rates were calculated from O_2 absorption. For the enzyme lifetimes, continuous experiments with enzyme immobilized on octyl Sepharose CL-4B of >88 μm were carried out in small agitated-aerated vessels with ca. 50 ml working volume. The concentration of the product in the outlet was detected spectrophotometrically at 234 nm. The gels were retained in the reaction mixture by filtering the outlet through #280 stainless steel wiremesh (53 μm opening).

Wetted inorganic particles were easily dispersed but showed low catalytic activities, whereas fully wetted amphiphilic gels (octyl- and phenyl Sepharose CL-4B) were less dispersed but had remarkably high activities. Octyl Sepharose CL-4B was investigated further. The octyl residue had dual effects: enzyme immobilization through hydrophobic interaction and imparting the gels with amphiphilic nature. The optimal batch reaction conditions were pH 10.5 for the immobilization, reaction mixture of 5% linoleic acid and 5% lecithin in n-decane, and 20% (v/v) gels. Continuous experiments using enzyme a) adsorbed through hydrophobic Interaction, b) covalently bound by CNBr, and c) physically adsorbed-crosslinked by glutaraldehyde revealed that the enzymes were still labile (half lives of several hours), suggesting the necessity of further investigations to increase the stability.

The author expresses his appreciation to T. Hibi, M. Hashimoto and K. Hatanaka for their technical assistance.

REFERENCES

1. MARTINEK, K. & BEREZIN, I. V. *J. Solid-Phase Biochem.* 2: 343 (1977).
2. KLIBANOV, A. M. *Anal. Biochem. 93:* 1 (1979).
3. CARREA, A. M., COLOMBI, F., MAZZOLA, G., CREMONESI, P. & Antonini, E. *Biotechnol. Bioeng. 21:* 39 (1979).
4. EMKEN, E. A. *J. Amer. Oil Chemts. Soc. 55:* 416 (1978).
5. YAMANE, T., NAKATANI, H., SADA, E., OMATA, T., TANAKA, A. & FUKUI, S. *Biotechnol. Bioeng. 21:* 2133 (1979).
6. FUKUI, S., OMATA, T., YAMANE, T. & TANAKA, A. in "Enzyme Engineering," vol. 5 (H. Weetall and G. Royer, eds.) Plenum, New York (1980) p. 347.

PROGRESS TOWARD ARTIFICIAL PHOTOSYNTHESIS

D. J. Graves

University of Pennsylvania,
Philadelphia, Pennsylvania, USA

Several years ago we proposed an artificial scheme to store chemically the energy of sunlight as a reactive chemical species (1). Its salient features included immobilized dyes, at least one immobilized enzyme, and a series of election transfer agents. The various species and chemical reaction events were postulated to be coupled cyclically so that the energies of two photons could be combined and trapped in a single molecule of some chemical species. This series of events is based on the major features of the "z scheme" of photosynthesis as it is presently understood.

We felt that analyzing this overall highly complex process and then examining small pieces of it in detail would lead to a better understanding of photosynthesis and of artificial photo-chemical energy storage systems. We began by tackling what appeared to be the most difficult aspects of the problem. By 1976 (2) we had shown that hydrogenase, previously reported to be a highly unstable and oxygen-sensitive enzyme, could be immobilized as a highly stable and reactive catalyst. This was only possible when the enzyme was isolated from *Alcaligenes eutropha* and the temperature was kept low. We also (3,4) studied photoelectrochemical reactions at an organic semiconductor-solution interface. Electrochemical techniques offer the advantage of coupling individual events with wires rather than by circulating electron transfer agents in a solution phase, but they introduce additional problems of their own.

More recently, we have turned our attention to the problem of immobilizing dyes which could carry out photochemical reactions or act as electron exchangers (which would be useful in many enzymatic reactions). Realizing such a goal has been more diffi-

143

Fig. 1. Hydrophobic immobilization of an electron carrier.

cult than anticipated. In summary, we have found that chemical
bonds added to a dye molecule usually cause a large change in its
redox properties. The most successful technique (Fig. 1) involves several steps: a) a dye capable of dimerizing is covalently coupled to a support phase (this usually results in
redox properties very different from those of the parent molecule); b) additional dye is allowed to adsorb on the covalently
bound dye. This second dye layer, which is hydrophobically
bonded, will have properties quite similar to those of the parent
molecule. An example is the dye thionine, which approximately re-
tains its original redox properties. Unfortunately, to date we
have not succeeded in simultaneously retaining photochemical
activity.

REFERENCES

1. GRAVES, D. J. & STRAMONDO, J. G. *Am. Inst. Chem. Eng. Symp.*
     Ser. No. 158 72: 43 (1976).
2. DELOGGIO, T., thesis University of Pennsylvania (1976).
3. AYERS, W. M., thesis University of Pennsylvania (1980).
4. AYERS, W. M. & GRAVES, D. J., *Am. Inst. Chem. Eng. Symp.*
     Ser. No. 198 76: 107 (1980).

LIGHT-SENSITIZATION OF A MICROBIAL PROTEASE

Y. Y. Lee, K. N. Kuan, and P. Melius*

Departments of Chemical Engineering and Chemistry*
Auburn University
Auburn, Alabama, USA

A trypsin-like enzyme (1-4) has been isolated from a commercial preparation of *Streptomyces griseus* pronase. The purification procedure involved successive affinity chromatography using carbo-benzoxy-L-phenylalanyl-triethylenetetraminyl-Sepharose 4B (cPTS) (5) and CM ion-exchanger columns. Starting with B grade pronase in 0.02 M Tris-HCl buffer at pH 7.2, the first step was affinity chromatography with the cPTS column with elution by 1 M guanidine HCl. The second step was ion-exchange chromatography with CM-52 (Whatman) ion-exchange cellulose and sodium acetate buffer at pH 4.2; elution was by a linear gradient of 0-0.2 M NaCl in

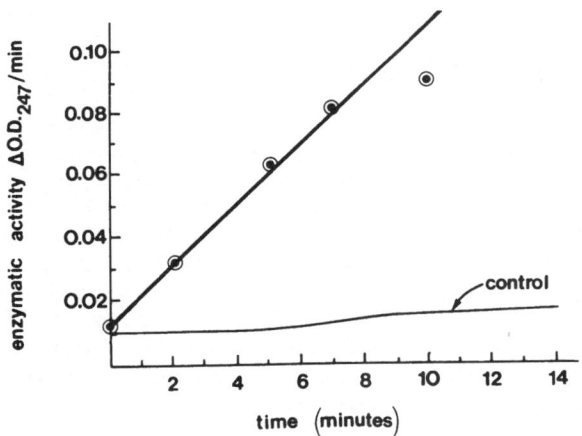

Fig. 1. Activation of *cis*-cinnamoyl-trypsin-like enzyme by UV-light.

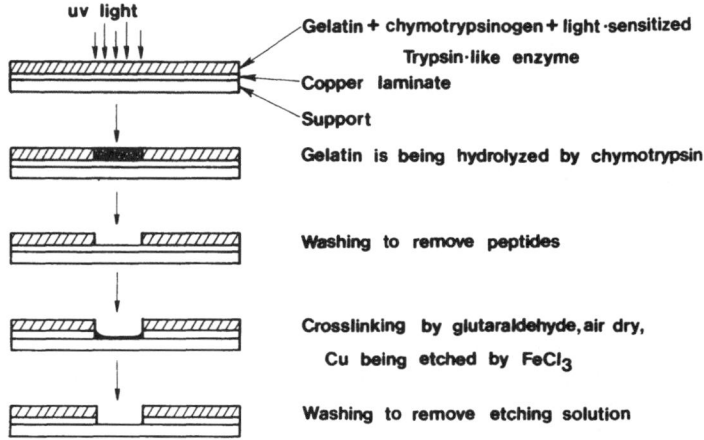

Fig. 2.   Image transfer onto a printed circuit board.

acetate buffer.   The third step involved concentration of the enzyme by dialysis against aquacide in sodium acetate-acetic acid
buffer at pH 7.8.

This microbial enzyme possessed a unique characteristic in
that it could be artificially light-sensitized by acylation with
various effector molecules.   The light-sensitization procedure was
devised so that the enzyme became inactive in the acylated form,
but regained its activity under UV-light.   The methods for inhibition and light reactivation were quite similar to those reported
for α-chymotrypsin (6).   Fig. 1 shows data on the reactivation of
the *cis* cinnamoyl-trypsin-like enzyme.   Furthermore, the degree
of reactivation was inversely proportional to the distance of the
light source from the enzyme.   We have investigated this system for
transferring an image onto a printed circuit board (Fig. 2).   By
using a stencil as shield, a certain portion of the board was exposed to UV-light.   In the light exposed area, the trypsin-like
enzyme regained its activity which in turn activated chymotrypsino-
gen to chymotrypsin.   These activated enzymes hydrolyzed gelatin,
while in the non-exposed area the enzymes remained inactive.   The
non-hydrolyzed gelatin area was cross-linked by glutaraldehyde and
air dried.   The copper surface was etched by dipping the plate
in ferric chloride solution.   After rinsing, a negative gelatin
image was obtained on the circuit board.

REFERENCES

1.   NARAHASHI, Y. *Meth. Enzymol. 19:*   651 (1970).

2. WAHLBY, S. *Biochim. Biophys. Acta 151:* 394 (1968).

3. AWAD JR., W. M., SOTTO, A. R., SIEGEL, S., SKIBA, W. E.,
    BERNSTROM, G. G., & OCHOA, M. S. *J. Biol. Chem. 247:*
    4144 (1972).

4. JURASEK, L., JOHNSON, P., OLAFSON, R. W., & SMILLIE, L. B.
    Can. J. Biochem. 49: 1195 (1971).

5. FUJIWARA, K., OSUE, K., & TSURU, D. *J. Biochem. (Tokyo) 77:*
    739 (1975).

6. KUAN, K., LEE, Y. Y., TEBBETTS, L., & MELIUS, P. *Biotechnol.
    Bioeng. 21:* 443 (1979).

ENZYMATIC PHOTOSENSITIVE MATERIALS

N. F. Kazanskaya

Chemistry Department, Moscow State University
Moscow, USSR

In up-to-date photographic materials, the principle of catalytic development is widely used. This permits a large quantum
yield increase of the primary light signal. The principal difficulty consists in the choice of pre-catalytic systems that are sufficiently stable and form active and specific catalysts on exposure to light. Some years ago we proposed to use as cis-cinnamoyl
derivatives of proteinases as photomaterials (1,2). However, the
chemical modification of protein active sites to make the protein
photosensitive is a difficult problem that decreases the number
of enzymes that can be used in photography.

The available methods of photo-immobilization allow one to
obtain the same results without premodification of the enzymes.
It is sufficient to have a photosensitive matrix to which any protein can be bound by the action of light. The binding is effected
on light-sensitive groups or on groups formed during exposure.
In such a system, the nature of the enzyme is of no consequence,
especially if the reaction of binding has a nonspecific free-
radical pattern. Table 1 presents data on the maximal photoimmobilization of α-chymotrypsin for three variants of photomaterial,
whose photosensitivity is endowed by the $-N_3$ group.

In such a process the concentration of the bound enzyme is
proportional to the exposure time. The same dependence is true
of the amount of product accumulated during dark development. If,
therefore, the reaction product is an insoluble dye, the linear
ratio between the dye density and exposure may be effected to give
half-tone image.

   The above-described photographic system enhances the primary
light signal by at least a factor of 10^3-10^4 for 10-20 min (usual
time for development). We have obtained successful images using
the following enzyme-substrate pairs: proteinases-indoxyl esters,
alkaline phosphatase-naphthyl esters, and dye diazocomponent,
peroxidase - 3,3'-benzaminobenzidine.

   Of some difficulty, in the described process, is the non-
specific sorbtion of protein in a carrier. So, the exposed material should be carefully washed using high concentrations of salts
and detergents.

REFERENCES

1. KAZANSKAYA, N. F. in "Enzyme Engineering", vol. 4, (G. Brown,
       G. Manecke, and L. B. Wingard Jr., eds.) Plenum, New York
       (1978) p. 401 and vol. 5 (H. Weetall and G. Royer, eds.)
       Plenum, New York (1980) p. 213.
2. BEREZIN, I. V., KAZANSKAYA, N. F., & MARTINEK, K. in "Enzyme
       Engineering: Future Directions", (L. B. Wingard Jr., I.V.
       Berezin, and A. A. Klyosov eds.) Plenum, New York (1980)
       p. 357.

TABLE 1

PHOTO-IMMOBILIZATION OF α-CHYMOTRYPSIN

Chromatography Paper*	Enzyme on Carrier $(mg/g \times 10^2)$
Modified with N-methyl, N-(2-chloroethyl)-n-azido- benzylamine and impregnated with enzyme solution	2.5
Impregnated with enzyme solutions and 4,4'-diazi- dostylbene-2,2'-disulfate natrium salt	1.7
Impregnated with p-azido- benzaldehydediacetal and treated with enzyme after exposure	5.0

*10-30 sec, = 370 nm, DPS-250 Hg lamp, power 1.5×10^{15}
quant/cek (at 313 nm); 30 cm.

ESTER EXCHANGE OF TRIGLYCERIDE BY ENTRAPPED LIPASE IN ORGANIC SOLVENT

K. Yokozeki, T. Tanaka, S. Yamanaka, T. Takinami,
Y. Hirose, K. Sonomoto*, A. Tanaka* and S. Fukui*

Central Research Laboratories, Ajinomoto Co., Inc.
Kawasaki and Department of Industrial Chemistry*
Faculty of Engineering, Kyoto University, Kyoto
Japan.

Production of fat with desired physical and chemical properties by exchanging fatty acid moieties of triglyceride with other
fatty acid(s) is of great industrial interest. Although triglyceride can be reformed chemically by hydrogenation or interesterification, both reactions occur at random positions and some cis
unsaturated fatty acyl moieties are converted to the trans
forms. Enzymes, such as lipase, will provide new types of triglyceride, depending on the substrate and position specificities.
Interesterification with lipase from hog pancreas has been reported in an aqueous system (1); however, the yield was not so
good. The very low solubility of substrates and products in the
aqueous system and the excess water in the reaction mixture tend
toward hydrolysis of triglyceride rather than interesterification.
Introduction of an appropriate organic solvent is required to set
up a homogeneous reaction system and to shift the reaction equilibrium to the desired direction.

To obtain interesterification by lipase, n-hexane was
selected as the best reaction medium. When a small amount of
buffer was used, interesterification of triglyceride did not
occur, probably because the n-hexane appeared to be in the form
of droplets. On the other hand, lipase catalyzed the reaction in
the presence of dispersers, such as celite. The relationships
among the amount of buffer or glycerol, the degree of fatty acid
incorporated, and the recovery of triglyceride were investigated.
Use of glycerol instead of buffer, as the enzyme activator, had
a tendency to promote the reaction toward interesterification.

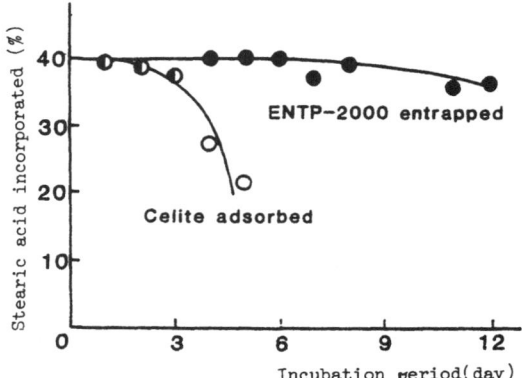

Fig. 1.   Use of lipase preparations for interesterification over
          24 hr at 40°C with shaking. Reaction mix: celite-
          adsorbed lipase entrapped in photo-crosslinked gels
          (ENTP-2000 entrapped) or celite-adsorbed lipase (equal
          to 5 mg of lipase), 0.25 g olive oil, and 0.25 g steric
          acid in 10 ml water-saturated n-hexane.

This may have occurred from a decrease in the hydrolysis of triglyceride (2).

     Acyl exchange of olive oil with palmitic acid and stearic
acid was studied using lipase from *Rhizopus delemar* under optimized conditions for obtaining cacao butter-like fat. The lipase
also was immobilized by adsorption, ionic binding, covalent binding, or entrapment using various types of supports. Celite-
adsorbed lipase entrapped with a hydrophobic photo-crosslinkable
resin prepolymer (3), ENTP-2000, exhibited the best activity among
supports tested (activity yield 75%). This suggests that the hydrophobicity of gel entrapping lipase would markedly affect the
diffusion of hydrophobic substrates, as in steroid bioconversion
in organic solvents (4). The stability of the lipase is shown in
Fig. 1.

REFERENCES

1.   STEVENSON, R. W. *et al*. *J. Am. Chem. Soc. 56:* 676 (1979).
2.   TANAKA, T. *et al*. *Agric. Biol. Chem. 45:* 2387 (1981).
3.   SONOMOTO, K., *et al*. *Eur. J. Appl. Microbiol. Biotechnol. 6:*
       325 (1979).
4.   FUKUI, S. *et al*. *Eur. J. Appl. Microbiol. Biotechnol. 10:*
       289 (1980).

SOLVENT PRODUCTION BY *CLOSTRIDIUM ACETOBUTYLICUM*

IN AQUEOUS TWO-PHASE SYSTEMS

B. Mattiasson, M. Suominen**, E. Andersson,
L. Haggstrom*, P.-Å. Albertsson and
B. Hahn-Hagerdal

University of Lund, Sweden, Royal Institute of
Technology*, Stockholm, Sweden, and University of
Helsinki**, Finland

One way to overcome product inhibition in biological conversions is by means of extraction. This has proven successful, with the aid of aqueous two-phase systems, in the bioconversion of cellulose (1). Aqueous two-phase systems are especially suitable for biological conversion processes because the high water content of both phases makes them compatible with biological material. This paper concerns the production of acetone and butanol in aqueous two-phase systems using *Clostridium aceto-butylicum*. This system was chosen because at a total solvent concentration of less than 2% the products have a strong inhibitory effect on the microorganism.

Fig. 1 compares product formation in a batch fermentation with that in an aqueous two-phase system. In both cases the same media were used: 40 g/L glucose, 10 g/L peptone, 10 g/L yeast extract, 0.8 g/L NH_4Cl, 0.6 g/L Na_2HPO_4, 0.4 g/L KH_2PO_4, 0.2 g/L $MgSO_4 \cdot 7 H_2O$ and traces of Fe^{3+}, Ca^{2+}, Zn^{2+}, Co^{2+}, Cu^{2+} and Mn^{2+}. The phase system consisted of medium supplied with 6% (w/w) Dextrane T-40 (Pharmacia Fine Chemicals, Uppsala, Sweden) and 25% (w/w) carbowax peg 8000 (Union Carbide, N.Y., USA), which resulted in a top to bottom phase ratio of 6:1. The partition coefficients of butanol, acetone, and ethanol in such a system without medium added were 2.0, 1.9, and 1.9, respectively. The microorganisms were completely partitioned to the bottom phase. The solvent concentration was measured by gas chromatography. In the phase system only the top phase was analyzed bacause the bottom phase holding dextran was too viscous. Solvent production

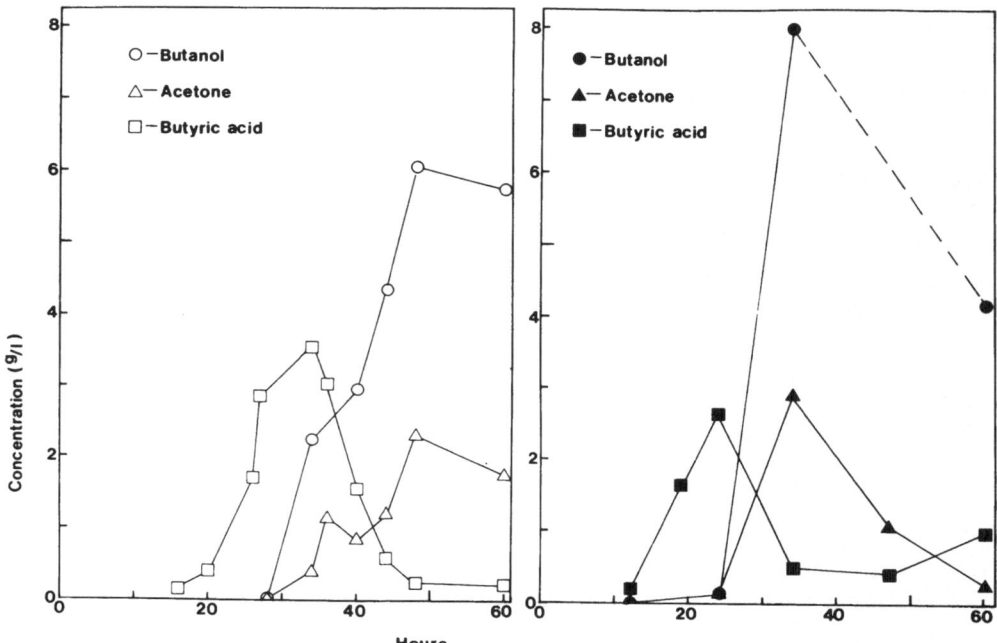

Fig. 1. Product formation by *Clostridium acetobutylicum* in batch
(open symbols) and aqueous two-phase (closed symbols)
systems. Concentrations refer to top phase.

appeared to occur earlier and be faster in the phase system than
in the batch system, indicating that the aqueous two-phase system
does not have any undesired effect on the metabolism of *Clostridium
acetobutylicum*. The mean productivity in the aqueous two-phase
system was estimated at 0.24 g/L/hr, which compares well with the
mean productivity of an ordinary batch process, 0.26 g/L/hr, with
13 g/L butanol produced after 50 hr.

Fig. 2 shows product formation in the same two-phase system
with the solvents stripped off by distillation when the total
solvent concentration in the top phase had reached 11 g/L. The
top phase was supplied with more substrate, 40 g/L glucose; and
solvent production was started again with approximately the same
yield as in the first run. The same result was obtained when the
top phase was replaced by fresh top phase without added nutrients.
C. acetobutylicum requires minute amounts of growth factors, which
means that the fraction of the medium left in the bottom phase
might very well support growth even if nutrients are excluded from
the top phase. However, high productivity of solvents was reached
with alginate immobilized *C. acetobutylicum* also when the
nutrients were excluded from the substrate medium (2), indicating

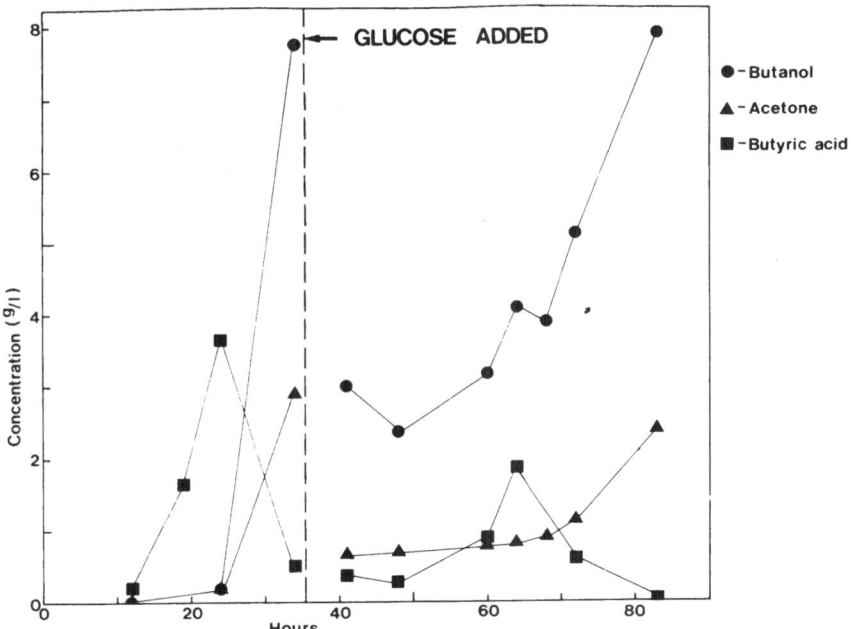

Fig. 2.   Product formation by Clostridum acetobutylicum in
          aqueous two-phase system with removal of solvents from
          the top phase. Concentrations refer to top phase.

that solvent production by *C. acetobutylicum* was not growth-
related. Similarly, once the cells reached the stationary growth
phase they could be immobilized in the bottom phase of an aqueous
two-phase system and produce solvents from a substrate without
further supply of nutrients. The results suggest continuous
solvent production with *C. acetobutylicum* and an aqueous two phase
reactor and solvent stripping (3).

This study was supported by the Swedish Board for Technical
Development.

REFERENCES

1. HAHN-HÄGERDAL, B., MATTIASSON, B. & ALBERTSSON, P.-Å.
        Biotechnol. Lett. 3: 53 (1981).
2. HÄGGSTRÖM, L. *Abst. Symp. Tech. Mikrobiol.* (H. Dellweg, ed.)
        Difodruck Schmacht, Bamberg (1979) p. 271.
3. HAHN-HÄGERDAL, B., MATTIASSON, B., ANDERSSON, E. & ALBERTSSON,
        P.-Å. *J. Chem. Technol. Biotechnol.* 32: 157 (1982).

ENZYME KINETICS AND MASS-TRANSFER ON TWO LIQUID PHASE

HETEROGENEOUS SYSTEMS

J. M. C. Duarte

QUATRUM, Empresa Nicional de Quimica Organica, SARL
Lisboa, Portugal

The rate of cholesterol to cholestenone conversion by the cholesterol oxidase of *Nocardia rhodochrous* is greatly increased when an apolar organic solvent is used to solubilize the cholesterol. The resulting emulsion is of the water-in-oil type. The system consists of: a) a water immiscible organic liquid phase where the substrate is solubilized and to where the product of the reaction must diffuse, b) an aqueous phase containing the *Nocardia* cells which may still be considered as another (solid) phase, and c) a gas phase which supplies the oxygen needed for the reaction. In such a system the enzyme kinetics are strongly dependent on the mass-transfer effectiveness and on the nature of the organic solvent (1).

The influence of cell concentration, X, on the rate of reaction, v, was evaluated. At practically constant cholesterol concentrations the enzymic rate of reaction must be balanced by the oxygen supply.

As the apparent Michaelis constant for oxygen, K_{O_2}, of cholesterol oxidase in the cells is higher than the air saturation, the enzymic rate equation may be simplified to yield a steady state oxygen concentration at the site of reaction as $K^*/(k_c x + K_L a)$, where $K^* = K_L a (O_2^*)$; (O_2^*) is the oxygen concentration in the gas phase; $K_L a$ is the overall mass-transfer coefficient for oxygen; and $K_c = V/K_{O_2}$. Thus, the specific rate of reaction can be described as:

$$v_{O_2} = \frac{K^*}{X(1 + K_L a/k_c X)} \tag{Eq. 1}$$

Therefore the rate of reaction is controlled by the magnitude of the complex parameter $(k_c/K_L a)X$; this parameter involves kinetics and mass-transfer constants and the cell concentration, similar to what happens in the case of immobilized enzyme kinetics (2, 3). When the value of this parameter is much larger than one, e.g. when the cell concentration is high and/or the value of $K_L a$ is small, the rate of reaction V_{O_2} is simply given by $K*/X$. This was the case when no forced O_2 aeration was used in small reactors (500 ml) at cell concentrations larger than 52 mg wet cells per ml of organic solvent. With these conditions the $K_L a$ obtained was about 4/hr.

When the supply of oxygen was not rate limiting, it was observed that the reciprocal plot of the rates against cell concentrations was linear. This fact was attributed to the limitation of the surface area of the dispersed phase for cholesterol transfer. This limitation became more severe when the cells were entrapped in hydrophilic gels (1) with rates tending to the rate of the completely aqueous system (4).

This work was done at the Chemical and Biochemical Engineering Department of University College London with a grant from the Calouste Gulbenkian Foundation of Lisboa.

REFERENCES

1. DUARTE, J. M. C. & LILLY, M. D. in "Enzyme Engineering" vol. 5 (H. Weetall and G. Royer, eds.) Plenum, New York (1980) p. 363.
2. REGAN, D. L., LILLY, M. D. & DUNNILL, P. D. *Biotechnol. Bioeng. 16:* 1081 (1974).
3. ATKINSON, B. & LESTER, D. E. *Biotechnol. Bioeng. 16:* 1299 (1974).
4. DUARTE, J. M. C. *Aabst. Sec. Cong. Mediterraneo Ingen. Quim.,* Barcelona, Spain (1981).

PHYSICO-CHEMICAL MEANS OF INCREASING THE YIELD OF END PRODUCTS IN BIOCATALYSIS

I. V. Berezin

Chemistry Department, Moscow State University
Moscow, USSR

Physicochemical approaches, which enable an increase in the yields of end products in enzyme-catalyzed reactions, have been analyzed comprehensively (1). These are: I) equilibrium approaches where higher yields are due to a shift in the thermo-dynamic equilibrium of the reacton; e.g., a) a method based on low solubility of the end product; b) a method of consecutive re-actions; c) enzymatic synthesis in aqueous-organic mixtures, and d) water-water-immiscible organic solvent biphasic systems; and II) nonequilibrium (kinetic) approaches, where the yield of the end product considerably exceeds the equilibrium level. These are, first and foremost, hydrolytic enzyme-catalyzed transfer reactions and kinetically controlled equilibrium. Some recently estab-lished unexplained facts are reported on the regulation of enzy-matic reaction by light.

In our laboratory regularities of enzymatic synthesis in a biphasic water-organic systems have been studied in detail. If water as a reaction medium is replaced by the biphasic system, a considerable (up to several orders of magnitude) shift of chemical equilibrium of the synthetic reaction toward the end product occurs (2). Also, a shift of the acid-base equilibrium (up to 5 and more pH units) of the reacting compounds occurs (3). The latter allows the pH optimum for catalyst (enzyme) effectiveness to coincide with the pH optimum for the catalyzed reaction equi-librium (4); it also enables a shift to be made in the equilibrium of the reaction of peptide bond hydrolysis toward synthesis (5). Up to now biphasic water-organic systems were used successfully for the enzymatic syntheses of N-acyl amino acid esters(4, 6), peptides (5 and unpublished results), benzyl penicillin (7) and gylcerophosphate (6).

The detail theoretical analysis and prospects of the biphasic method in biotechnology are given elsewhere (1-3).

REFERENCES

1. MARTINEK, K., et al. J. Appl. Biochem., in press.
2. MARTINEK, K., et al. Biochim. Biophys. Acta 658: 76 (1981).
3. MARTINEK, K., et al. Biochim. Biophys. Acta 658: 90 (1981).
4. MARTINEK, K., et al. Biotechnol. Bioeng. 23: 1115 (1981).
5. SEMENOV, A. N., et al. Biotechnol. Bioeng. 23: 355 (1981).
6. MARTINEK, K., et al. Bioorg. Khim. (Russ.) 3: 696 (1977).
7. SEMENOV, A. N., et al. Dokl. Akad. Nauk SSSR (Russ.) 258: 1124 (1981).

HEAT-STABLE L-LACTATE DEHYDROGENASE AND ITS APPLICATION TO AN ENZYME REACTOR.

T. Ohta, H. Taguchi and H. Matsuzaswa

Department of Agricultural Chemistry, University of Tokyo, Tokyo, Japan

In recent years, many stable enzymes have been isolated from thermophilic bacteria. These thermostable enzymes may have tremendous potential in enzyme technology. We isolated thermophilic bacteria from the Kawamata Hot Springs in Japan. One of the isolated bacteria, named *Thermus caldophilus* GK-24, was chosen as a good source of stable enzymes, including an extracellular protease. The purification and properties of lactate dehydrogenases have been reported for some thermophilic bacteria (1-3), but little is known about the enzyme from *Thermus* strains. Extremely heat-stable L-lactate dehydrogenase (EC 1.1.1.27) has been purified from *T. caldophilus* GK-24.

The enzyme was purified using ion-exchange and affinity chromatography. The specific activity of the most highly purified sample measured at 52°C was 73.5 U/mg. The purified enzyme preparation migrated as a single protein band having a MW of 31,000 on SDS/ polyacrylamide gel electrophoresis. The native MW was estimated as approximately 120,000 by gel filtration.

The optimum enzyme reaction for the reduction of pyruvate was found at pH 4.5; and almost no activity was observed in the neutral pH region. In the presence of 0.2 mM fructose-1,6-diphosphate (FDP), however, the enzyme reaction was remarkably activated in the pH range from 5.8 to 8.0 and was not practically changed in the pH region from 4 to 5. Therefore, two peaks were found at pH 4.2 and 7.1. The maximum activation was obtained with 0.2 mM FDP (about 30-fold activation). Inorganic phosphate partly activated the enzyme reaction. D-lactate was not a substrate.

The optimal temperature of the enzyme reactions was 80° for the reduction of pyruvate and $95^\circ C$ (highest temperature tested) for the oxidation of lactate. By heating at $90^\circ C$ for 60 min and with 0.5 mM FDP, only 15% of the activity was lost as compared to the same treatment in the absence of FBP.

.This enzyme can be used as a useful stable element in a bioreactor system to convert pyruvate into lactate, though its activity at 30°-40° is much lower as compared with that at the optimal temperature.

ACKNOWLEDGMENT

This work was supported partially by a grant for scientific research from the Ministry of Education, Science and Culture of Japan.

REFERENCES

1. WEERKAMP, A. & MACELROY, D. *Arch. Microbiol.* 85: 113 (1972).
2. SCHAR, H-P & ZUBER, H. *Hoppe-Seyler's Z. Physiol. Chem.* 360: 795 (1979)
3. LAMED, R & ZEIKUS, J. G. *J. Bacteriol.* 141: 1251 (1980).

MULTIENZYMES AND COFACTORS IMMOBILIZED WITHIN LIPID-POLYAMIDE MEMBRANE MICROCAPSULES FOR SEQUENTIAL SUBSTRATE CONVERSION

Y. T. Yu and T. M. S. Chang

Artificial Cells and Organs Research Centre
McGill University, Montreal, Canada

Microencapsulation is one of the methods of enzyme immobilization in which an unlimited number of enzymes can be immobilized within each microcapsule (1, 2). For multistep reactions with multienzyme systems it would be advantageous to immobilize cofactors within the microcapsules. Lipid-polyamide membrane microcapsules have been prepared, which can restrict the passage of hydrophilic molecules like KCl and NaCl (2, 3). Recently by using an extension and modification of this approach, it was possible to retain free cofactors within the microcapsules (4). Some substrates, e.g., α-ketoglutarate, could also be retained within such microcapsules. This way lipophilic substrate can diffuse into the microcapsules, in which sequential conversion and cofactor recycling can take place (4).

A typical example is the study in which lipid-polyamide membrane microcapsules were prepared to contain glutamate dehydrogenase, alcohol dehydrogenase, NAD^+, α-ketoglutarate, ADP, KCl, and $MgCl_2$. After two washings with buffer solution there was no leakage of enzymes, cofactor, or α-ketoglutarate. Furthermore, the microcapsules after two washings still retained the same rate of enzyme activity and cofactor recycling rate, as compared to the microcapsules before washing.

Various amounts of NAD^+ (0.25-10.0 μmoles) were used; and the conversion rate of ammonia into glutamate was studied (Table 1). Even with a low amount of NAD^+ (0.25 μmoles) it was possible to convert 10 μmoles of ammonia into glutamate with cofactor recycling. Without cofactor recycling 10 μmoles of NADH must be microencapsulated in order to achieve the same extent of conversion.

163

Permeability studies showed that there was negligible permeation of hydrophilic substrates of even small molecules like urea
through the lipid-polyamide membrane microcapsules. By decreasing
the proportions of cholesterol in the cholesterol-lecithin lipid,
it was possible to obtain permeable membranes for urea and at the
same time with no leakage of NAD$^+$ and α-ketoglutarate. This way
urea permeating through the lipid polyamide membrane was sequentially converted into glutamate.

ACKNOWLEDGMET

The support of the Medical Research Council of Canada, MRC-
SP-4, is gratefully acknowledged.

REFERENCES

1. CHANG, T. M. S. *Science 146:* 524 (1964).
2. CHANG, T. M. S., "Artificial Cells," Charles C. Thomas,
      Springfield, Illinois (1972).
3. ROSENTHAL, A. M. & CHANG, T. M. S. *J. Memb. Science 6:* 329
      (1980).
4. YU, Y. T. & CHANG, T. M. S. *FEBS Lett. 125:* 94 (1981).

TABLE I

CONVERSION OF AMMONIA INTO GLUTAMATE

Cofactor (μmoles)	Amount of Ammonia Converted (μmoles)		
	45 min	90 min	180 min
NAD$^+$ 0.25	2.4 ± 2.8	6.6 ± 2.8	10.0 ± 2.4
NAD$^+$ 0.50	8.0 ± 1.6	9.5 ± 2.0	11.0 ± 2.0
NAD$^+$ 1.00	8.0 ± 1.6	9.6 ± 2.0	13.0 ± 1.2
NAD$^+$ 10.00	8.8 ± 2.4	12.0 ± 2.8	15.0 ± 2.0
NADH 10.00*	10.6 ± 2.4	10.8 ± 2.4	10.8 ± 1.2

*No recycling of NADH

TYROSINE AND 3,4-DIHYDROXYPHENYL-ALANINE SYNTHESIS BY

CITROBACTER FREUNDII

N. S. Egorov, M. B. Koupletskaja, and E. N. Kondratieva

Department of Microbiology, Moscow State University
Moscow, USSR

Tyrosine phenol lyase catalyzes the splitting of L-tyrosine and L-3,4-dihydroxyphenyl-alanine (DOPA). The reversibility of both reactions allows this enzyme to be used for the synthesis of L-tyrosine and L-DOPA (1,2).

L-tyrosine $\rightleftharpoons$ phenol + pyruvate + NH_3

L-3,4-dihydroxyphenyl-alanine $\rightleftharpoons$ pyrocatechol + pyruvate + NH_3

Among the strains of bacteria of different genera studied by us (Erwinia, Escherichia, Citrobacter, Proteus, Pseudomonas, Xanthomonas), only three have revealed the presence of tyrosine phenol lyase. All these bacteria are facultative anaerobes and can oxidize formate (3). Omeliansky named such microorganisms Bacterium formicum; but in Bergey's Manual (4) they are identified as Citrobacter freundii.

As result of the use of medium containing peptone and formiate (3), we succeeded in isolating another strain of bacteria producing tyrosine phenol lyase. Thus several strains appear to be capable of oxidizing formate and synthetizing this enzyme. However, there were no strains forming syrosine phenol lyase among the methylotrophic bacteria using methanol and methylamine or among the bacteria oxidizing phenols.

Some strains of C. freundii forming tyrosine phenol lyase are capable of producing about 19 g/L L-tyrosine per 2 hr. With periodic addition of phenol to the medium, cells of the most active strains synthetize up to 75 g/L tyrosine per 12 hr and up to

95 g/L per 24 hr. The use of pyrocatechol and pyruvic acid
neutralized by ammonia allows these bacteria to synthetize up to
45 g/L DOPA per 48 hr; but the addition of ammonium acetate decreases the yield of DOPA.

In contrast to earlier work (1,2,5), *C. freundii* strains
investigated by us synthetized very little tyrosine and produced
no DOPA if pyruvate was replaced by serine. In a reaction mixture
with lactate instead of pyruvate, tyrosine was not produced; and
DOPA was synthetized at a slow rate (3-6 g/L per 48 hr). In the
presence of other organic acids or glycerol, neither tyrosine nor
DOPA was formed. We succeeded in increasing DOPA synthesis from
lactate by growing certain strains of *C. freundii* on medium containing lactic acid. Cells grown on such a medium produced up to
40 g/L DOPA per 48 hr from lactate and pyrocatechol.   .

REFERENCES

1. YAMADA, H., KUMAGAI, H., ENIE, H., MATSUI, H. & OKUMURA, S.
       Proc. IV IFS: Ferment. Technol. Today 445 (1972).
2. ENEI, H., MATSUI, H., YAMASHITA, K., OKUMURA, S. & YAMADA, H.
       Agr. Biol. Chem. 36: 1861 (1972).
3. OMELIANSKY, V. L. *Ctrbl. Bakt. 2:* 1903.
4. "Bergey's Manual of Determinative Bacteriology," Williams and
       Wilkins Co., Baltimore (1974).
5. OGATA, K., YAMADA, H., ENEI, H.& OKUMURA, S., United States
       Patent 3,791,924 (1974).

ENZYME CATALYSTS FOR THE STEREO AND REGIO SELECTIVE

OXYFUNCTIONALIZATION OF ORGANIC SUBSTRATES

S. W. May

School of Chemistry
Georgia Institute of Technology
Atlanta, Georgia, USA

From the viewpoint of enzyme technology, oxygenases, which incorporate molecular oxygen directly into organic substrates, are particularly intriguing since selective, direct oxygenation of organic substrates has traditionally been an unresolved challenge to organic chemistry. We have focused on oxygenase-catalyzed heteroatom oxygenation, epoxidation, hydroxylation, and ketonization of simple organic substrates (Fig. 1). These represent basic, synthetically important reactions. In all cases, the reactions involve highly stereo and regio selective insertions of molecular oxygen into organic substrates. Although we illustrate these three oxyfunctionalization reaction types for a few specific cases, work in progress indicates that several other enzymes, heretofore considered simple hydroxylases, also readily carry out these three processes with regio and stereo specificity.

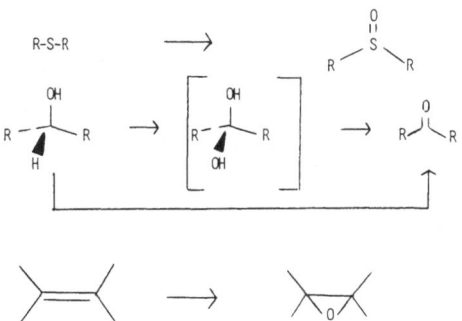

Fig. 1. Basic types of enzymatic oxygenation reactions investigated.

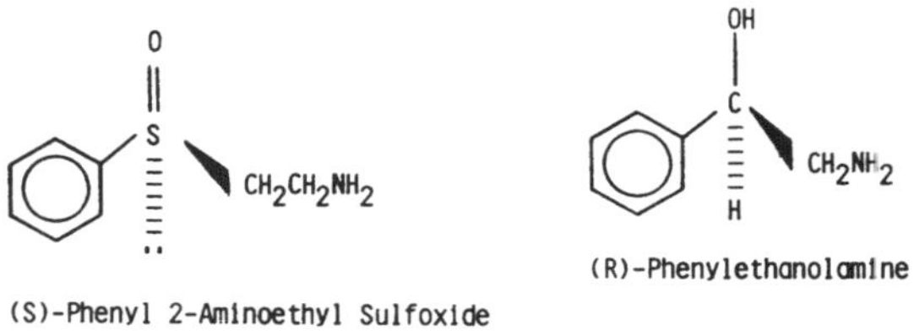

(S)-Phenyl 2-Aminoethyl Sulfoxide   (R)-Phenylethanolamine

Fig. 2.   Stereochemical consistency of sulfoxidation and hydroxy-
          lation products produced by DBH.

        Working with the copper-containing monooxygenase, dopamine-
B-hydroxylase (DBH), we demonstrated for the first time sulfoxidation by an enzyme heretofore considered to be a hydroxylase (1,2).
Sulfoxidation exhibits all the characteristics diagnostic of a
monooxygenase reaction.   For example, oxygen, substrate, and electron consumptions are stoichiometric with sulfoxide production.
Product identities were established unequivocally by direct isolation and uv, ir, nmr, and mass spectral examination.   Sulfoxidation
is stereospecific; and the stereochemistry of oxygen attack is
fully consistent with that established for DBH-catalyzed methylene
hydroxylation (Fig. 2)   On the basis of detailed kinetic experiments, we have established that sulfoxidation proceeds much more
readily than hydroxylation for comparable substrates.   In fact,
DHB catalyzes sulfoxidation more rapidly than any other reaction;
and designation of this enzyme as a hydroxylase is obviously misleading.   Thus, preparative scale production of mg/ml quantities
of essentially optically pure sulfoxides is easily accomplished,
even with crude enzyme preparations.   In sharp contrast, microsomal
sulfoxidation has been reported to result in only the slightest
(1%) enantiomeric enrichment.

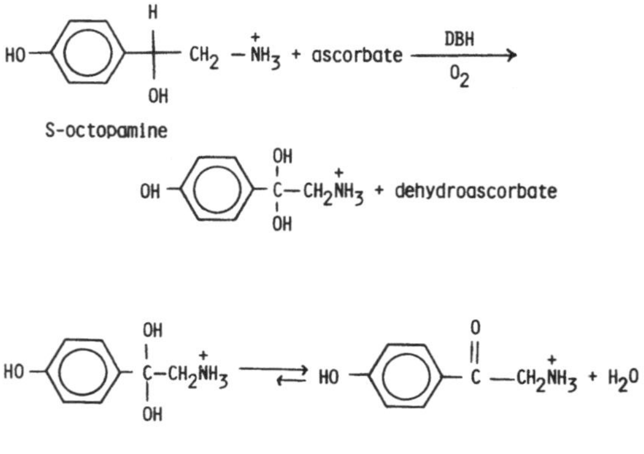

S-octopamine

Fig. 3.   Pathway for ketonization.

We have utilized two approaches to the enzymatic generation of ketones with controlled stereo and regio specificity. First, we have demonstrated that enzymatic hydroxylation of the enantiomers of normal products can generate ketones via an oxygenase pathway, a process which is totally distinct from the familiar alcohol degydrogenase reaction. Secondly, we have identified and studied a prototype alcohol dehydrogenase which is specific for only secondary alcohols, but which is sufficiently tolerant of substrate secondary structure so as to be of potential interest from a synthetic viewpoint. The following two examples illustrate these approaches. S-octopamine, the enantiomer of the product generated by DBH hydroxylation of tyramine, is converted by DBH to the corresponding ketone (3). Ketonization is a kinetically facile process; and a body of evidence clearly indicates that it occurs through an oxygenase pathway (Fig. 3). Preparative scale experiments easily produce isolated yields in the mg/ml range, even with crude enzyme preparations. We are currently investigating whether this approach to stereo selective ketonization will be applicable to other oxygenases. The secondary alcohol dehydrogenase, which we have isolated from *Pseudomonas*, readily produces ketones from a variety of straight chain and alicyclic secondary alcohols of moderate chain length; while primary alcohols are unreactive (4). The dehydrogenase has a broader specificity than similar enzymes recently described in the literature. Preparative scale generation of ketones using both crude and purified enzyme preparations have been successfully demonstrated.

An area of longstanding interest in our laboratory has been the highly stereo and regio selective generation of epoxides from simple olefins by the ω-hydroxylation system isolated from *Pseudomonas oleovorans* (5). Aside from detailed mechanistic and biochemical analyses, our studies have established that enzymatic epoxidation exhibits a highly unusual specificity, far different from that of non-enzymatic epoxidizing agents, such as peracids. For example, terminal olefins are epoxidated exclusively, even in the presence of more highly substituted double bonds. Simple internal cycloalkenes are hydroxylated but not epoxidated; whereas exocyclic double bonds are epoxidized. Other experiments have established that hydrophobic interactions with substrates critically affect reactivity, a circumstance which might well be exploited to advantage. Thus, the enzymatic epoxidation system clearly possesses a specificity of synthetic interest, e.g. where selective epoxidation of a terminal double bond in the presence of a more highly substituted double bond is desired. Stereochemical studies have uncovered unique and exciting aspects of this enzymatic epoxidation reaction. First, epoxidation of olefins is stereo specific, e.g. only R(+)-7,8-epoxy-1-oxtene is produced from octadiene. Note that stereo selective epoxidations of simple alkenes to give products of high optical purity are impossible to accomplish with any known chemical agent, including even chiral

peracids. Second, we have found that the presence of a preformed
asymmetric group on the substrate can alter the stereochemical
course of epoxidation. This finding may establish a basis for control of the stereochemistry of this enzymatic reaction. Results
we have obtained with specifically deuterated substrates support
a radical mechanism for epoxidation (6).

In summary, we are identifying, characterizing, and developing
for possible synthetic use, oxygenase systems which carry out oxygenation of heteroatoms, oxygenative ketonization of alcohols,
and epoxidation of olefins. All of these reactions represent basic,
synthetically important transformations. In all cases, the reactions are highly regio and stereo selective, and have been carried
out on a preparative scale. Our results have established for the
first time that enzymes previously considered to be simple hydroxylases readily carry out these three reaction types. In our view,
oxygenases possess great potential for future applications in enzyme technology.

REFERENCES

1. MAY, S. W. & PHILLIPS, R. S. *J. Am. Chem. Soc. 102:* 5981
      (1980).
2. PHILLIPS, R. S. & MAY, S. W. *Enzyme Microb. Technol. 3:* 9
      (1981).
3. MAY, S. W., PHILLIPS, R. S., MUELLER, P. W. & HERMAN, H. H.
      J. Biol. Chem. 256: 2258 (1981); *256:* 8470 (1981).
4. MAY, S. W., STELTENKAMP, M. S., BORAH, K. R., KATAPODIS, A.
      G., & THOWSEN, J. R. *J. Chem. Soc. Chem. Commun.* 845 (1979).
5. MAY, S. W. *Enzyme Microb. Technol. 1:* 15 (1979).
6. MAY, S. W., GORDON, S. L. & STELTENKAMP, M. S. *J. Am. Chem.
      Soc. 99:* 2017 (1977).

Session III
NEW IMMOBILIZATION TECHNIQUES OF BIOMATERIALS AND THEIR APPLICATIONS
Chairmen: D. Thomas and N. Ise

HIGH-RATE, CONTINUOUS WASTE PROCESSOR FOR THE PRODUCTION OF HIGH BTU GAS USING IMMOBILIZED MICROBES

R. A. Messing

Research & Development Division Corning Glass Works
Corning, New York, USA

Earlier studies showed the relationship between the dimensions of a microbe and the accumulation of that microbe in a porous, inorganic structure (1-3). The results indicated that cells that reproduced by fission could be loaded to levels between 10^8 and 10^9 cells/g of ceramic if the ceramic pore diameters were in the range of 1 through 5 times the major dimension of the cell. The rationale for this pore diameter criterion, when growth is required, is diagrammatically represented in Fig. 1. It is as-

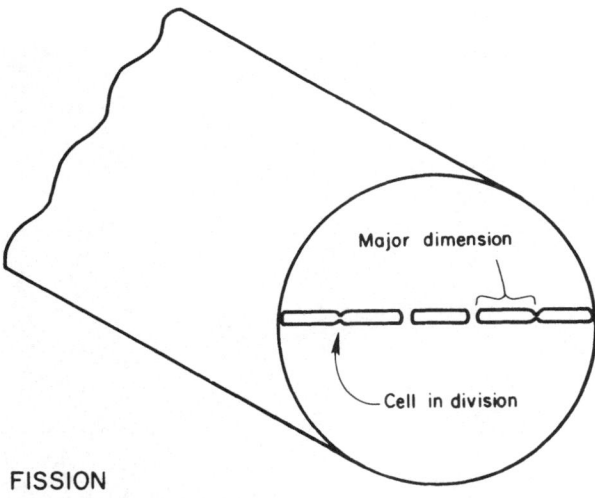

FISSION

Fig. 1. Diagramatic representation of reproduction of cells within an optimized pore.

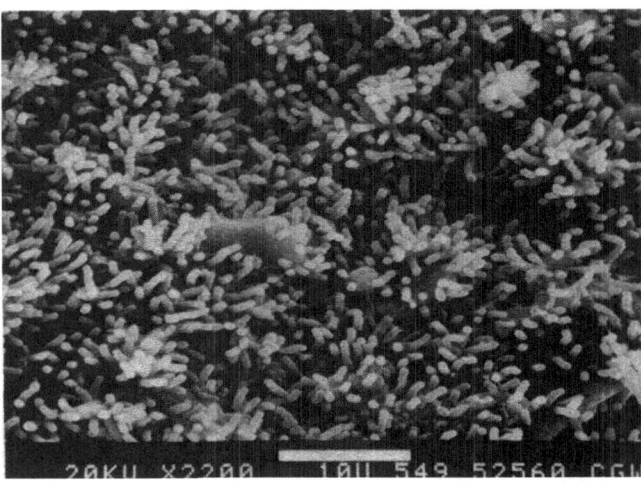

Fig. 2.   Scanning electron micrograph of cells from sewage
          immobilized on alumina.

sumed that the cells are immobilized on their minor dimension;
they thus must protrude toward the opposite wall of the pore.   If
another cell is immobilized on the opposite wall of the pore and
the cells undergo fission, then a minimum of 4 times the major
dimension of the cell is required to nearly fill the pore.   When
a cell wants to exit from that pore, an additional cell length is

Fig. 3.   Scanning electron micrograph of cells from sewage
          immobilized on cordierite.

required, is diagrammatically represented in Fig. 1. It is assumed that the cells are immobilized on their minor dimension;
they thus must protrude toward the opposite wall of the pore. If
another cell is immobilized on the opposite wall of the pore and
the cells undergo fission, then a minimum of 4 times the major
dimension of the cell is required to nearly fill the pore. When
a cell wants to exit from that pore, an additional cell length is
required, giving a pore diameter of 5 times the major dimension
of the cell. Scanning electron micrographs of bacteria from aero-
bically treated sewage, immobilized on two different porous ceramic materials, indicate that the cells were actually immobilized
on their minor dimension (Fig. 2 and 3). Note that all of the
cells were immobilized on their minor dimension and that the cell
packing was very close. It should also be noted that the cells
were packed tightly into the depth of the pore. A logical reason
for cell immobilization on the minor dimension might be that maximal cell surface still remains available for the transfer of
metabolites.

The system selected for demonstrating the value of this immobilized cell approach was one of waste conversion and energy
production (4,5). The goals of our studies were to determine the
functionality of the immobilized cells and also to design a reactor which would more efficiently convert the carbon in the feed
to methane. In order to accomplish this goal, an additional horizontal stage was added to the anaerobic filter (4).

MATERIALS AND METHODS

The reactor is shown schematically in Fig. 4 (6). The liquid
level controller activates the gas pump when the liquid level
falls too low. A feed pump is mounted at the base of the hydrolytic-redox stage. A check valve is inserted into the exit tubing
of the anaerobic stage. Both stages contain controlled-pore
ceramic material for immobilizing the microbes. The hydrolytic-
redox stage used a Pharmacia 16/20 water-jacketed column, while
the anaerobic stage used a modified 250 x 15 mm Lab Crest column
with a water jacket.

The feed material was sewage from the Corning Municipal
Sewage Waste Treatment Plant. It was stored at 4 - 6°C. Before
use, the pH was adjusted to 8.6 - 8.9 with sodium hydroxide and
filtered through cheesecloth and glass wool to remove coarse par-
ticulates. An extruded cordierite about 2 mm in diameter and
2 - 6 mm long was used as carrier; the average pore diameter was
3μ (pore diameter distribution 2 - 9μ, pore volume 0.44 cc/g, and
porosity 56.5%). The hydrolytic-redox stage received 24.5 g
while the anaerobic stage had 51 g of cordierite. Another support used was extruded brick containing carbon, (tradename Dura-

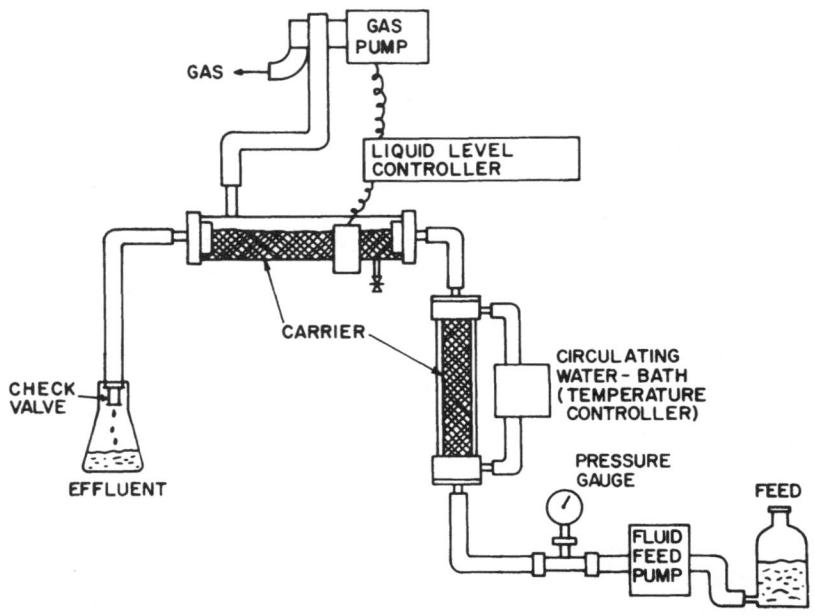

Fig. 4.   Test reactor system. The first carrier is a hydrolytic-
redox stage; the second carrier is an anaerobic stage.

lite Noire), produced by F. Guery, Rambervillers, France. The
brick had an average pore diameter of 6 (pore diameter distribution 0.8-30 , pore volume 0.29cc/g, and porosity of 34%). The
hydrolytic-redox stage received 20 g, while the anaerobic stage
had 50 g of the brick.

Gas determinations were done with a CEC 104 mass spectrometer, total carbon with a Dohrmann Model DC-50 carbon analyzer, and
chemical oxygen demand (COD) with the E. P. A. Certified ampule
method. The latter procedure is a modified Jirka, Carter colorimetric determination (7). Seeding of the reactors was allowed to
proceed for 7 - 10 days, at which time the reactors were operational. The reactors were then operated with continuous upflow
feeding and no recycling.

RESULTS AND DISCUSSION

Typical results are shown in Tables 1 and 2 for the cordierite
and brick supports, respectively. No dramatic trends in COD reductions, in conversions of delivered total carbon to methane, or in
recovery of reduced total carbon as methane were noted when the
residence time was varied from 1.9 - 5.5 hr. When the temperature

TABLE 1

PERFORMANCE OF REACTOR USING CORDIERITE AS SUPPORT

Day #	Residence Time (hr)	T (°C)	Gas Composition (%)				
			CH_4	CO_2	O_2	H_2	A
57	2.5	30	93.1	4.2	0	2.1	0.1
62	1.9	30	92.4	2.5	0.1	4.9	0.1
111	3.2	40	92.0	2.8	0.2	4.5	0.2
114	3.8	40	92.0	4.2	0 2	3.5	0.1

TABLE 1 (Contd)

COD (BOD*) (mg/l)				Delivered Total C Converted to CH_4 (%)	Reduced Total C Recovered as CH_4 (%)
Feed	Effluent	% Reduced	Psi		
1600	330	79.4	1.6	39.1	82.4
9900	270	70.0	2.4	37.5	110.0
1510	380	74.8	1.0	38.4	81.9
1600	320	80.0	1.0	41.4	81.4

*Biological oxygen demand

was varied from $20° - 40°C$ and the COD loading was varied from
810 - 2600 mg/L, a number of modest performance advantages were
identified. The single BOD reduction shown in Table 2 on the 62nd
day of operation indicated that greater than 97% reduction was
achieved at a residence time of 5.1 hours. Under conditions of
comparable residence times and COD loadings, the same reactor on
day 81 showed an increase in COD reduction from 64.2 to 77.7% and
conversion of delivered total carbon to methane from 31.7 to 42.1%
for a temperature increase from 20 to 30°C. At low levels of COD
(Table 1) on day 62 and at 30°C there was an increase of COD reduction from 70% to 86.4% and conversion of total carbon delivered
to methane from 37.5% to 45.9% with an increase of residence time
from 1.9 to 3.9 hours.

TABLE 2

PERFORMANCE OF REACTOR USING BRICK AS SUPPORT

Day #	Residence Time (hr)	T (°C)	Gas Composition (%)				
			CH_4	CO_2	O_2	H_2	A
60	4.9	30	92.7	3.1	0.5	3.6	0.1
62	5.1	30	92.5	3.2	0.6	3.7	0.1
67	3.5	30	90.1	2.4	0.7	6.8	0.1
69	3.9	30	91.7	2.2	0.7	5.3	0.1
74	4.7	30	91.7	5.1	0.1	3.0	0
76	4.2	30	93.2	3.7	0.5	2.7	0
81	3.3	30	93.7	3.8	0.2	2.4	0

TABLE 2   (Contd)

COD (BOD*) (mg/l)				Delivered Total C Converted to CH_4 (%)	Reduced Total C Recovered as CH_4 (%)
Feed	Effluent	% Reduced	Psi		
1750	230	86.9	0.5	44.9	83.4
905*	25*	97.2*	0.5	51.1	95.4
1120	140	87.5	0.5	45.9	105.0
810	110	86.4	0.5	45.9	108.0
1600	330	79.4	0.7	54.3	116.0
1700	320	81.2	0.5	49.2	88.4
2600	580	77.7	0.5	42.1	90.6

*Biological oxygen demand

COD reduction is a measure of the microbe performance, while the conversion of total carbon delivered to methane is a measure of both the microbe and reactor performance. The recovery of reduced total carbon as methane is a measure primarily of the gas transfer efficiency of the system. Since the most prolific production of methane by methanobacter occurs in the logarithmic growth phase, a major portion of the carbon delivered to the cells

would be required for growth and maintenance. It has been estimated that 40 - 60% of the delivered carbon would be required to
fulfill the needs for growth and maintenance due to the continual
removal of cells from the reactor. From the data labeled % of
Delivered Total C Converted to CH_4, it may be noted that at residence times greater than 3.3 hr and temperatures of at least 30°,
we have converted greater than 90% of the carbon not utilized for
growth to methane. This is further supported by the fact that less
than 5.1% of the evolved gas was removed as carbon dioxide.

Although the reactors were seeded from operational reactors
and then operated as packed-bed continuous reactors, other reactors were initiated by seeding solely from the sewage feed.
With the raw sewage for seeding, it required approximately 17 days
for the reactor to evolve significant quantities of gas and to
achieve COD reductions approximately one-half of those observed
with the fully operational reactors. By the 29th day all reactors
were fully operational.

The cordierite reactor was operated at feed rates of 28 - 48
ml/hr; and the gas evolved at rates of 12 - 16 ml/hr. The brick
reactor was operated at 18 - 27 ml/hr with gas evolution rates of
8 - 22 ml/hr. Under these conditions no serious problems were
encountered either with plugging of reactors or removal of gas.

The gas compositions in Tables 1 and 2, were greater than 90%
methane and less than 5.1% carbon dioxide under certain conditions.
In other words high BTU gas of pipe-line quality was evolved directly from these reactors without scrubbing, when sewage feed was
employed over a wide range of COD concentrations. The traditional
anaerobic waste treatment processes normally evolve a gas which
contains 50 - 75% methane and 25 - 50% carbon dioxide. The residence time required for traditional reactors is 10 - 30 days. It
should be noted that no hydrogen sulfide could be found in the
gas when a sewage feed was employed; however, with other waste
feed hydrogen sulfide did appear in the evolved gas.

CONCLUSION

The hydrolytic-redox stage reduces the size of waste cellulose and proteins by both hydrolytic and redox mechanisms in the
presence of facultative microbial populations. At the same time
oxygen is removed from the system; and carbon dioxide is produced.
Large quantities of carbon dioxide are dissolved due to vertical
operation under pressure of the hydrolytic-redox stage. In the
anaerobic stage immobilized methanobacter populations are used to
convert small molecules, such as acetic acid, ethanol, propionic
acid, propanol, and carbon dioxide to methane. Since this stage
is also under pressure, high quantities of carbon dioxide remain

dissolved so that larger quantities of carbon dioxide get reduced to methane. Since methane is hydrophobic and carbon dioxide is readily hydrated, the methane is liberated more rapidly from solution. Thus, the more complete conversion of carbon dioxide to methane and the greater transfer of methane from the liquid to gaseous phases is due to a combination of the high biomass accumulated in the stage and the horizontal assembly of the stage with maximized fluid to gas interface.

ACKNOWLEDGMENTS

M. Takeguchi and L. Simpson prepared the scanning electron micrograph photographs of the immobilized microbes. Analytical support was provided by L. Morse, D. Gardner, L. D. Kinney, A. Kacyon, V. Altemose, and E. H. Fontana. B. Sharma provided an economic evaluation and technical discussions; and G. Durand of Toulouse, France, identified the producer of Duralite Noire. And finally, A. Wise and D. Burkhead assisted in the preparation of this manuscript.

REFERENCES

1. MESSING, R. A. & OPPERMANN, R. A. *Biotechnol Bioeng.* 21: 49 (1979).
2. MESSING, R. A., OPPERMANN, R. A., & KOLOT, F. B. *Biotechnol. Bioeng.* 21: 59 (1979).
3. MESSING, R. A., OPPERMANN, R. A., & KOLOT, F. B. *Am. Chem. Soc. Symp. Ser.* 106: 13 (1979).
4. YOUNG, J. C. & MCCARTY, P. L. *J. Water Poll. Control Fed.* 41: R160 (1969).
5. JENNETT, J. C. & DENNIS JR., N. D. *J. Water Poll. Control Fed.* 47: 104 (1975).
6. MESSING, R. A. *Biotechnol. Bioeng.*, in press.
7. JIRKA, A. & CARTER, M. *Anal. Chem.* 47: 1397 (1975).

NEW DEVELOPMENTS IN THE PREPARATION AND CHARACTERIZATION OF

POLYMERBOUND BIOCATALYSTS

J. Klein and G. Manecke*

Institute of Chemical Technology, Technical University
of Braunschweig, and Institute of Organic Chemistry,*
Free University of Berlin, Federal Republic of Germany

Within the last two years (1,2) various new methods have been
developed related to the preparation and characterization of biocatalysts obtained by polymer entrapment of whole cells. It is
the intention to discuss briefly the concepts and some results with
reference to more detailed data published elsewhere.

PREPARATION OF POLYMERIC NETWORKS FOR CELL ENTRAPMENT

Ionotropic gelation: The ionotropic gelation process, especially in combination with Ca^{2+} alginates, has become one of the
most widely used entrapment methods for whole cells of different
origin. It should be kept in mind, that the name alginate stands
for a large variety of products, differing in molecular size and
chemical microstructure. These are important parameters with respect to the mechanical stability and the salt tolerance; for example, the Ca^{2+}/Na^{+} ratio is usually between 0.2 and 0.05 mole/mole.
While the Ca-alginate matrix generally is liable to redissolution,
a series of buffer systems can be presented, which would allow the
use of stable gels in the pH range 3 - 9.

A practical solution to the phosphate buffer instability of
Ca-alginate gels is the ionotropic gelation of polycations with
multivalent anions. This has been achieved by the use of chitosan
in combination with $K_4(Fe(CN)_6)$ or polyphosphate. As in the case
of alginates, macroporous, mechanically stable polymer networks
can be formed under rather mild physiological conditions (3,4).
A summary of presently available ionotropic gelation systems is
given in Fig. 1.

POLYELECTROLYTES	MULTIVALENT COUNTERIONS
POLYANIONS ALGINATE $COO^{\ominus}$ X$^{\ominus}$ CARBOXYMETHYL- CELLULOSE CARBOXY-GUAR- GUM COPOLY-STYRENE- MALEIC ACID	Ca^{2+}, Fe^{2+}, Zn^{2+}... Al^{3+}, Fe^{3+}.....
POLYCATIONS CHITOSAN $NH_3^{\oplus}$ X$^{\ominus}$	$Fe(CN)_6^{4-}$, $Fe(CN)_6^{3-}$ POLY-PHOSPHATE POLY-ALDEHYDO-CARBONIC ACID POLY-1-HYDROXY-1- SULFONATE-PROPEN-2 ALGINATE

Fig. 1.   Gel forming polyions and possible counterions.

Covalent Crosslinking of Prepolymers: in light of the reversibility of the ionotropic networks new possibilities for the formation of covalent, irreversible crosslinked gels, especially from
polymeric precursers, are of interest.   Such systems have been developed starting from polyvinylalcohol (PVA) and polyethylenimine
(PEI).   In a first approach, PVA was esterified with 3-mercapto-
propionic acid to form polymers containing pendant thiol groups.
Simple air oxidation led to crosslinks via disulfide bonds (PVA I)
(5).   Secondly, PVA was esterified with acrylic acid to lead to
photocrosslinkable water soluble polymers; crosslinking was achieved
with visible light in the presence of riboflavin-5'-phosphate as
photosensitizer.   Based on PEI, crosslinking can be achieved either
by reaction with glutaraldehyde (25% w/w) (PEI GA) or with the
watersoluble polyacrolein-bisulfite-addition compound (PEI PA).
The optimum concentration of the aldehyde groups has been determined
to be 0.6 (for PA) and 0.8 (for GA) aldehyde groups/PEI repeat unit.

The properties of these matrices were examined with *Arthrobacter simplex* cells (A.T.C.C. 6949) used for the steroid transformation of cortisol to prednisolone the Δ^1-dehydrogenase activity of the entrapped cells was studied at 30°C in the presence of
the cofactor menadione.   Typical reaction mixtures were as follows:
100 mg polymer cell conjugate, 100 ml substrate solution containing
2.0 mmol/L cortisol, 0.2 mmol/L menadione and 1% DMF.   Prior to
the conversion reaction, the polymer cell conjugates were incubated
with 0.2 mM cortisol solution for a period of 24 hr.   Comparing
the different immobilization methods, the PVA based conjugates

TABLE 1

COMPARISON OF Δ^1-DEHYDROGENASE ACTIVITY OF A. *SIMPLEX*
CELLS FOR DIFFERENT POLYMERIC CARRIERS

Catalysts	Cortisol Incubation	Cell Concentration (g/g)	Rel. Activity (%)
PVA I	No	0.50	82
PVA I	Yes	0.50	294
PVA II	Yes	0.79	245
PEI GA	Yes	0.67	54.6
PEI PA	Yes	0.67	34.5

showed a performance considerably superior to the PEI preparations
(Table 1). The optimum cell concentration was determined in each
case as examplified for PVA I gels in Table 2. Initial cortisol
incubation was responsible for a considerable increase in relative
activity; this was due to an increased enzyme level, possibly with

TABLE 2

DEPENDENCE OF THE ACTIVITY OF PVA I-*ARTHROBACTER SIMPLEX*
CONJUGATES ON THEIR CELL CONCENTRATION

Cell Concentration* ($g_{cells}/g_{conj.}$)	Activity (μmol/g/hr)		Rel. Act. ** (%)
	Per g Conjugate	Per g Cells	
0.17	0.0	0.0	0
0.30	97.0	323.3	294
0.46	141.4	307.4	279
0.50	161.5	323.0	294
0.67	98.7	147.3	134
0.74	22.0	29.7	27
0.80	8.0	10.0	9

*Determined gravimetrically; concentration in polymer-cell-
conjugate matrix.
**Activity of the free cells: 110 μmol/g/hr.

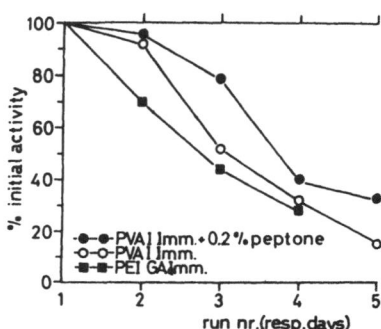

Fig. 2.   Repeated batchwise transformation of cortisol with immobi-
          lized *A. simplex* cells. Reaction conditions: PVA I -
          A. simplex conjugate 100mg; PEI GA- *A. simplex* conjugate
          200mg; substrate solution: 100ml resp. 50ml; cortisol
          2,0 mmol/l; menadione: 0.2 mmol/l; 1% DMF 30°C; after
          each run (24 hrs.) the substrate solution was renewed.

an increased cell level. If the cell concentration reached too
high a value, diffusion limitations obviously became so dominant
that the activity dropped to very low values. As will be seen
later this seemed to be a typical feature for immobilized cell
catalysts. With respect to kinetic properties, the apparent K_M
values increased from 1.3 mmol/L for free cells to 4.4 mmol/L for
PVA I; and the optimum menadione concentration changed from 0.3
mmol/L for free cells to 0.5 mmol/L for PVA I. Finally the time
dependent activity decay is compared in Fig. 2.

    Polycondensation of Reactive Oligomers: Polyurethanes (6)
and epoxy resins (7) are the most important systems in this group.
While the mechanical and chemical stability of these matrices have
been excellent, problems have been observed with respect to toxicity and enzyme deactivation. For example, with the systems described in the literature, so far no living yeast cells could be
immobilized. In the case of epoxy systems considerable progress
could be achieved by a) choosing different precurser compounds
and b) introducing a pregelling time on the order of 15 min before
mixing the cells with the condensing oligomers. With these changes
the viability of the immobilized yeast cells could be demonstrated
by the formation of etanol from glucose under limited growth conditions. As can be seen from Fig. 3, the decaying catalyst activity
of *Saccharomyces c.* cells could be restored repeatedly by intra-
particle cell growth. In addition, the continuous steady state
production of ethanol could be achieved (Fig. 4).

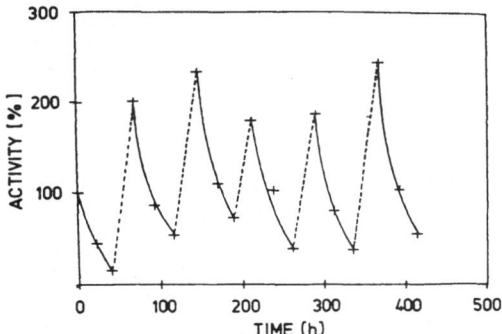

Fig. 3. Behavior of yeast cells immobilized in epoxy carriers
        with repeated discontinuous incubation.

PHYSICAL CHARACTERIZATION

        The first attempts to obtain quantitative data for the charac-
terization of the mechanical properties of polymer cells composits
were reported previously (2). A more complete description of the
methods of single particle compression, column pressure drop, and
abrasion also has been published (8). The column pressure drop
method not only can be applied to packed bed columns but also to
integrated,openpore foams (6), e.g. polyurethanes. Very few data,
however, are available on transport properties, as characterized
by the diffusivity and matrix porosity.

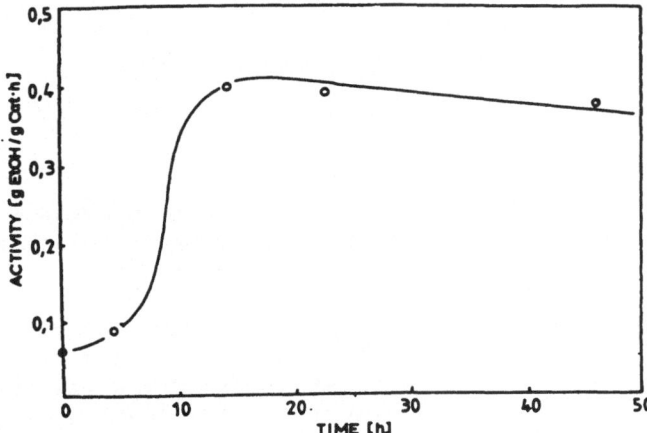

Fig. 4. Behavior of yeast cells immobilized in epoxy with con-
        tinuous incubation; results shown as ethanol production.

Effective Diffusivity: It is obvious that the porous polymeric carrier as well as the immobilized cells give rise to transport resistants for the diffusion of small molecules.   For such
a diffusional process

$$D_{eff} = D_o \, e^{-av_p} \qquad \text{(Eq. 1)}$$

Eq. 1 has been proposed (9) to relate the effective diffusivity,
D_{eff}, to the coefficient of free diffusion, D_o, and the volume
fraction of the polymer, v_p.   The exponential coefficient, a, is
a constant related to molecular size, with a = 4 for small molecules like oxygen (10).   To account for the immobilized cells,
Eq. 1 has to be modified, to include the volume fraction of cells,
v_B.

$$D_{eff} = D_o \, e^{-a(v_p + v_B)} \qquad \text{(Eq. 2)}$$

To check the quantitative validity of Eq. 1 or 2 and to obtain exact information on the relation of the coefficient a to
molar size, some diffusion experiments were carried out.   By mixing
a known number of catalyst beads of homogeneous substrate concentration, C_o (g/L), with a known volume of fresh solvent, the diffusion process can be followed by the concentration increase in the
supernatant liquid.   Let C_∞ be the substrate concentration in the
particle at long times, $\bar{C}$ the concentration at time t, r the
particle radius in cm, and $D = D_{eff}$ in cm^2/sec.   The diffusion coefficient can be obtained (8) from the slope by plotting ln
$[(\bar{C} - C_\infty)/(C - C_\infty)]$ vs t, as per Eq. 3.

$$\frac{\bar{C} - C_\infty}{C_o - C_\infty} = \frac{6}{\pi^2} \sum_{n-1} \frac{1}{n^2} \exp\left(\frac{-\pi^2 \, D_{eff} \, t}{r^2}\right) \qquad \text{(Eq. 3)}$$

Fig. 5 gives two examples for such diffusional processes in a
chitosan matrix.   The relation of parameter a to molecular weight
for some typical substrates is given in Table 3.

The exact determination of D_{eff} is necessary for the quantitative computation of catalytic effectiveness factors for immobilized
cell catalysts (11).   A typical example for the application of such
diffusivity data is the optimization of cell loading with respect
to the maximum absolute activity of a biocatalyst.   The dependence
of the relative and absolute activity of immobilized *Acetobacter*
sp. cells for the production of gluconic acid by air oxidation of

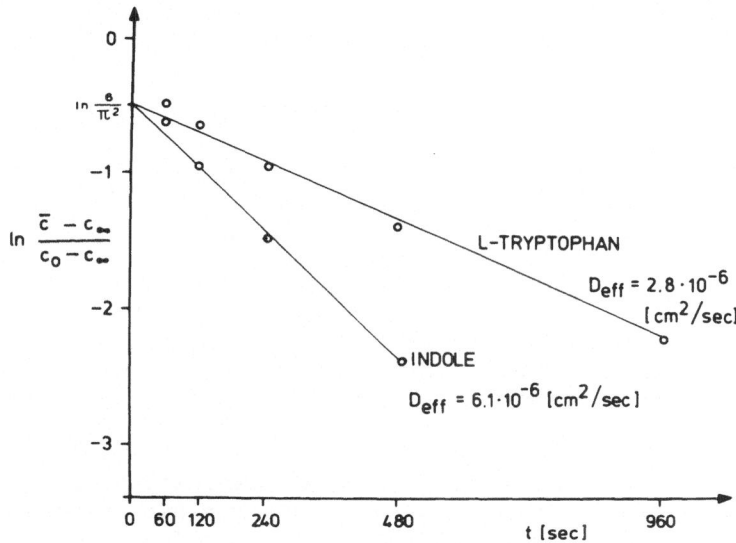

Fig. 5. Determination of the effective diffusivity D_{eff} in polymer matrices.

glucose (12) is shown in Fig. 6. The calculated values were based on the Thiele Modulus concept (8), using diffusivity data from Eq. 2. The maximum in the activity curve is explained by the increased blocking of pores for oxygen transport with increasing number of immobilized cells.

Matrix Porosity: The determination of diffusivity coefficients provides some qualitative information on the relative pore sizes of different polymeric carriers. With regard to inclusion (enzyme entrapment) or exclusion (substrate selectivity), however, it is

TABLE 3

VALUES FOR PARAMETER "a" FOR SOME RELEVANT SUBSTRATE MOLECULES

Substrate	Matrix	"a"
O_2		4
Indol	chitosan	6
L-Tryptophan	chitosan	8
Pen G	chitosan	12

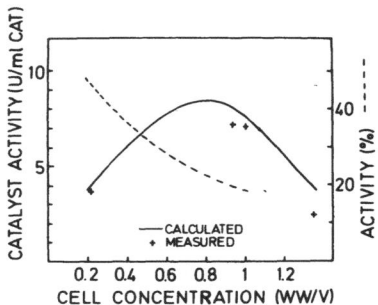

Fig. 6. Relative and absolute activity of immobilized *Acetobacter*
         sp cells as a function of cell loading in the production
         of gluconic acid from glucose.

of interest to obtain direct experimental information of the pore
sizes. Size exclusion chromatography (SEC) can easily be adapted
to this problem (13). If polymeric probe molecules of different
molecular size can be found, then chromatographic runs through a
column packed with a porous stationary phase will give information
about partial or total exclusion from the pores. If, as in the
case of dextran in water, the coil dimensions of the probe molecules
are available, direct information on the limiting pore size for
complete exclusion can be obtained from elution volume data.

     Such experiments have been performed, using porous alginate
beads of 70 μm medium particle diameter as the stationary phase.
As can be seen from Fig. 7 all of the dextran standards of
MW $\geq$ 160,000 were totally excluded from the matrix, while all molecules of MW < 110,000 diffused through the carrier with certain re-

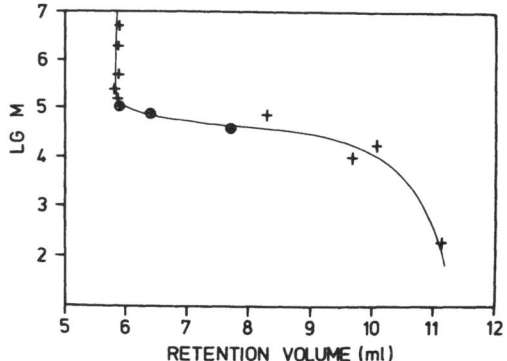

Fig. 7. Pore size determination of polymer matrices by steric
         exclusion chromatography with dextran standards; M means
         MW.

strictions. Based on coil size determination (5) the maximum pore size of this special alginate preparation was in the region of 170 Å. In principle it is possible to relate pore size directly to pore diffusion data; and such attempts are presently under way in our laboratory. In the case mentioned above (14) a good correlation between activity loss due to enzyme leakage from a polymeric carrier and carrier structure could be obtained.

ACKNOWLEDGMENT

Support of this work by the Federal Government (BMFT) is gratefully acknowledged.

REFERENCES

1. KLEIN, J. & WAGNER, F. in "Enzyme Engineering," vol. 5, (H. Weetall and G. Royer, eds.) Plenum, New York (1980) p. 335.

2. KLEIN, J., WASHAUSEN, P., KLUGE, M. & ENG, H. in "Enzyme Engineering," vol. 5 (H. Weetall and G. Royer, eds.) Plenum, New York (1980) p. 359.

3. VORLOP, K.-D. & KLEIN, J. *Biotechnol. Lett. 3:* 65 (1981).

4. WAGNER, F., LANG, S., BANG, W.-G., VORLOP, K.-D. & KLEIN, J. this volume.

5. MANECKE, G. & BEIER, W. *Angew. Makromol. Chem.*, in press.

6. KLEIN, J. & KLUGE, M. *Biotechnol. Lett. 3:* 65 (1981).

7. KLEIN, J. & ENG, H. *Biotechnol. Lett. 1:* (1979).

8. KLEIN, J. & WAGNER, F. *Dechema Monogr. 84:* 265 (1979).

9. WHITE, M. L. & DORION, G. H. *J. Polym. Sci. 55:* 731 (1961).

10. KLEIN, J. & SCHARA, P. *Appl. Biochem. Biotechnol. 6:* (1 (1980).

11. ENG, H., Thesis, Techn. University Braunschweig, (1980).

12. VORLON, K.-D., KLEIN, J., & WAGNER, F. *Abst. 6th Intern. Ferment. Symp. London (Canada)* F 12.1.12. (1980).

13. SNYDER, L. R. & KIRKLAND, I. J., "Introduction to Modern Liquid Chromatography", J. Wiley & Sons, New York (1974).

14. STOCK, J., Thesis, Techn. University Braunschweig, (1981).

15. BRANDRUP, J., IMMERGUT, G., eds., "Polymer Handbood", J. Wiley & Sons, New York (1974).

BIOCONVERSION OF LIPOPHILIC OR WATER INSOLUBLE COMPOUNDS BY IMMOBILIZED BIOCATALYSTS IN ORGANIC SOLVENT SYSTEMS

S. Fukui and A. Tanaka

Laboratory of Industrial Biochemistry
Department of Industrial Chemistry, Faculty of
Engineering, Kyoto University, Kyoto, Japan

Enzymes and microbial cells were entrapped inside gels of optionally hydrophobic or hydrophilic character as well as network structure by using prepolymers of photo-crosslinkable resins or urethane resins. Biocatalysts so immobilized were used for successful bioconversion of various highly lipophilic or water-insoluble compounds in water-organic co-solvent systems or water-saturated organic solvents. The influence of the hydrophobicity and the net work structure of the gels and the polarity of the solvents were studied in terms of the catalytic activity and operational stability of the gel-entrapped biocatalysts. In a less-polar solvent system, the hydrophobicity of the gels used to entrap microbial cells or enzyme affected markedly the catalytic activity of the gel-entrapped biocatalysts. The use of hydrophobic gels is preferable for bioconversion of highly lipophilic compounds. The activity of gel-entrapped cells correlated closely with the partition of substrates between the gels and external solvents. Bioconversion of steroids and terpenoid in organic solvents could be achieved successfully by gel-entrapped microbial cells whose stability was improved by immobilization. The selection of suitable gel hydrophobicity makes it possible to control conversion routes in reactions involving two or more reactants of different hydrophobicity. Synthesis of water-insoluble compounds, such as adenine arabinoside from uracil arabinoside and adenine, was catalyzed by entrapped bacterial cells which showed an excellent stability in a water-organic co-solvent system. Furthermore, lipase adsorbed on Celite and then entrapped in a hydrophobic gel, revealed high activity and good stability in the ester exchange reaction of triglyceride.

In the case of the bioconversion of lipophilic as well as water-insoluble or only slightly soluble compounds, it will be desirable to carry out enzyme-mediated reactions in solutions containing an organic co-solvent or in a suitable organic solvent system if the enzyme activity is maintained in such a reaction system. The use of organic solvents will overcome the poor aqueous solubility of the substrate or other components of lipophilic or hydrophobic nature and, when necessary, can shift an unfavorable thermodynamic equilibrium in the desired direction. However, in principle, the catalytic activity of the native enzyme decreases rapidly; and the substrate specificity disappears in organic solvents. Generally speaking, the immobilization of enzymes or cells onto a suitable support or inside an appropriate gel seems to be one of the most promising approaches for enzyme stabilization.

In vivo, many enzymes, especially enzymes which catalyze the transformation of lipophilic substrates, function in membrane-associated states. Entrapment of enzymes inside suitable gels will give a microenvironment analogous to that of membrane-associated enzymes. Multi-point interaction between entrapped enzymes and gel matrices should stabilize the active conformation of the biocatalysts and render them more resistant against the action of organic solvents.

However, the physical and chemical properties of immobilized biocatalysts used for conversion of lipophilic compounds in organic solvent systems are more complicated. For instance, the affinity of lipophilic substrates for the gels and the diffusion of reactants through the gel matrices are important aspects. A low affinity of hydrophilic gels for hydrophobic substrates will lower the apparent activity of the gel-entrapped biocatalyst. Thus, the use of gels with a suitable balance of hydrophilicity and hydrophobicity and net-work structures is necessary. For preparing such gels, we have developed new immobilization methods by the use of photo-crosslinkable resin prepolymers and urethane prepolymers. The advantages of the prepolymer methods are a) the inclusion of the biocatalyst in the gel can be achieved by a very simple and mild technique and b) tailor-made gels of desired physical and chemical properties can be obtained by the use of appropriate prepolymers.

Fig. 1 shows the structures of the prepolymers of photo-crosslinkable resins (ENT and ENTP) (1-4) and the prepolymers of urethane resins (PU) (4-7), which have been used in our studies. The photo-crosslinkable resin prepolymers have photo-sensitive functional groups, such as acryloyl, at the terminals of the linear chain. The chain length of prepolymer can be adjusted by using poly(ethylene glycol) or poly(propylene glycol) of optional

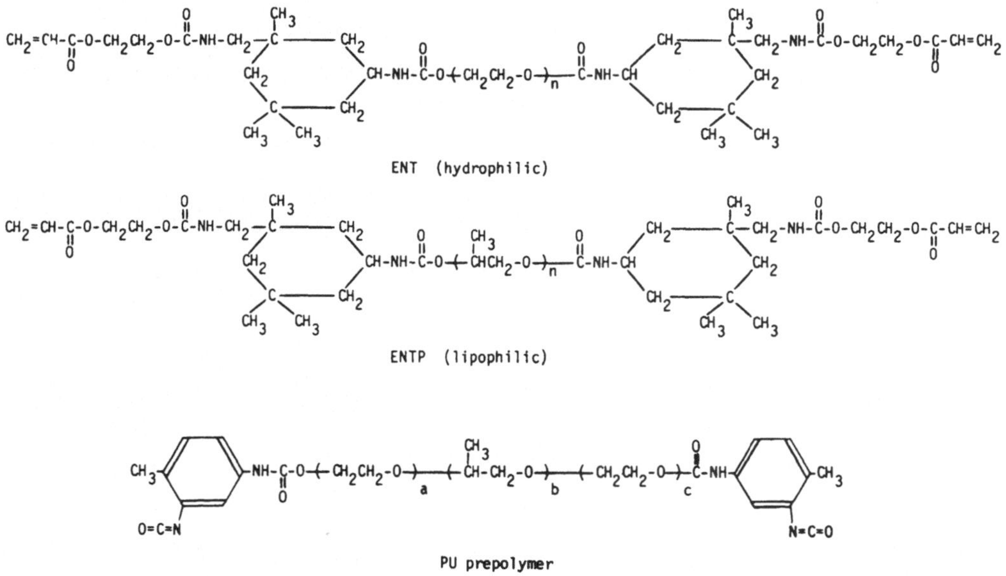

Fig. 1.   Structures of photo-crosslinkable resin prepolymers (ENT and ENTP) and urethane prepolymers (PU). ENT is hydrophilic, while ENTP is hydrophobic. PU-3 and PU-6 are water-miscible prepolymers with diol MWs of 2529 and 2627, NCO contents of 4.2% and 4.0%, and ethylene glycol contents of 57% and 91%, respectively. PU-3 gives a hydrophobic gel, and PU-6 a hydrophilic gel.

chain length in the linear main chain. Illumination of a mixture of photo-crosslinkable resin prepolymer and biocatalyst with near-UV rays for several minutes gives the gel-entrapped biocatalyst. The chain length of prepolymer correlates closely to the net-work structure of the gels so formed. The hydrophilicity or hydrophobicity and ionic character of the gels are also adjusted by selecting the main skeleton of the prepolymers.

In the case of urethane propolymers having isocyanate groups at both terminals of the linear skeleton of the polyether diol made from poly(ethylene glycol) and poly(propylene glycol), the prepolymers can be mixed with an aqueous solution of the enzymes or a water suspension of the cells or organelles and proceed to react with each other and become readily crosslinked by forming urea linkages with liberation of carbon dioxide. Prepolymers with different hydrophilicity or hydrophobicity can be obtained by changing the ratio of poly(ethylene glycol) to poly(propylene glycol) in the polyether moiety. The net-work of gel matrices

TABLE 1

BIOCONVERSION OF LIPOPHILIC COMPOUNDS BY IMMOBILIZED BIOCATALYSTS
IN ORGANIC SOLVENTS

Reaction	Substrate	Product
steroid Δ^1-dehydro-genation*	4-androsterene-3,17-dione	1,4-androstadine-3,17-dione
17 β-hydroxy steroid dehydrogena-tion*	β-stradiol	estrone
3-β hydroxysteroid dehydrogenation*	cholesterol	cholestenone
menthyl ester hydrolysis**	dl-menthyl succinate	l-menthol + d-menthyl suc-cinate
interesterification of triglyceride	olive oil + stearic acid	cacao butter-like fat + oleic acid

*Nocardia rhodocrous; **Rhodotorula minuta ***Rhizopus delemar lipase.

also can be controlled by changing the chain length of the pre-polymers.

Table 1 shows several examples of the bioconversion of lipo-philic compounds in organic solvent systems. Δ^1-Dehydrogenation of steroids and dehydrogenation of the 3 β-hydroxy group and the 17 β-hydroxy group of steroids could be carried out successfully with immobilized cells of Nocardia rhodocrous (1,3,8-11). The stereo-selective hydrolysis of the menthyl ester was also carried out with immobilized cells of Rhodotorula minuta (12). Immo-bilized lipase of Rhizopus delemar was found to be useful for the ester exchange reaction of triglyceride (13).

N. rhodocrous can convert testosterone (TS) to 1,4-androstadiene-3,17-dione (ADD) via Δ^1-dehydrotestosterone (DTS) or 4-androstene-3,17-dione (4-AD) in the presence of phenazine methosulfate (PMS). The latter material is an artificial elec-tron acceptor, which is absolutely required for the Δ^1-dehydrogen-

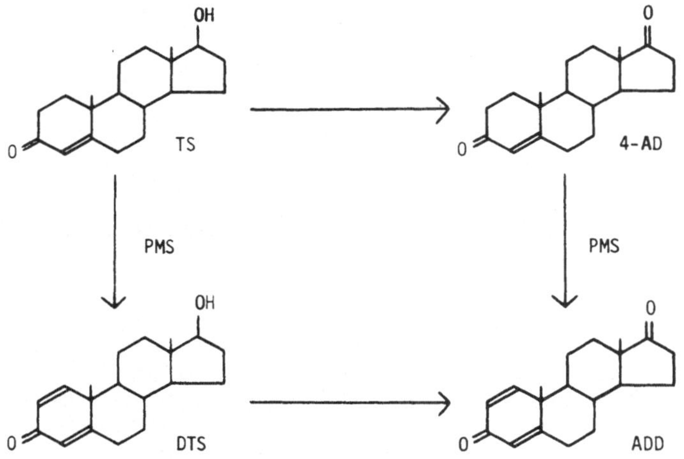

Fig. 2.   Bioconversion of testosterone by *Nocardia rhodocrous*.

ation reaction and which stimulates the dehydrogenation of 17 β-
hydroxysteroids. When the bacterial cells are entrapped in gels of
different hydrophilicity or hydrophobicity, the properties of the
gels have a major influence on the reaction routes (9). In the
case of TS conversion in a water-saturated mixture of benzene and
n-heptane (Fig. 2), hydrophobic gel-entrapped cells formed 4-AD as
the major reaction product. On the other hand, DTS was the main
product with hydrophilic gel-entrapped cells. This different profile in dehydrogenation products can be explained by a low affinity
of PMS for hydrophobic gel-entrapped cells. With hydrophilic gels
the stimulated uptake of PMS as well as the inhibitory action of PMS
toward 17 β- hydroxysteroid dehydrogenase will be responsible for
the accumulation of DTS.

The stereo-specific hydrolysis of *dl*-menthyl succinate has
also been carried out by gel-entrapped cells of *Rhodotorula
minuta* in water-saturated n-heptane (Fig. 3) (12). The stability
of the cells was markedly improved by immobilization. The flow
sheet for *l*-menthol production in large quantity is illustrated
in Fig. 4.

Not only microbial cells but also enzymes can be used in
organic solvent systems. The ester exchange reaction of triglyceride was mediated by lipase to produce cacao butter-like fat from
olive oil in water-saturated n-hexane (13). In this case, the
oleic acid moiety at positions 1 and 3 in olive oil should be substituted regio-specifically by a saturated fatty acid, such as
stearic acid (Fig. 5). This reaction requires a small amount of
water; a large amount of water leads to hydrolysis of triglyceride
rather than ester exchange. When the reaction was carried out in

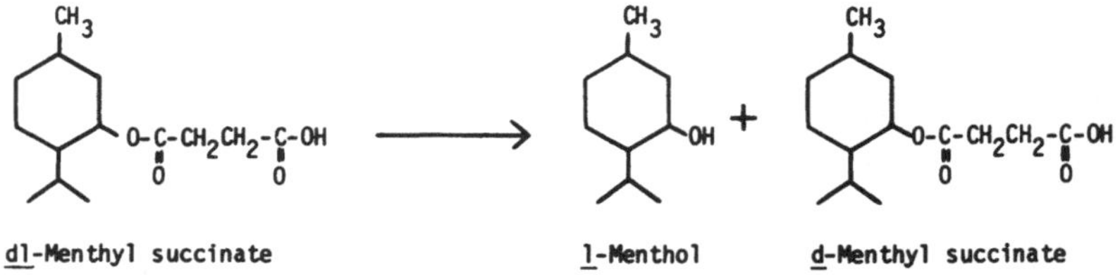

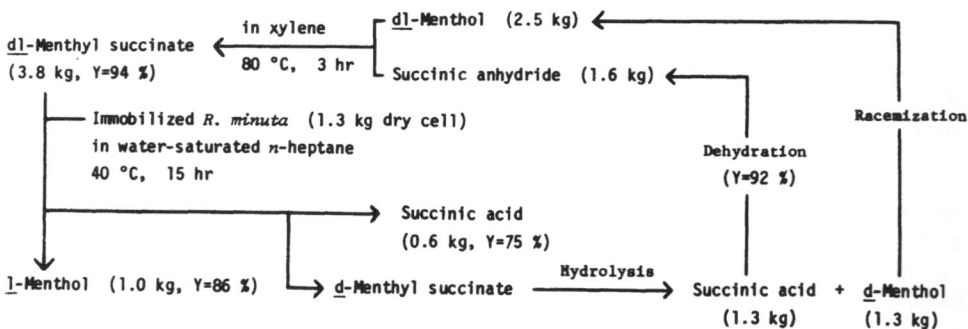

Fig. 3. Stereo-selective hydrolysis of *dl*-methyl succinate by
         Rhodotorula minuta.

n-hexane, lipase adsorbed on Celite showed good activity, as reported previously (14,15); while the enzyme suspended in the solvent did not show any activity. The activity of the Celite-
adsorbed lipase was quickly decreased during repeated use, probably
due to desorption of enzyme from Celite. Among various immobilization methods, the enzyme entrapped in a hydrophobic gel, ENTP-2000,
was found to have a high activity (Table 2) and a markedly improved
operational stability.

     Table 3 shows the bioconversion by immobilized biocatalysts
in mixtures of water and water-miscible organic co-solvents.
Dehydrogenation of hydrocortisone to form prednisolone was done
by immobilized acetone-dried cells of *Arthrobacter simplex* in a
reaction mixture containing 10% methanol (2,7). Immobilized
growing cells of *Curvularia lunata* hydroxylated Reichstein's
Substance S to hydrocortisone in the presence of 2.5% dimethyl
sulfoxide or methanol (16,17). Immobilized cells of *Enterobacter
aerogenes* catalyzed the transglycosylation of uracil arabinoside
to form an anti-viral drug, adenine arabinoside (18).

Fig. 4. Flow sheet for *l*-methol production by *Rhodotorula
         minuta*.

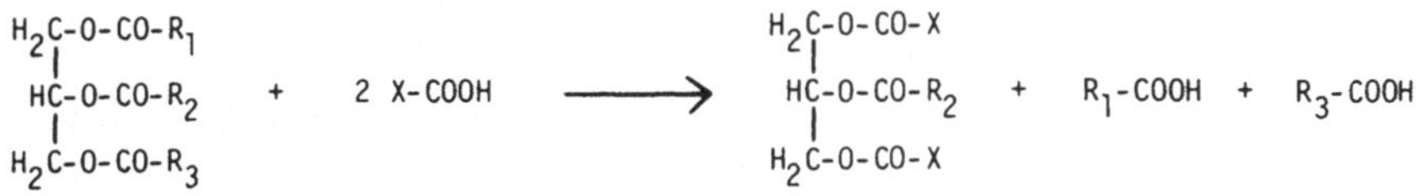

$$
\begin{array}{l}
H_2C\text{-}O\text{-}CO\text{-}R_1 \\
| \\
HC\text{-}O\text{-}CO\text{-}R_2 \\
| \\
H_2C\text{-}O\text{-}CO\text{-}R_3
\end{array}
\quad + \quad 2 \; X\text{-}COOH \quad \longrightarrow \quad
\begin{array}{l}
H_2C\text{-}O\text{-}CO\text{-}X \\
| \\
HC\text{-}O\text{-}CO\text{-}R_2 \\
| \\
H_2C\text{-}O\text{-}CO\text{-}X
\end{array}
\quad + \quad R_1\text{-}COOH \quad + \quad R_3\text{-}COOH
$$

Fig. 5.   Ester exchange reaction of triglyceride mediated by
          Rhizopus delemar lipase.

An antibiotic effective against DNA viruses, adenine arabinoside (AraA), can be synthesized by *E. aerogenes* from chemically
prepared uracil arabinoside (AraU) via transglycosylation (19).
However, the solubility of the substrates and products in water
is very low; so it is difficult to carry out the reaction at high
concentrations of the substrates. We have tried to introduce an
organic solvent into the reaction system; and we found that
dimethyl sulfoxide was the best from the criteria of product solubility and enzyme stability. The reaction should be carried out
at $60^{\circ}C$ to avoid the degradation of adenine, one of the substrates, by the deaminase (Fig. 6). When the reaction was done
with 40% dimethyl sulfoxide, free bacterial cells gradually lost
activity because of the leakage of the enzyme from the cells.
By entrapment of the cells with either photo-crosslinkable resin
prepolymer or urethane prepolymer, the transglycosylation activity
of the cells was maintained for at least 35 reaction batches over
35 days, even when the reaction was carried out at $60^{\circ}C$ in 40%

TABLE 2

INTER-ESTERIFICATION ACTIVITY OF LIPASE ENTRAPPED WITH
DIFFERENT PREPOLYMERS

Prepolymer	Adsorption on Celite	Activity Yield (%)
None	+	100
None	−	0
ENT-1000	−	13
ENT-2000	−	18
ENT-4000	−	20
ENT-6000	−	14
ENTP-2000	−	29 − 82
ENTP-2000	+	75
PU-3	−	19
PU-6	−	15

TABLE 3

BIOCONVERSION BY IMMOBILIZED BIOCATALYSTS IN WATER-ORGANIC
CO-SOLVENT SYSTEMS

Reaction	Substrate	Product	Cell Type
steroid Δ^1-dehydrogenation	hydrocortisone	prednisolone	*Arthrobacter simplex*
steroid 11 β-hydroxylation	Reichstein's substance S	hydrocorti-sone	*Curvularia lunata*
steroid 11α-hydroxylation	progesterone	11α-hydroxy-progesterone	*Rhizopus stolonifer*
transglycosyla-tion	uracil arabino-side + adenine	adenine arabinoside + uracil	*Enterobacter aerogenes*

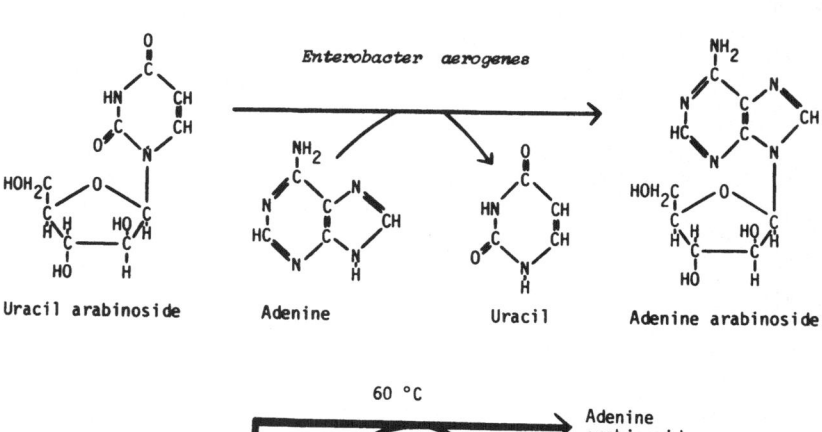

Fig. 6. Main and side reactions of adenine arabinoside synthesis by *Enterobacter aerogenes*.

dimethyl sulfoxide.   The yield of the product was almost 100%, based on the amount of adenine added (18).

As mentioned here, enzymes and microbial cells (non-living treated cells, resting cells, spores and growing cells) can be entrapped with appropriate prepolymers.   The entrapped biocatalysts can be used successfully for the bioconversion of water-insoluble compounds in organic solvent systems and also in water-organic co-solvent systems.   The prepolymer methods allow selection of adequate hydrophobicity and net-work size of the gels for application to different types of bioconversion.   The easy selection of gel properties for immobilization of biocatalysts will be very important in order to extend the use of immobilized biocatalysts in various fields.

REFERENCES

1. OMATA, T., TANAKA, A., YAMANE, T. & FUKUI, S. *Eur. J. Appl. Microbiol. Biotechnol. 6:* 207 (1979).
2. SONOMOTO, K., TANAKA, A., OMATA, T., YAMANE, T. & FUKUI, S. *Eur. J. Appl. Microbiol. Biotechnol. 6:* 325 (1979).
3. OMATA, T., IIDA, T., TANAKA, A. & FUKUI, S. *Eur. J. Appl. Microbiol. Biotechnol. 8:* 143 (1979).
4. FUKUI, S., SONOMOTO, K., ITOH, N. & TANAKA, A. *Biochimie 62:* 381 (1980).
5. FUKUSHIMA, S., NAGAI, T., FUJITA, K., TANAKA, A. & FUKUI, S. *Biotechnol. Bioeng. 20:* 1465 (1978).
6. TANAKA, A., JIN, I.-N., KAWAMOTO, S. & FUKUI, S. *Eur. J. Appl. Microbiol. Biotechnol. 7:* 351 (1979).
7. SONOMOTO, K., JIN, I.-N., TANAKA, A. & FUKUI, S. *Agric. Biol. Chem. 44:* 1119 (1980).
8. YAMANE, T., NAKATANI, H., SADA, E., OMATA, T., TANAKA, A. & FUKUI, S. *Biotechnol. Bioeng. 21:* 2133 (1979).
9. FUKUI, S., AHMED, S. A., OMATA, T. & TANAKA, A. *Eur. J. Appl. Microbiol. Biotechnol. 10:* 289 (1980).
10. FUKUI, S., OMATA, T., YAMANE, T. & TANAKA, A. in "Enzyme Engineering," vol. 5 (H. Weetall and G. Royer, eds.) Plenum Press, New York (1980) p. 347.
11. OMATA, T., TANAKA, A. & FUKUI, S. *J. Ferment. Technol. 58:* 339 (1980).
12. OMATA, T., IWAMOTO, N., KIMURA, T., TANAKA, A. & FUKUI, S. *Eur. J. Appl. Microbiol. Biotechnol. 11:* 199 (1981).
13. YOKOZEKI, K., YAMANAKA, S., TAKINAMI, K., HIROSE, Y., TANAKA, A., SONOMOTO, K. & FUKUI, S. *Eur. J. Appl. Microbiol. Biotechnol.*, in press.
14. TANAKA, T., ONO, E., ISHIHARA, M., YAMANAKA, S. & TAKINAMI, K. *Agric. Biol. Chem. 45:* 2387 (1981).

15.   MACRAE, A. R. *Abstr. 2nd Eur. Congr. Biotechnol.* (East-
      bourne, England), p. 43 (1981).
16.   SONOMOTO, K., HOQ, M. M., TANAKA, A. & FUKUI, S. *J. Ferment.
      Technol. 59:* 465 (1981).
17.   SONOMOTO, K., HOQ, M. M., TANAKA, A. & FUKUI, S. *Appl.
      Environ. Microbiol.*, in press.
18.   YOKOZEKI, K., YAMANAKA, S., VTAGAWA, T., TAKINAMI, K., HIROSE,
      Y., TANAKA, A., SONOMOTO, K. & FUKUI, S. *Eur. J. Appl.
      Microbiol. Biotechnol.*, in press.
19.   UTAGAWA, T., MORISAWA, H., MIYOSHI, T., YOSHINAGA, F.,
      YAMAZAKI, A. & MITSUGI, K. *FEBS Lett. 109:* 261 (1980).

MULTIPLE CARRIER REGENERATION AND ENZYME IMMOBILIZATION *IN SITU*

J. E. Prenosil and E. Stuker

Chemical Engineering Department ETH
Zurich, Switzerland

Traditionally, enzyme immobilization is a batch operation performed outside of an enzyme reactor. Enzyme immobilization *in situ* means fixing the enzyme to a carrier in the reactor in order to produce the biocatalyst directly where it is going to be used. The advantages over the traditional procedure are: a) no need for a special immobilization plant; b) additional enzyme immobilization is possible so as to increase the catalytic activity; c) re-immobilization is possible after the carrier is regenerated; d) regeneration and re-immobilization can be part of a general sanitation cycle and as such easily automated; and 3) reduced enzyme consumption should result.

The immobilization of β-galactosidase and trypsin on an ion exchange resin served as the model processes. β-Galactosidase from *A. Niger* (Lactase AN, Societe Rapidase, France-Seclin) and trypsin from bovin pancreas (Fluka A G, Buchs, Switzerland, *ca.* 9000 BAEE-U/mg.) were adsorbed on phenol-formaldehyde ion exchange resin (Duolite S-761, Diaprosim, France-Vitry). The carrier was packed in a column reactor; and the enzyme solution was recirculated through the reactor. Samples of the solution were taken during the adsorption. Catalyst samples were taken after the immobilization to determine the enzyme activity. Crosslinking by glutaraldehyde followed the enzyme adsorption.

The results showed that the initial rate of adsorption was higher in the column reactor so that the equilibrium was attained earlier. For both β-galactosidase and trypsin the activity of the composite prepared *in situ* was higher or the same, as compared with the one from the batch procedure. The catalyst particle position in the column did not seem to affect its final activity

in a systematical way; and the flow rate did not produce an effect.
A significant increase in the adsorption rate was observed at
higher temperature (42°C). To establish the cumulative effect of
the adsorbed residue and carrier stability on the catalyst activity,
multiple regeneration and reimmobilization cycles were performed.
Two column reactors with biocatalyst prepared *in situ* were operated at 60°C, pH 4.5, and 5% lactose substrate. The catalyst was
regenerated monthly either by 0.5 M CH_3COOH or 0.1 M NaOCl, followed by enzyme re-immobilization. The results are shown in Fig. 1.

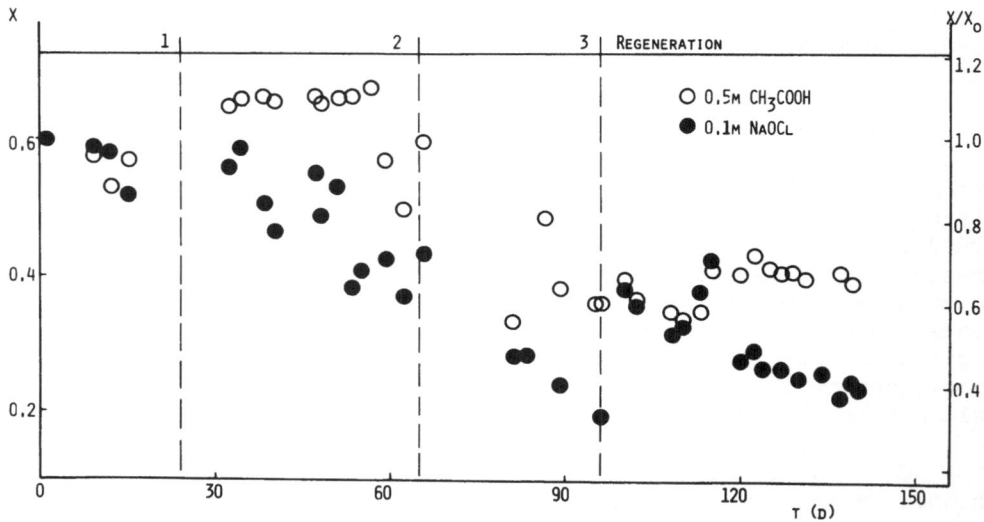

Fig. 1. Conversions from reactors regenerated and re-immobilized
         in situ as a function of time.

VIABILITY AND BIOSYNTHETIC CAPACITY OF IMMOBILIZED PLANT CELLS

P. Brodelius,* F. Constabel and W. G. W. Kurz

Pure and Applied Biochemistry,* University of Lund
Lund, Sweden and National Research Council of Canada
Prairie Regional Laboratory, Saskatoon,
Saskatchewan, Canada

Cells of the alkaloid-producing plant *Catharanthus roseus*, grown as suspension cultures, were immobilized by entrapment in various gels (1) and studied by monitoring plasmolysis, respiration, cell growth, and cell division. Cells entrapped in alginate, carrageenan, agarose and agar were found to be fully viable (1). Agarose-entrapped cells appeared to be essentially unaffected by immobilization as indicated by unrestricted respiration, cell growth, and cell division (Fig. 1A); while cells entrapped in alginate were restricted in respiration, cell growth, and cell division (Fig. 1B). Viable preparations of immobilized plant cells were also biosynthetically active as shown by synthesis of the indole alkaloid ajmalicine (1). Both *de novo* synthesis and synthesis from added precursors took place. Enhanced alkaloid synthesis observed with alginate-entrapped cells in this and other (2) studies may result from secondary metabolite formation being proportional to the reciprocal of the growth rate.

REFERENCES

1. BRODELIUS, P. & NILSSON, K. *FEBS Lett.* *122:* 312 (1980).
2. BRODELIUS, P., DEUS, B., MOSBACH, K. & ZENK, M. H. in "Enzyme Engineering," vol. 5 (H. H. Weetall and G. P. Royer, eds.) Plenum Press, New York (1980) p. 373.

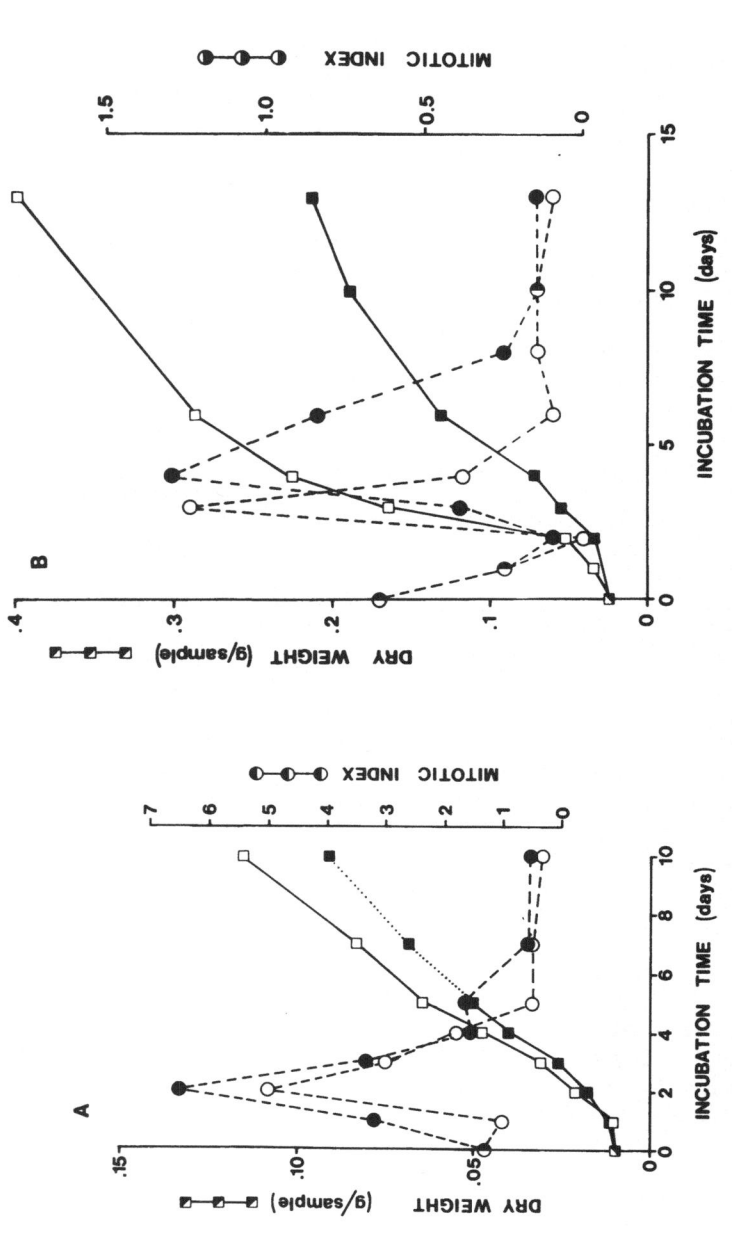

Fig. 1. Dry weight and mitotic index of freely suspended (open symbols) and immobilized (solid symbols) cells of *C. roseus* as a function of incubation time. A: Agarose-entrapped cells (The dotted line indicates that significant amounts of freely suspended cells were found in the samples). B: Alginate-entrapped cells.

IMMOBILIZED ENZYMES OF HETEROGENEOUS STRUCTURE

K. Nakamura, K. Hibino and Y. Yano

Department of Agricultural Chemistry
University of Tokyo, Tokyo, Japan

Immobilized enzymes of heterogeneous structure were prepared by covalent binding and entrapment to improve the catalytic and mechanical properties. The effect of mass transfer diffusivity and partitioning on the observed reaction rate was analyzed by three structural models of active and inactive metrices: uniform, parallel, and serial.

Glucose isomerase was extracted from lyophilized cells of *Streptomyces phaeochromogenes*, and used for immobilization after partial purification or crystallization. First the enzyme was covalently bound on aminoethylated polyacrylamide gel particles by one of three methods: diazo coupling, cross-linking using glutaraldehyde, and peptide formation. The immobilized enzyme particles were suspended in an acrylamide monomer solution, and the mixture was extruded through a small tube into an organic solvent. An aqueous solution of the initiator was continuously added near the exit of the tube. The activity and the stability of the free as well as the immobilized enzymes were measured using 20 mM Tris buffer, pH 8.0, at 60° C in 10 mM MgSO4 and either glucose or fructose.

The addition of an adsorption step to the enzyme purification procedure increased the specific activity of the purified enzyme considerably. This highly purified enzyme gave increased specific rate constants for the immobilized enzyme when Bio-Gel P-6 was used as the carrier and gave increased specific rate constants and activity per unit volume of the immobilized enzyme particles when the more porous Bio-Gel P-300 was used as the carrier.

The effect of aminoethylation on the activity of the bound
glucose isomerase (glutaraldehyde binding method) was different
among fine, medium, and coarse carrier particles of Bio-Gel P-300.
The activity expressed per unit surface area was largest for the
coarse particles, probably because the matrix shrinkage had a
greater influence on the fine particles. The glutaraldehyde binding method resulted in higher activity of immobilized enzyme as
compared to the other two methods (Table 1). The yield of activity
was 69-75% for primary immobilization by the glutaraldehyde method,
while it could be more than 80% in the secondary step of immobilization. The covalently bound and then entrapped enzyme kept 80%
of its initial activity for more than 30 days in a packed bed reactor, while the primary immobilized enzyme lost its activity in
20 days with a concurrent increase of the pressure drop.

The effect of mass transfer on the reaction rate was so
small that there was no marked difference in effectiveness factor
estimated by the three structural models of plate-shaped immobilized enzyme. Either the uniform or serial models, however,
seemed to fit better, probably because the reaction rate constant
could be estimated without much change under the different conditions of reaction.

TABLE 1

COMPARISON BETWEEN COVALENT BINDING METHODS USING BIO-GEL P-300

Immobilization	Activity*	
	(unit/mg protein)	(unit/cm^3 carrier)
glutaraldehyde binding	2.33	5.50
diazo coupling	0.53	1.00
paptide bond formation	2.58	1.33
native	3.67	

*carrier 50-100 mesh; aminoethyl concentration 1.70×10^{-5} mole/cm^3
  carrier; 50% of initial activity remained after the enzyme was
  held 3 hr at a pH in the range 5-10 for 30 min at 85-90°C.

FIBROUS SUPPORT FOR IMMOBILIZATION OF ENZYMES

H. Ichijo, T. Suehiro, A. Yamauchi,
S. Ogawa, M. Sakurai* and N. Fujii*

Research Institute for Polymers and Textiles
Yatabehigashi, Tsukuba, and Nitivy Co., Ltd.*
Chuoh-ku, Tokyo, Japan

There are methods, such as adsorption, trapping, cross-linking and covalent-bonding, available for the immobilization of enzymes. No one method seems to have a clear advantage over the others. Therefore, an attempt was made to develop a poly(vinyl alcohol) (PVA) fibrous support for enzyme immobilization.

PVA is considered to be a good material for an immobilization support because it shows hydrophilic properties and can easily incorporate various functional groups by acetalization. Since aminated PVA fiber adsorbs enzymes by electrostatic attraction, the immobilization procedure is simple; and the reaction between immobilized enzymes and substrates is nearly independent of the size of the substrates. This newly developed type of fiber, which is formed as a bundle of super fine filaments (SFF) each measuring 1 μm or less in diameter, permits much more surface area than the conventional enzyme immobilization supports. The fibrous support is expected to have low flow resistance in a column and to be applicable in continuous enzymic reaction systems, because it permits a variety of fibrous shapes such as knitted fabric, string, and filter paper, suitable for intended applications.

The diameter of the SFF, measured directly with scanning electron microscopy, was in the range of 0.1 to 0.6 μm, even finer than ordinary fibers of which the diameter was about 20 μm. The surface area of a SFF sample, measured with methanol (1, 2), was about 15% of that of activated charcoal. Large size pores formed a larger proportion of the pore size distribution for SFF than for activated charcoal. Therefore, SFF was considered to have good properties for adsorption of large enzyme molecules.

The yield on adsorption of invertase by SFF was 93% in 5 min
and 97% in 240 min, while the yield by an ordinary fiber was only
9% in 5 min and 26% in 240 min. As the quantity of invertase added
in this experiment was small, the maximum capacity for adsorbed
invertase was not found from these results. However, it is obvious
that SFF has even greater ability for immobilizing invertase than
does an ordinary fiber. The amount of adsorbed enzyme increased
linearly with nitrogen content, ranging from 0.2 to 0.5 wt% and
reached almost a maximum at 1.0 wt%. It is considered to be
reasonable that invertase is immobilized on SFF by ionic bonds.
The nitrogen content of 1.0 wt% was found to be enough for immobilizing invertase. The amount of adsorbed invertase was almost
proportional to the initial enzyme concentration. One gram of SFF
adsorbed over 800 mg of invertase at 200 mg/dl and was expected
to immobilize more enzyme at higher enzyme concentration. The
observed activity of immobilized invertase increased linearly below
400 mg/g fiber; and increased gradually even above 400 mg/g fiber.
The relative activity was found to be as
high as 10%, even at 800 mg of adsorbed enzyme/g fiber.

When 0.5 M sucrose solution was passed through the SFF filter
paper containing adsorbed invertase at a space velocity 60 hr^{-1},
the yield of glucose was almost 100% for 7 days.

It is concluded that the aminated SFF is an excellent support for the immobilization of invertase by ionic bonds, because
the amount of invertase adsorbed on SFF was over 800 mg/g fiber;
and the immobilized invertase showed high observed activity. The
results obtained from the continuous flow experiment with SFF
filter paper and adsorbed invertase suggest that SFF has a high
possibility to be applied to an industrial flow enzymatic reaction
system.

REFERENCES

1. URANO, K. *Hyomen 13:* 738 (1975).
2. URANO, K., SUGIYAMA, K., IIZUKA, T. & KOMORI, M. *Mizushori
    Gijutsu 15:* 1159 (1974).

HIGH TEMPERATURE CELL-TRAPPED ULTRAFILTRATION MEMBRANES

E. Drioli, G. Iorio, M. De Rosa,* A. Gambacorta,*
and B. Nicolaus*

Istituto di Principi di Ingegneria Chimica,
Universite di Napoli and Istituto di Chimica di
Molecole di Interesse Biologico,* CNR, Napoli
Italy

The possibility of using ultrafiltration (UF) and reverse
osmosis (RO) membranes filled with biocatalyst might represent a
significant improvement to the development of membrane technology
and enzyme engineering (1,2). However the possibility of preparing industrial level UF and RO membranes filled with enzymes or
whole cells, using traditional techniques, has been limited until
now by the need for non-aqueous solvents in the casting solutions
and high temperature annealing which destroys the catalytic
properties. The recent isolation of *Caldariella acidophila* (3),
an extreme thermophile growing optimally at 87°C and whose
enzymes are generally stable to protein denaturating agents,
offers an interesting opportunity for using the phase inversion
technique for the preparation of UF and RO membranes filled with
whole cells as the enzyme source. Moreover *C. acidophila* contains enzymes of industrial interest, such as β-galactosidase.

Artificial membranes filled with *C. acidophila* have been
prepared by several methods. Cellulose acetate and polysulfone
were used to obtain asymmetric membranes by the phase inversion
technique (1,2). Albumin and glutaraldehyde were used for cell
immobilization in membranes by the co-crosslinking method (4).
And finally, a hydrophilic polyisocyanate was used in preparing
porous polyurethane structured foams in thin films (5). The
physico-chemical properties of trapped-cell β-galactosidase
activity in the above models were similar to those shown by the
enzyme in the free cells (3). At the optimal pH the β-galactosidase exhibited maximal activivy at about 100°C and appeared
stable for up to 24 hr at room temperature and pH of 3-8. Incu-

bation of trapped-cell β-galactosidase for up to 24 hr at room temperature with organic solvents did not cause any loss of activity. After 8-9 months of wet storage at 4° C no decrease of enzymatic activity was observed. Cell entrapment emparted a significant increase in enzymatic activity in comparison with intact free cells. This effect may be a consequence of cytoplasmic membrane permeabilization of the microorganism caused by the entrapment procedures. The increase of enzymatic activity differed with the membrane preparation and was 35 fold greater for the polyurethane system. All the membranes had a flat sheet configuration and was tested in standard ultrafiltration systems. The permeate flow rate, β-galactose degree of conversion, and stability of the system were studied in the range of 70-85° C.

REFERENCES

1. DE ROSA, M., GAMBACORTA, A., ESPOSITO, E., DRIOLI, E, & GAETA, S. *Biochemie 62*: 517 (1980).
2. DRIOLI, E., IORIO, G., MOLINARI, R., DE ROSA, M., GAMBACORTA, A. & ESPOSITO, E. *Biotechnol. Bioeng. 23*: 221 (1981).
3. BUONOCORE, V., SGAMBATI, O., DE ROSA, M., ESPOSITO, E. & GAMBACORTA, A. *J. Appl. Biochem. 2*: 390 (1980).
4. DRIOLI, E., GAETA, S., CARFAGNA, C., DE ROSA, M., GAMBACORTA, A. & NICOLAUS, B. *J. Membr. Science 6*: 345 (1980).
5. DRIOLI, E., IORIO, G., SANTORO, R., DE ROSA, M., GAMBACORTA, A. & NICHOLAUS, B. *J. Mol. Cat.*, in press.

IMMOBILIZED MODIFIERS FOR PROTEINS (IMPs)

W. H. Scouten, C. Lewis, A. Barnett, R. Haller
and W. Iobst

Department of Chemistry, Bucknell University
Lewisburg, Pennsylvania, USA

A wide series of insolubilized protein modification reagents have been synthesized and applied to the modification and/or purification of a variety of amino acids, model proteins, and cell membranes. These materials possess sterically hindered protein modification reagents which have potential applicaiton as protein topology probes.

Peroxyacid resins, cyanogen bromide activated polysaccharides, e.g. agarose, and cleavable bifunctional reagents immobilized on various matrices were prepared and employed as protein modifiers (1-3). One idealized scheme of protein modification reagents is shown below (4). These reagents all possess the advantages of ease of handling and speed of application; while many of them were also useful topology probes. The chief disadvantage in their use lies in the limited capacity of the reagents in terms of the concentration of protein modification reagent/g matrix.

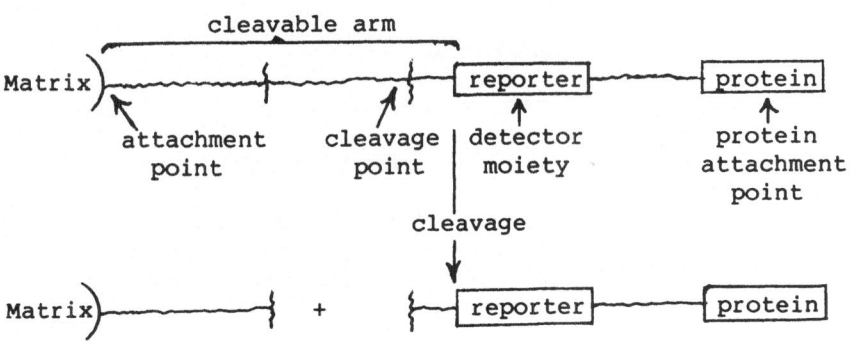

Cyanogen bromide activated agarose was employed as an insoluble reagent for converting protein lysyl amine residues into guanidine analogs. Cell membranes were also isolated by adding whole cells in isotonic solution to CNBr activated agarose, washing away excess cells, and then lysing the resulting immobilized cells with the hypotonic buffers. The immobilized ghosts that resulted could be eluted from the agarose by treatment with nucleophiles, e.g. 1 M NH_2OH. Similarly, amine functions of lysozyme (employed as a model protein) were modified to form substituted guanidines by treating lysozyme, bound to CNBr activated agarose, with nucleophiles bearing radioactive, fluorescent or chromophoric groups.

(R = chromophore, fluorophore or radio label)

A third series of immobilized modifiers for proteins (Imps) were prepared from cleavable binfunctional cross linking reagents which were immobilized on reactive matrices, e.g. Sepharose AH, to form a cleavable insolubilized protein reagent. For example, excess dithiobis(succinimidyl proprionate) (DTSP) when reacted with Sepharose AH yielded an excellent reagent for isolating cell membranes or for modifying proteins to thioacylate lysyl side chains, effectively converting the NH_2 group into a sulfhydryl residue. This is particularly effective using ^{35}S or ^{14}C radio labels as reporter groups.

REFERENCES

1. WILCHEK, M., OKA, T. & TOPPER, Y. *Proc. Nat. Acad. Sci. USA* 72: 1055 (1975).
2. PATCHORNICK, A. & KRAUS, M. A. *Encycl. Polym. Sci. Tech. Suppl. 1:* 468 (1976).
3. SCOUTEN, W. H. *Enzyme Engineering Symposium 4:* 368 (1978).
4. SCOUTEN, W. H., "Affinity Chromatography: Bioselective Adsorption on Inert Matrices," *Wiley-Interscience, New York (1981).*

ENZYME IMMOBILIZATION IN POROUS SUPPORTS

J. E. Bailey

Department of Chemical Engineering, California
Institute of Technology, Pasadena, California,
USA

Enzyme immobilization on the internal surfaces of porous insoluble materials provides high volumetric activity and minimal
diffusional resistance for substrates and products. Preparation
of these biocatalysts requires impregnation of the support with
enzyme, a complex transient diffusion-surface attachment process.
The influences of enzyme and support properties and immobilization
conditions on enzyme loading and on internal distribution of
enzymes in the catalyst particles have been studied.

The effect of particle size was investigated by immobilizing
glucoamylase under stagnant conditions to porous activated carbon
that had been pretreated with carbodiimide (1). Smaller particles
loaded more enzyme and exhibited higher specific activities than
larger particles (Table 1), strongly suggesting that enzyme was

TABLE 1

PARTICLE SIZE EFFECTS ON IMMOBILIZATION

Particle radius (μm)	Final loading (mg protein/g carbon)	Activity (μmole glucose/ min/g carbon)
125–297	35.1	49.6
490–598	29.7	27.2
847–1180	25.9	15.2

preferentially attached near the exterior pellet surface. The final loading was calculated from enzyme depletion from the immobilizing solution.

Different experiments explored contacting effects by circulating enzyme solution through a fixed bed of support particles at a controlled flow rate. Higher flow rates gave more rapid initial immobilization rates but lower final loadings. To explore this phenomenon further, two different immobilization experiments in the recirculation reactor were terminated when approximately the same loading had been achieved. The catalyst prepared at the higher recirculation flow rate was more than twice as active as the other preparation. One hypothesis consistent with this result is more nonuniform distribution of enzyme within the pellet at the higher recirculation flow rate (2,3).

In order to visualize the internal distribution of enzyme directly, glucoamylase labelled with fluorescein isothiocyanate was immobilized on CNBr-activated Sepharose, which was sectioned to 10 μm and photographed with an incidence fluorescence microscope at 480X. Enzyme was localized near the outer pellet surface for high recirculation rates during immobilization, but the distribution of enzyme was uniform for immobilization using a lower recirculation rate. For ordinary substrate reaction kinetics the nonuniformly loaded pellets will exhibit higher activity (2). On the other hand, the smaller protein bovine serum albumin was found to be uniformly distributed even when immobilized at the high recirculation rate.

This research demonstrates that the activity of a biocatalyst prepared by attaching enzyme to the internal surfaces of a porous solid depends upon the size of the porous particles, the contacting pattern between enzyme solution and the porous support, and the relationship between protein and particle pore dimensions. Careful design of the immobilization process is therefore very important in determining process characteristics of the resulting catalyst.

Y. K. Cho, H. Chain, J. Mehner, and Y. H. Park obtained the experimental data. Support for this research was provided by the National Science Foundation.

REFERENCES

1. CHO, Y. K. & BAILEY, J. E. *Biotechnol. Bioeng. 20:* 1651 (1978).
2. CORBETT JR., W. E. & LUSS, D. *Chem. Eng. Sci. 29:* 1473 (1974).
3. LASCH, J. *Fed. Europ. Biochem. Soc. Proc. 52:* 495 (1979).

IMMOBILIZATION OF MICROBIAL CELLS USING GELATIN AND GLUTARALDEHYDE

Q. Wang, X. Ji and Z. Yuan

Shanghai Institute of Biochemistry, Chinese
Academy of Sciences, Shanghai, China

Immobilized enzymes and cells (1,2) should be of considerable economical and practical importance. Therefore, since 1976 we have been working on techniques for immobilization of microbial whole cells (3,4). Entrapment in gelatin and crosslinking with glutaraldehyde has been employed successfully for the immobilization of *Echerichia coli* D816 cells for penicillin acylase activity and *Streptomyces roseofulus* Kc 13-575 cells for glucose isomerase.

Wet cells (20 g) were mixed with 10 ml 10% gelatin and 1 ml 25% glutaraldehyde, layered, and stored in the refrigerator. The gel films were cut into 2 mm particles. If needed, the gel was tanned with 0.25% glutaraldehyde. Electron micrographs of the immobilized cells showed that large quantities of cells were entrapped within the gel. The ratio of cells to gelatin and the amount of glutaraldehyde used were critical for the apparent activity. The activity recoveries were 30-40% for immobilized glucose isomerase and 30-35% for immobilized penicillin acylase, which could be increased one fold by crushing the gel particles. Some of the enzymatic properties of the immobilized and native cells are listed in TABLE 1.

Enzyme stability was considerably enhanced by immobilization. The immobilized penicillin acylase possessed higher resistance to heat than did the native cells and was more stable at extreme pH values of <4 and >9. However, the resistance to heat and extremes of pH were essentially the same for the immobilized and native glucose isomerase.

Good operational stability was shown in the case of immobilized *E. coli*. After continuously passing a solution of 2.5%

TABLE 1

SOME BASIC PARAMETERS

Enzyme Properties	Penicillin acylase (E. coli)		Glucose Isomerase (Str. roseofulus)	
	Native	Immobilized	Native	Immobilized
Optimum pH	8.4	9.0	7-8	7-8
Optimum Temperature ($^{\circ}$C)	50	55	85	85
K_m (mM)	6.3	11.4	0.5	1.33 M

penicillin G in 0.2 M phosphate buffer over the cells for 103 days at 37°C, the immobilized *E. coli* column showed no loss of penicillin acylase activity. The immobilized glucose isomerase column maintained a conversion of 42% during continuous isomerization of 45% glucose solution at 65°C for up to nearly 20 days. An industrial process based on immobilized *E. coli* penicillin acylase has been in use since 1978. After 285 cycles of 6-amino penicillanic acid production over 7 1/2 months, the hydrolytic rate was maintained unchanged. This is one of the first industrial applications of immobilized enzymes in China. The immobilized glucose isomerase process has been employed on a pilot plant scale (5). Immobilized aspartase in *E. coli* and α-galactosidase in *Thermophilic bacilli* also have been satisfactorily prepared by this same method.

REFERENCES

1. YUAN, Z., LIU, S. & YUAN, J., "Immobilized Enzymes and Affinity Chromatography", Science Press, Beijing, China (1975).
2. MOSBACH, K. *Meth. Enzymol.* 44: (1976).
3. *Prog. Biochem. Biophys.* 2: 65 (1980).
4. WANG, Q., YE, X., ZHAO, F., LIU, G., LIU, G., OU, G., SHEN, C. & XU, J. *Acta Biochimica Biophysica Sinica 12:* 305 (1980).
5. JI, X., LI, H., ZENG, Y., BAI, P., GU, G., WANG, L. & XU, S. *Progress Biochem. Biophys.*, in press.

IMMOBILIZATION OF β-GALACTOSIDASE FROM *E. COLI* CSH36 AND ITS

MICROBIAL CELLS USING CELLULOSE BEADS

Y. Hong, S. Kwon, M. Chun and M. Sernetz*

Department of Agricultural Chemistry, Korea
University, Seoul Korea and Institut fur Biochemie*
Justus-Liebig-Universitat, Giessen
Federal Republic of Germany

Using porous cellulose beads as a suitable matrix, Chen *et al.* (1) and Chun *et al.* (2) immobilized some enzymes by covalent or ionic binding, respectively; and Linko *et al.* (3) entrapped *S. cerevisiae* yeast cells within the beads. The retention of activity and the kinetic properties of β-galactosidase and *E. coli* mutant cells immobilized respectively on a) cellulose beads which were treated with p-benzoquinone and b) tannin by covalent binding and adsorption coupling were investigated in this work.

Constitutive β-galactosidase from *E. coli* CSH36 was used as the enzyme source. Porous cellulose beads (water content 91.5%, 230 μm dia) were prepared by a modification of the method of Chen *et al.* (1). Cellulose acetate powder (12.5g) was completely dissolved in 100 ml of dimethylsulfoxide and acetone in a ratio of 6:4. Thereafter, the cellulose-solvent mixture was dropped into 30% acetone solution with stirring. The high surface area, caused by the porous structure, has made this a good matrix for attachment via covalent and adsorption coupling. The dried porous cellulose beads (50 mg) were activated by 3% tannic acid, a mixture of the same volume of 0.1 M p-benzoquinone and 3% tannic acid, or 0.1 M p-benzoquinone (4). The pH values of the activation reagents were adjusted to 11 with 1 N NaOH. The activation was performed with shaking for 6 hrs. β-Galactosidase was immobilized on the activated cellulose beads by covalent binding and adsorption with shaking for 16 hrs. In the case of whole cell immobilization, lyophilized cells were used.

The optimum pH and temperature of the microbial β-balactosidase and the extracted enzyme were negligibly affected by the im-

mobilization. The Km values of the native and immobilized enzyme
were 4.0 X 10^{-4} M and 7.5 X 10^{-4} M, respectively. About 75% of the
native enzyme activity was immobilized. The stability and operational reuse of the immobilized materials are shown in Fig. 1.
In the tannin-p-benzoquinone activated cellulose beads, the remaining activity was over 80% of the initial enzyme activity after 20
runs. This kind of method is easy to apply and gives high coupling yields with enzymes. We foresee that this cell-matrix binding
method can be applicable in immobilizing cells with relatively
good enzyme activity.

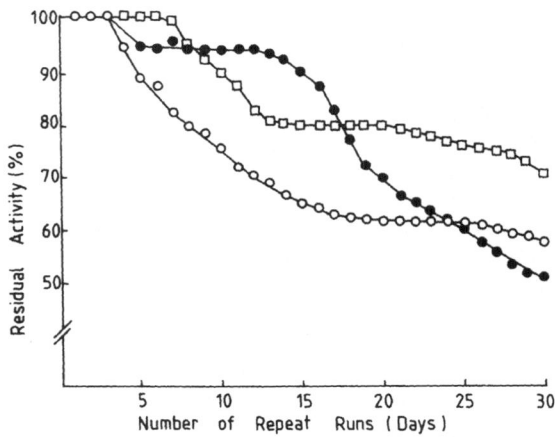

Fig. 1.   Retention of β-D-galactosidase activity of immobilized
          enzyme on reuse. Activator: O tannin; ● benzoquinone
          (BQ); □ BQ-tannin mix.

REFERENCES

1. CHEN, L. F. & TSAO, G. T. *Biotechnol. Bioeng. 18:* 1507
    (1976).
2. CHUN, M., DICKOPP, G. & SERNETZ, M. *J. Solid-Phase Biochem.*
    5: 211 (1980).
3. LINKO, Y-Y, POUTANEN, K., WECKSTROM, L. & LINKO, P. *Enzyme*
    Microb. Technol. 1: 26 (1979).
4. WATANABE, T., FUJIMURA, M., MORI, T., TOSA, T. & CHIBATA, I.
    J. Appl. Biochem. 1: 28 (1979).

HYDROPHOBIC IMMOBILIZATION OF ENZYMES AND POLYNUCLEOTIDES

ON TRITYL AGAROSE

P. Cashion, A. Javed, V. Lentini, D. Harrison,
J. Seeley and G. Sathe

Department of Biology, University of New Brunswick
Fredericton, New Brunswick, Canada

It has been reported that agarose can be chemically derivatized to form tritylated [$(C_6H_5)_3$-C-] agarose (TA) (1-3). The
trityl group is attached by a stable ether bond. Depending upon
the concentration of trityl groups (10-100 µmol/ml), the resin
binds a variety of polynucleotides such as poly A terminated mRNA
and denatured DNA as well as most enzymes. Of the 30 or so
enzymes tested to date, only pancreatic RNase failed to bind. To
maintain specificity polynucleotide binding is best done at lower
degrees of tritylation (10-40 µmol/ml) while enzymes are generally
immobilized at higher levels (50-100 µmol/ml). From the structures of the respective macromolecules which bind, and from their
sensitivity to low dielectric constants, to low ionic strengths,
and to chaotropic salts, it is probable that the binding is hydrophobic in nature.

There are striking resemblances between the binding of both
polynucleotides and enzymes to TA and the reported binding of
these macromolecules to a variety of other matrices (1-3). The
polynucleotide binding specificity and the conditions for optimal
binding and release by lignin containing celluloses, benzoylated
celluloses, and detergent impregnated NO_2-celluloses all strongly
resemble that of TA. Likewise occurs for the reported binding of
proteins to NO_2-cellulose, phenoxyacetyl cellulose, and alkylated
or tannin derivatized polysaccharides. Recently Butler has shown
that polyphenolic condensed tannins, proanthocyanidins, interact
strongly with proline enriched proteins (4). A feasible mechanism
encompassing all of these associations can be suggested (1-3).
It involves simple hydrophobic stacking type interactions between,
in most cases, polyaromatic groups on the resin and accessible

219

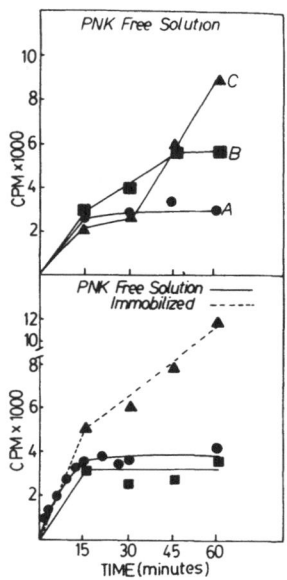

Fig. 1. Enhanced stability of polynucleotide ligase when immo-
         bilized on trityl agarose. T$_4$ polynucleotide kinase
         immobilization and assay procedure as in (2). Top: A,
         free solution assay containing 2 mg bapped poly C in
         200 µL buffer and 5 µL ^{32}PATP + 2 unit PNK incubated at
         37°C for 60 min.; B, same as A plus 2 unit PNK added
         after 15 min.; C, same as A plus 2 unit PNK added after
         30 min. Bottom: Free solution assays same as A above;
         aliquots taken at times shown and activity checked.
         Immobilized assay involved 4 identical columns, at 15,
         30, 45 and 60 min.; one column was assayed for activity.

aromatic or aliphatic groups on the particular macromolecule, be
it a protein or nucleic acid.

     It has been reported (2,3) that TA-immobilized polynucleotide
kinase (T$_4$) exhibits 3-4 times more activity than the corresponding free enzyme in solution during a 60-min. incubation. Recent
work shows that this apparent higher activity is due to an enhanced stability during incubation, not to a higher specific
activity (Fig. 1). The enzyme free in solution becomes inactive
after 20 minutes, while the TA-bound enzyme is fully active during a 60 min. incubation. Certain enzymes currently being studied,
such as Butler's 5' phosphodiesterase and bacterial and intestinal
alkaline phosphatases, remain fully active for at least 8 months
in TA-immobilized form with repeated use (5).

Reusability of an enzyme immobilization matrix is a desirable feature (6). It has recently been found that TA resins are easily and quickly regenerated and hence fully reusable (7). Furthermore, given the mild conditions involved in eluting most enzymes from TA columns, it is, in fact, possible to recover and hence reuse the originally immobilized enzyme as well.

Given the frequent reports in the current literature of catalytically important interactions between enzymes and membrane phospholipids (8), our current hypothesis is that the efficacy and simplicity of immobilizing enzymes to TA derives from the fact that the hydrophobic TA fibers sufficiently resemble the *in vivo* phospholipid milieu of biomembranes, that enzymes are predisposed to bind to TA fibers *in vitro* as well. This is supported by the strong resemblance between the binding of alkaline phosphatase to natural membranes and its binding to either TA alone or to TA-phosphatidyl choline adducts (5).

REFERENCES

1. CASHION, P., *et al. Nucl. Acids Res. 8:* 1167 (1980).
2. CASHION, P., *et al. Nucl. Acids Res. Symp. Ser. 7:* 173 (1980).
3. CASHION, P., *et al.* in "Gene Amplification and Analysis," vol. 2 (J. G. Chirikjian and T. S. Papas, eds.) Elslevier-North Holland, New York (1981) p. 551.
4. HAGERMAN, A. E. & BUTLER, L. G. *J. Biol. Chem. 256:* 4494 (1981).
5. CASHION, P., *et al. Biotechnol. Bioeng.,* in press.
6. COULET, P., CARLSSON, J. & PORATH, J. *Biotechnol. Bioeng. 23:* 663 (1981).
7. CASHION, P., *et al. Biotechnol. Bioeng.,* in press.
8. DUFOUR, J. & GOFFEAU, A., *J. Biol. Chem. 255:* 19591 (1980).

INFLUENCE OF THE ACTIVATION DEGREE OF THE SUPPORT ON THE

PROPERTIES OF AGAROSE-NUCLEASE

A. Ballesteros, J. M. Guisan, and J. Serrano

Instituto de Catalisis y Petroleoquimica, C.S.I.C.
Madrid, Spain

(presented by M. Engasser)

When an enzyme to be covalently insolubilized has many residues which are able to react with the activated support, an increase in the degree of support activation can favor insolubilization by multiple-point attachment. The degree of activation of the matrix also will influence the speed of the insolubilization reaction and the maximum capacity for bound protein.

Micrococcal nuclease (E.C.3.1.31.1), containing 149 amino acid residues of which 23 were lysines, was insolubilized on agarose using CNBr activation of the support. Two agarose gels with very different porous structures were used: Se 2 (Sepharose 2B from Pharmacia Fine Chemicals) and Ag 22 (prepared in our laboratory by reducing the porosity of commercial Sepharose 6B). The mean pore radius and specific surface areas were as follows: Se 2, 1250 $\overset{\circ}{A}$ and 8.2 m^2/cm^3; Ag 22, 30 $\overset{\circ}{A}$ and 240 m^2/cm^3, respectively (2). The dimensions of the enzyme were 30 x 30 x 40 $\overset{\circ}{A}$ (1). By varying the CNBr concentration we obtained agaroses with very different concentrations of reactive cyanates (3), as determined according to Kohn and Wilchek. From the specific surface area, the density of cyanate groups on the surface (external and internal) of the agarose gels was calculated and expressed as the No. of cyanates/1000 $\overset{\circ}{A}^2$ (1000 $\overset{\circ}{A}^2$ is approximately the area occupied by a molecule of nuclease). This represents the maximum theoretical number of covalent bonds between the matrix and the amino residues in the enzyme molecule. Ag 22 gave 2.0 and 0.5 cyanate/1000 $\overset{\circ}{A}^2$ for 80 and 20 mM cyanate, respectively; while Se 2 gave 11.0, 3.7, and 1.1 cyanate/1000 $\overset{\circ}{A}^2$ for 15, 5.0, and 1.5 mM cyanate, respectively (all with 151 µg nuclease/ml of gel).

We demonstrated previously (4) that nuclease, insolubilized on Sepharose 2B yields derivatives which, when assayed using thymidine 5'-(p-nitrophenyl phosphate) 3'-phosphate as substrate, were not affected by diffusional limitations. In the present report when the Se 2 activation was increased the k'_{cat} of the insoluble derivatives decreased, but the K'_m also decreased (5). Eadie-Hofstee plots for the least porous (Ag 22) showed mixed enzymic reaction-internal diffusion kinetics (6); and in contrast to the derivatives obtained with the Se 2 matrix, the Ag 22 with the higher support activation had a higher specific activity. These values of specific activity may be the consequence of different distribution of enzyme within the support. In derivatives obtained using a support of pore size intermediate between that of Se 2 and Ag 22, no influence of the extent of activation of the matrix on the specific activity was found, indicating the superposition of opposite effects, i.e. different conformational changes of the enzyme (single or multiple point attachment) and different distribution of the enzyme inside the matrix.

This work has been supported by the Spanish Commission Asesora de Investigacion Cientifica y Tenica, and by the Caja de Ahorros y Monte de Piedad de Madrid.

REFERENCES

1. ANFINSEN, C. B., CUATRECASAS, P. & TANIUCHI, H. in "The Enzymes," 3rd Edit., vol. 4 (P. D. Boyer, ed.), Academic Press, New York (1971) p. 177.
2. SERRANO, J., thesis, Autonomous University of Madrid (1981).
3. KOHN, J. & WILCHEK, M. *Biochem. Biophys. Res. Comm. 84:* 7 (1978).
4. GUISAN, J. M., MELO, F. V. & BALLESTEROS, A. *Appl. Biochem. Biotechnol. 6:* 25 (1981).
5. GUISAN, J. M. & BALLESTEROS, A. *J. Solid-Phase Biochem. 4:* 245 (1979).
6. GUISAN, J. M., MELO, F. V. & BALLESTEROS, A. *Appl. Biochem. Biotechnol. 6:* 37 (1981).

IMMOBILIZATION OF BIOFUNCTIONAL COMPONENTS BY RADIATION

POLYMERIZATION AND APPLICATIONS

I. Kaetsu, M. Kumakura, T. Fujimura, M. Yoshida,
F. Yoshii, M. Asano, M. Tamada, and N. Kasai

Takasaki Radiation Chemistry Research Establishment
Japan Atomic Energy Research Institute
Watanuki, Takasaki, Gunma, Japan

The purposes of this study were a) to propose a new physical immobilization method using vinyl monomers and radiation polymerization at low temperatures and b) to show some applications and characteristics of the method.

Biocomponents, such as enzyme, drug, antigen, antibody, microbial cell, or tissue cell, were mixed with monomers. These were shaped into various forms, cooled to 0 to -78°C, and polymerized by radiation with γ-rays to form a composite of polymer and the biocomponent. The biological activity of the composite was measured and evaluated according to conventional analytical methods. The characteristics of the method can be summarized as follows: a) Freedom of choice of carrier polymer, for example hydrophilic or hydrophobic, rigid plastic or soft elastomer. b) Freedom to select shape, form, and structure of composite, for example membrane, emulsion, porous gel, or multiple-layer gel. c) Freedom to choose biocomponents to be immobilized, such as low or high MW enzyme, organism, or whole cell. d) Freedom to distribute biocomponent from physical attachment on carrier surface to inner entrapment within carrier matrix. 3) Freedom to specify mobility of biocomponent in polymer from low (tight fixation) to high (weak adsorption).

These factors can be changed and controlled easily by variation of the conditions. Denaturation of the biocomponent by radiation damage was avoidable, if the radiation conditions were limited to temperatures below -20°C and radiation doses less than 1 mega rad, which was sufficient for polymerization of the monomer.

The present method has been applied in several different areas. In applications involving surface reactions the composite fixing of an enzyme on the carrier surface was used to catalyze the hydrolysis of a high MW substrate, such as starch, protein, and cellulose. The immobilized enzyme also was used in the hydrolysis of solid substrates such as starch plant, cellulosic waste, saw dust, chaff, and bagasse. The hydrolyses were successful. Surface active composites containing antibody (IgG) were applied to immunological reactions with antigens such as α-feto protein; and the results enabled the surface reaction to be carried out with large size antigens.

A second area of application was that of controlled slow release of drugs. A composite with drug entrapped inside a carrier was used for the controlled release. The first application was carried out in chemotherapy with an anticancer composite embedded in the polymer. This was very effective in retardation of pain and tumor growth with no observed secondary reaction. The same technique also was applied successfully to release of steroid hormones, such as teststeron. A third area, that of cell immobilization was applied to the immobilization of erythrocytes on a carrier surface such as methoxypolyethyleneglycol methacrylate polymer. The composite was stable and did not cause hemolysis, yet allowed oxygen adsorption and desorption. Chloroplasts from spinach were stabilized markedly by immobilization in the whole cell state with the same polymer. It was found that immobilization with various vinyl polymers was effective with cultured and growing cells. For example, *S. formosensis* (yeast), cultivated in the polymer, had a 13 times higher rate than when grown in a suspension of intact cells.

IMMOBILIZATION OF WHOLE CELL GLUCOSE ISOMERASE WITHIN SOYBEAN PROTEIN

C. L. Lai

Department of By-product Utilization
Taiwan Sugar Research Institute,
Tainan, Taiwan, 700 China

An inexpensive natural polymer, soybean protein, was used as a matrix for immobilization of whole cells of *Streptomyces* sp. for glucose isomerase. The immobilization consisted of the following steps: a 5% soybean protein solution was boiled for a few minutes and cooled to 60°C (1). *Streptomyces cells were mixed* with the heated protein solution and then entrapped within the protein by coagulation with $MgCl_2$. After filtration the cell-protein mixture was extruded to form pellets and then dehydrated. The resulting dry preparation was treated with hexamethylenediamine and glutaraldehyde and finally washed and dried in vacuum to about 12% moisture.

The effect of coagulants on the activity of the immobilized cells was examined. Magnesium chloride was found to be the best among the coagulants tested. The activities of the immobilized preparations coagulated with calcium chloride and glucono-delta-lactone were 39% and 44% that of the preparation with magnesium chloride, respectively. The activity decay was presumably due to high calcium content in the matrix in the case of calcium chloride coagulation and to the low pH (5.1) of the matrix in the case of glucono-delta-lactone coagulation.

Dehydration of the extruded pellets before addition of hardening reagents was an important step for obtaining good mechanical strength. Hardening treatment of highly hydrated pellets resulted in a highly porous preparation of poor mechanical strength.

The immobilized cells with the best combination of activity and mechanical strength were obtained with a cell-to-carrier ratio of 1:4, on a dry weight basis. This ratio probably represented

227

a compromise between the activity and mechanical strength. The
activity of the immobilized cells increased linearly with increasing
cell-to-carrier ratio; but the mechanical strength declined sharply
above a ratio of 1:4. The optimum temperature and pH of the immobilized cells were almost identical with those of the free cells.

The operational stability of the immobilized cells was determined in a packed column (2.5 x 40 cm) system operated at 62°C and
pH 8.2 with 40% (w/w) glucose syrup at a flowrate of 0.5 space
velocity. The column retained 69% of the initial activity after
continuous operation for 23 days; and the half-life was estimated
to be 37 days. No apparent problem of back pressure was noticed
during the continuous operation.

REFERENCES

1. HASHIZUME, K., MAEDA, M. & WATANABE, T. *Nippon Shokuhin Kogyo
    Gakkaishi 250:* 21 (1978).

ENZYME STABILIZATION IN HIGH MACROMOLECULAR CONCENTRATION ENVIRONMENTS

G. Greco Jr., G. Marrucci, and L. Gianfreda*

Istituto di Principi di Ingegneria Chemica, and
Facolta di Farmicia,* Universita di Napoli, Italy

The joint use of completely polarized ultrafiltration (UF) membrane enzymatic reactors and of stabilizing, linear chain, soluble polymers has been proposed as an alternative to the more traditional enzyme immobilization techniques (1-3). When an enzyme is injected into an unstirred UF cell, it undergoes accumulation at the membrane surface (concentration polarization). Due to the high local concentrations, strong protein-protein interactions occur. These generally result in considerable improvement in enzyme stability often without reduction in activity. Further enhancement can be produced when a soluble, linear chain inert macromolecule is also injected into the system. A polymeric network is formed that drastically reduces the enzyme mobility, thus preventing to a considerable extent the unfolding of the proteic macromolecule. The polymeric structure is labile enough to be easily removed by rinsing the UF membrane to restore its permeability and rejection properties and not give rise to any appreciable resistances to substrate mass transfer.

A comparison was performed using a) stirred, b) polarized, and c) concentration polarized reactors with stabilizing macromolecule injection under the same general operating conditions (3). The reacting system was p-nitro-phenyl phosphate hydrolysis by acid phosphatase (E.C.3.1.3.2 from potato). The stabilizing macromolecule was SEPARAN MGL (polyacrylamide, Dow Chemical Co., Midland MI). The results are summarized in Fig. 1 in terms of the specific reaction rate. The proportionality between log(rate) and t suggests first order deactivation kinetics (4-5) with enzyme half-lives of 5.1 hr (stirred), 12.3 hr (polarized), and 26.9 hr (stabilized).

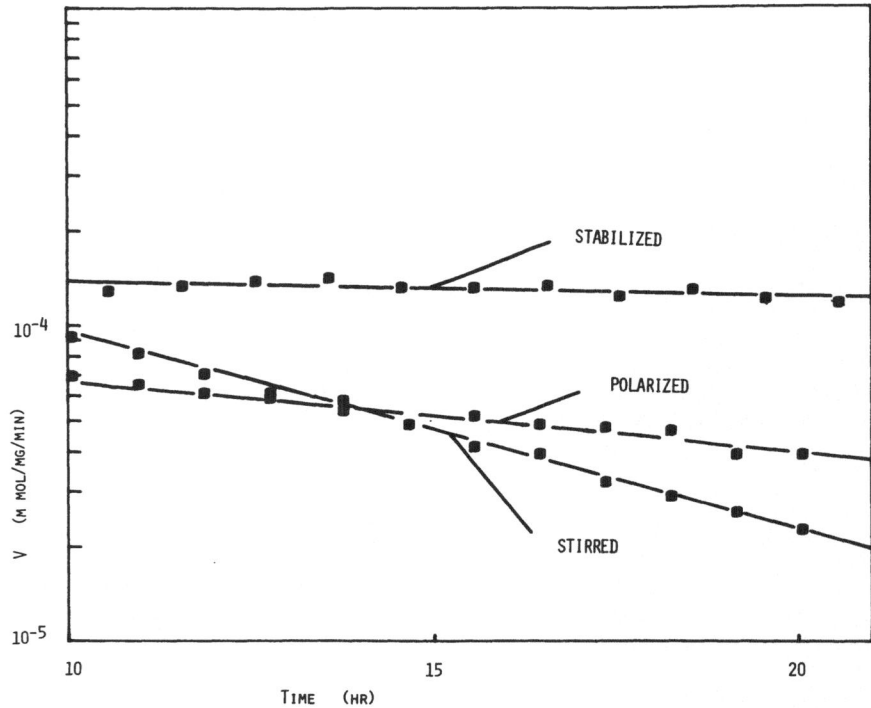

Fig. 1. P-nitrophenyl phosphate hydrolysis by acid phosphatase
         with substrate feed concentration of 5 mM (saturating)
         in 100 mM Na-Citrate buffer, pH 5.60, 55°C.

        A more detailed investigation was performed on cellobiose
hydrolysis by β-glucosidase (E.C.3.2.1.21 from sweet almonds)
stabilized with TYLOSE H4000 T (hydroxyethyl cellulose, Hoechst
A. G., Frankfurt) (6). The results are reported in Fig. 2 in terms
of enzyme half-life and the reciprocal absolute temperature. By
inspection of Fig. 2, one can conclude that stabilization by a
factor of 3.6 was obtained by switching to the polarized system
and by a factor of 20.2 when use was made of a stabilizing polymer.
Absolute activity as a function of pH showed that only minor, if
any, reductions in specific reaction rate took place due to
polarization and to stabilization, since the optimal pH was unaffected. Therefore, a true increase in enzyme stability was
achieved and not a spurious effect deriving from substrate mass
transfer resistances, since the activity levels were unchanged in
the three configurations and, furthermore, the activation energy
of the deactivation process was unaffected.

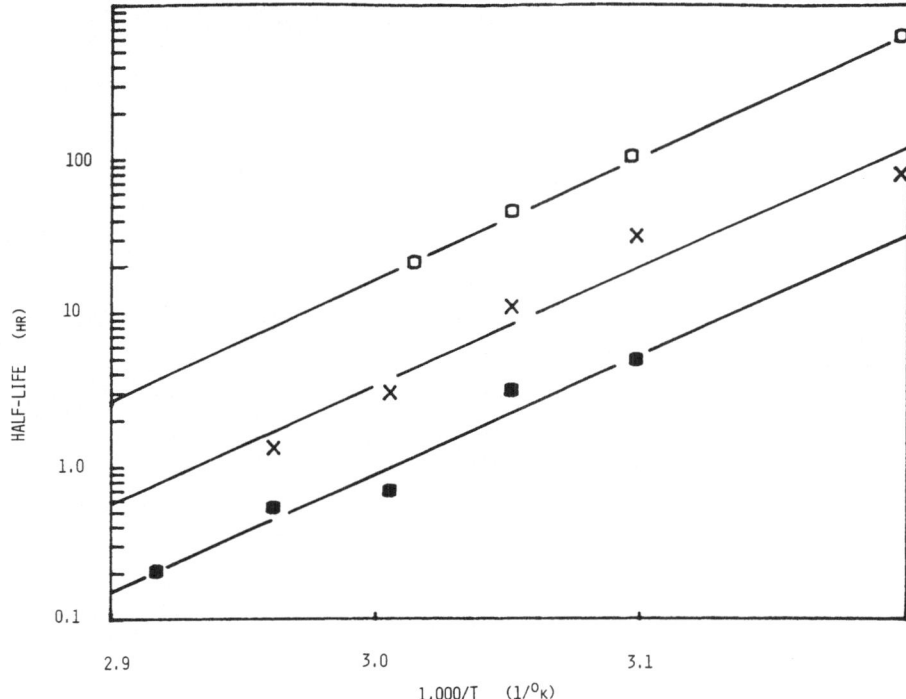

Fig. 2. Cellobiose hydrolysis by β-glucosidase. Enzyme thermal
inactivation: ■ native enzyme, X polarized reactor,
□ stabilized reactor.

REFERENCES

1. GIANFREDA, L. & GRECO JR., G. *Biotechnol. Lett. 33:* 3
    (1981).
2. GRECO JR., G. & GIANFREDA, L. *Biotechnol. Bioeng.*, in
    press.
3. GRECO JR., G., LIVOLSI, A. M., SCARFI, M. R., MANSI, F. R.
    & GIANFREDA, L. *Eur. J. Appl. Microbiol. Biotechnol.*, in
    press.
4. GRECO JR., G., ALBANESI, D., CANTARELLA, M., GIANFREDA, L.,
    PALESCANDOLO, R. & SCARDI, V. *Eur. J. Appl. Microbiol.
    Biotechnol. 8:* 249 (1979).
5. GRECO JR., G., GIANFREDA, L., ALBANESI, D. & CANTARELLA, M.
    J. Appl. Biochem., in press.
6. GIANFREDA, L., LIVOLSI, A. M., SCARFI, M. R. & GRECO JR.,
    G. *Enzyme Microb. Technol.*, in press.

INTRINSIC STABILITY OF THERMOPHILIC ENZYMES: 6-PHOSPHOGLUCONATE

DEHYDROGENASE FROM *BACILLUS STEAROTHERMOPHILUS* AND YEAST

F. M. Veronese, E. Boccu and A. Fontana*

Institutes of Pharmaceutical and Organic*
Chemistry, University of Padua, Padua, Italy

In this communication, it will be shown that a thermophilic enzyme is not only thermostable (1,2), but also quite resistant to most common protein denaturants.

6-Phosphogluconate dehydrogenase was isolated to homogeneity from thermophilic *B. stearothermophilus* NCA 1503 (1); and the stability was studied on a comparative basis with that of the mesophilic enzyme from yeast. It was shown that the thermophilic enzyme had a T m about 30°C higher than the mesophilic one (Fig. 1A). The stability extended over a wider pH range for the thermophilic enzyme, particularly in the alkaline region (Fig. 1B). Organic solvents such as acetone, dioxane, and dimethylformamide led to enzyme inactivation of 40-70% and 2-10% for the thermophilic and mesophilic enzymes, respectively (Fig. 1C). Noteworthy stability towards the proteolytic enzymes trypsin, chymotrypsin, and elastase was observed with the thermophilic enzyme. For example, the thermophilic enzyme showed chymotrypsin activity for at least 24 hr, while the mesophilic one was fully inactivated after only 20 min. (bottom, Fig. 1C).

This work was carried out with Italian C.N.R aid as a special project on fine chemicals.

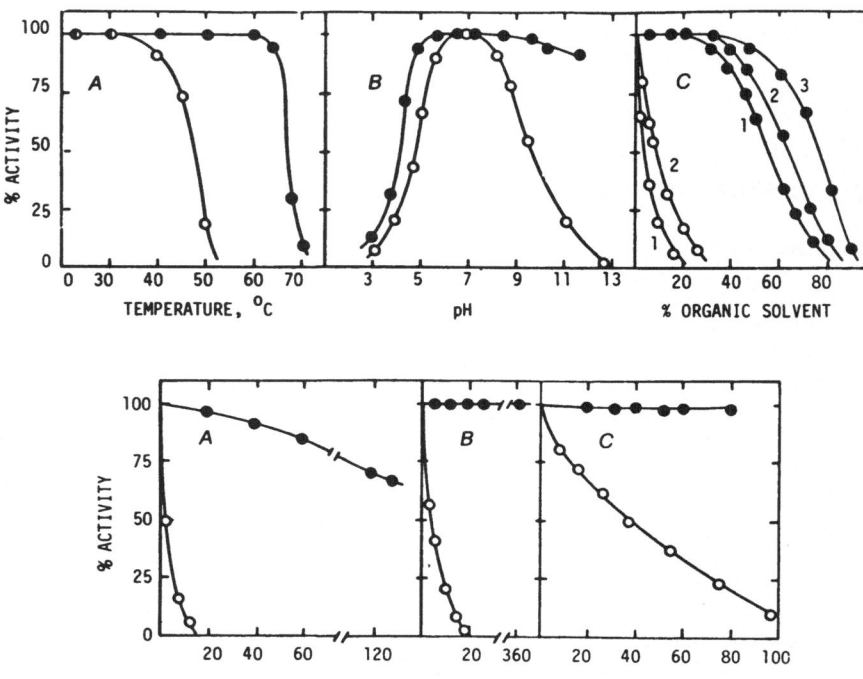

Fig. 1.   Effects of conditions on activity of *B. stearothermo-*
          philus ● and yeast O 6-phosphogluconate dehydrogenase in
          0.1 M potassium phosphate buffer, pH 7.2, containing
          1 mM EDTA and 1 mM dithiothreitol. Top: A) heated
          15 min. in buffer then assayed; B) incubated 30 min.
          at 20°C in buffer then assayed; C) incubated 90 min.
          in aqueous-acetone 1), -dioxane 2), and -dimethylforma-
          mide 3). Bottom: incubated in phosphate buffer, pH 7.2,
          at 37° plus trypsin (A), chymotrypsin (B), and elastase
          (C) at a protease:enzyme ratio of 2:100.

REFERENCES

1. VERONESE, F. M., BOCCU, E., FONTANA, A.,BENASSI, C. A. &
   SCOFFONE, E. *Biochim. Biophys. Acta 522:* 277 (1978).
2. VERONESE, F. M., BOCCU, E. & FONTANA, A. *Biochemistry 15:*
   4026 (1976).

STABILIZATION OF PENICILLIN AMIDOHYDROLASE IMMOBILIZED ON

EUPERGIT C

K. Sauber and D. M. Kramer*

Hoechst AG, Frankfurt and Rohm GmbH,* Darmstadt
Federal Republic of Germany

Carrier bound enzymes are playing an increasing important role in industrial production, in part through development of carriers with good filtration properties and high loading capacity. Economically it is important how often one can reuse carrier bound enzymes. Therefore, we have looked at higher stability under repeated use. Since dithiols are known stabilizers of enzymes (1), we have tried to combine this effect with solid phase enzymology.

Details of the immobilization and stabilization are described elsewhere (2). The stability under operating conditions was tested batch-wise in the following manner: 2 g immobilized enzyme, 20 ml 5% (w/v) substrate in 0.01 M phosphate buffer at pH 7.5, and NaN_3 were stirred at 37° to a substrate turnover of 95 - 98%. The enzyme beads were filtered by suction on fritted glass and added to fresh 20 ml substrate solution. The procedure was repeated sequentially 20 times. With penicillin acylase covalently immobilized on Eupergit C (Oxirane-acrylic-beads), the activity was 231 U/g wet weight. This product showed 58% loss of activity after 20 cycles of operation (5% penicillin G solution in a stirred batch reactor (Fig. 1). When treated with either dithioglycol or dithiothreitol, the immobilized acylase was significantly stabilized showing only 42% and 18% loss of activity, respectively.

We concluded that dithiols gave additional bonds between the enzyme and the carrier so that unfolding and hence deactivation was reduced.

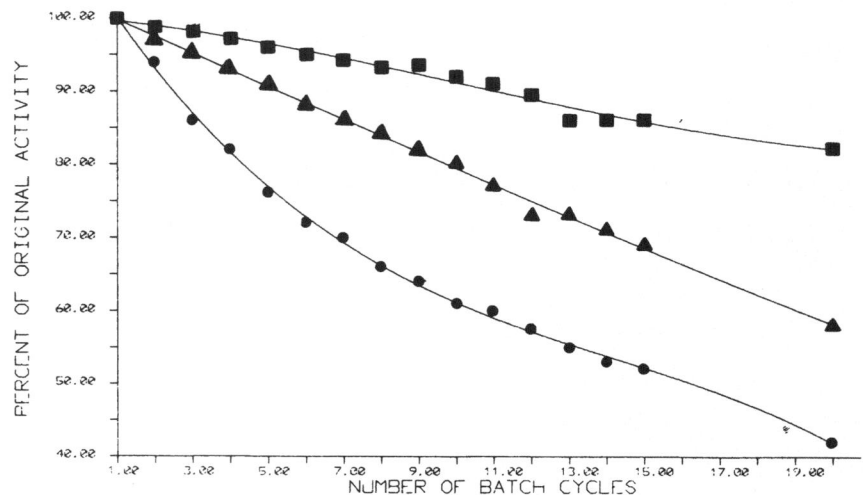

Fig. 1.   Loss of activity during repeated use in a batch reactor.
Code: ● untreated, ▲ treated with dithioglycol,■
treated with dithiothreitol.

REFERENCES

1.   CARLSSON, J., thesis, Uppsala University (1974).
2.   German Patent 2732301.

STABILIZATION OF FUMARASE ACTIVITY OF *BREVIBACTERIUM FLAVUM*

CELLS BY IMMOBILIZATION WITH κ-CARRAGEENAN AND POLYETHYLENEIMINE

T. Tosa, I. Takata, and I. Chibata

Research Laboratory of Applied Biochemistry
Tanabe Seiyaku Co. Ltd.
Yodogawa-ku, Osaka, Japan

In 1974 we succeeded in the industrial production of L-malic acid using *Brevibacterium ammoniagenes* cells immobilized with polyacrylamide gel (1), and in 1977 this method was changed to the new method using *Brevibacterium flavum* cells immobilized with κ-carrageenan (2). As a result, the productivity of L-malic acid using this new method was increased to 9 times that of the conventional polyacrylamide gel method (3). In this κ-carrageenan method, the stabilization of fumarase activity was considered to be enhanced by increased interaction between the microbial cells and the matrix. In order to obtain even more stable preparations and to improve the productivity of this system, we attempted to immobilize the microbial cells using κ-carrageenan along with some polymers.

The addition of polycationic polymers to the immobilization medium increased the operational stability of the fumarase activity of the immobilized cell column. This stabilization effect was particularly evident with polyethyleneimine. So, we investigated the optimal conditions for immobilization, using polyethyleneimine, and for continuous production of L-malic acid, using this improved immobilized cell column. The heat stability of the immobilized preparation increased so that the column could be operated at relatively high temperatures of 50-55°C for long times. Table 1 shows the relative productivities of L-malic acid, utilizing the various immobilized preparations. The productivity of immobilized with κ-carrageenan and polyethyleneimine increased to 21 times that of *B. ammoniagenes* immobilized with polyacrylamide; and the productivity per hour was enhanced to 1.8 times that of *B. flavum* immobilized only with κ-carrageenan. In

TABLE 1

COMPARISON OF PRODUCTIVITIES OF L-MALIC ACID
BY VARIOUS IMMOBILIZED PREPARATIONS

Microbial Cells and Immobilization Method	Operation Temperature ($^\circ$C)	Fumarase Activity (μmole/hr/ml of gel)	Operational Statility (Halflife; days)	Relative Productivity* (%)
B. ammoniagenes Polyacrylamide	37	530	53	100
B. flavum Polyacrylamide	37	610	94	273
κ-Carrageena	37	900	160	897
κ-Carrageenan + Polyethylene-imine	37	980	243	1587
+ Polyethylene-imine	45	1420	165	2073
+ Polyethylene-imine	50	1670	128	1992
+ Polyethylene-imine	55	2160	74	1730

*Each immobilized cell column was operated to the same activity
(265 μmole/hr/ml of gel) as that for the halflife of *B. ammoniagenes* immobilized with polyacrylamide; the productivity of *B.
ammoniagenes* immobilized with polyacrylamide was taken as 100.
Productivity = $\int_o E_o \exp(-kd\ t)dt$ where E_o = initial fumarase
activity, K_d = decay constant, and t = operation period.

1980 the industrial production system of L-malic acid was changed
to this improved method.

REFERENCES

1. YAMAMOTO, K., TOSA, T., YAMASHITA, K., & CHIBATA, I. *Eur. J.
    Appl. Microbiol. 3:* 169 (1976).
2. TAKATA, I., YAMAMOTO, K., TOSA, T., & CHIBATA, I. *Eur. J. Appl.
    Microbiol. Biotechnol. 7:* 161 (1979).
3. TAKATA, I., YAMAMOTO, K., TOSA, T., & CHIBATA, I. *Enzyme
    Microb. Technol. 2:* 30 (1980).

APPLICATION OF POLYETHYLENE GLYCOL-BOUND NAD DERIVATIVE AND

THERMOSTABLE DEHYDROGENASE IN A MODEL ENZYME REACTOR

I. Urabe, N. Katayama, and H. Okada

Department of Fermentation Technology, Osaka
University, Osaka, Japan

Enzyme reactors that utilize coupled dehydrogenase reactions with NAD recycling can employ macromolecular NAD derivatives of high cofactor activity with a wide variety of dehydrogenases. We have prepared polyethylene glycol-bound NAD derivatives (PEG-NAD) by coupling N^6-(2-carboxyethyl)-NAD (N^6-CE-NAD) to aminopolyethylene glycol (MW 3,000 - 3,700) with water soluble carbodiimide [1]. PEG-NAD had high cofactor activity and was used in a continuous enzyme reactor [1]. We also studied the applicability of using thermostable dehydrogenases.

Table 1 shows the initial rate of reduction or oxidation of N^6-CE-NAD(H) and PEG-NAD(H) with thermophile *(Bacillus stearotherthermophilus* and *Thermus thermophilus)* and mesophile (pig and yeast) malate dehydrogenases (MDH) relative to the rates of change of NAD(H). Thermophile MDH showed higher relative activity than mesophile MDH, while the NAD derivatives were better cofactors than the natire NAD for *T. thermophilus* MDH. The increase in the relative activity of *T. thermophilus* MDH for the NAD derivatives was due to the increase in Vm (Table 2).

The continuous production of L-lactate was carried out in an ultrafiltration apparatus fitted with a UM-2 ultrafiltration membrane at 30°C. PEG-NAD, rabbit muscle lactage dehydrogenase, and *T. thermophilus* MDH were kept in the reactor; and substrates of pyruvate and malate were fed continuously to the reactor. For reactor operation of 25 days, the enzyme half-life was 20 days. These results indicated that PEG-NAD and thermostable dehydrogenases are applicable for use in enzyme reactors.

TABLE 1

COFACTOR ACTIVITY OF NAD(H) DERIVATIVES FOR THERMOPHILE
AND MESOPHILE MALATE DEHYDROGENASES (NAD, NADH = 100)

MDH Source	N^6-CE-NAD	PEG-NAD	N^6-CE-NADH	PEG-NADH
Pig heart	59	48	93	75
Yeast	36	35	87	51
B. stearothermo philus	96	73	112	85
T. thermophilus	432	207	210	95

TABLE 2

KINETIC CONSTANTS FOR THE REDUCTION OR OXIDATION OF NAD(H)
AND THEIR DERIVATIVES WITH PIG HEART AND T. THERMOPHILUS MDH

MDH Source	NAD(H) derivative	Reduction		Oxidation	
		K_m (µM)	V_m (%)	K_m (µM)	V_m (%)
Pig heart	Native	33.6	100	9.5	100
	N^6-CE-	25.3	44	12.7	99
	PEG-	42.8	43	30.3	99
T. thermophilus	Native	26.4	100	5.1	100
	N^6-CE-	42.9	566	9.5	230
	PEG-	28.8	195	22.0	122

This work was partly supported by Japanese Ministry of Education grant No. 56850030.

REFERENCES

1. FURUKAWA, S., KATAYAMA, N., IIZUKA, T., URABE, I. & OKADA, H. FEBS Lett. 121: 239 (1980).

USE OF THE POROUS MINERAL SPHEROSIL AS A CARRIER FOR ENZYMES:

FIXATION AND PURIFICATION

B. Mirabel

Rhone-Poulenc Recherches, Centre de
Recherches de la Croix de Berny
Antony, cedex, France

Spherosil[R] is a form of silica beads with controlled porosity.
The principal methods for attachment of enzymes to it are: adsorption by ionic or hydrophobic interaction; adsorption followed
by crosslinking with gluraldehyde; and covalent bonding via silani-
zation, esterification, or crosslinking to a thin layer of functional polymer. The resulting enzyme-supports are suitable for
use in columns because of the non-swelling, rigid, pressure resistant nature of the support. Pilot scale facilities have been
used with acylase, amyloglucosidase, lipase, lactase, glucose isomerase, penicilline amidase, and invertase.

Spherosil also is very suitable for enzyme purification.
Examples include: a) extraction of whey protein (an industrial
plant with 1.5 m^3 of Spherosil is in use in France); b) extraction
of lysozyme from egg whites; c) purification of human blood albumin
(an industrial plant with about 1 m^3 of Spherosil is in use in
France); and d) purification of urokinase from human urine.

Session IV

INDUSTRIAL APPLICATIONS OF IMMOBILIZED BIOMATERIALS

Chairmen: A. Michaels and H. Samejina

POTENTIAL APPLICATION OF IMMOBILIZED VIABLE CELLS IN THE FOOD

INDUSTRY: MALOLACTIC FERMENTATION OF WINE

S. Gestrelius

NOVO INDUSTRI A/S
Bagsvaerd, Denmark

Malolactic fermentation of wine is a bacterial conversion of
L-malic acid to L-lactic acid and carbon dioxide. One of the proposed enzyme mechanisms is shown in Fig. 1. It has not been possible to demonstrate that pyruvate and NADH are intermediates in this
reaction, only that NAD is required for malolactic activity and
that NADH and pyruvate are also formed, but in minute amounts (1).

The malolactic conversion is considered a desirable manufacturing step for three reasons: a) to decrease the acidity of the
wine, b) to stabilize the wine by reassuring that the fermentation
will not take place in the bottle, and c) to increase the flavor
complexity of the wine. In cool areas with high contents of malate

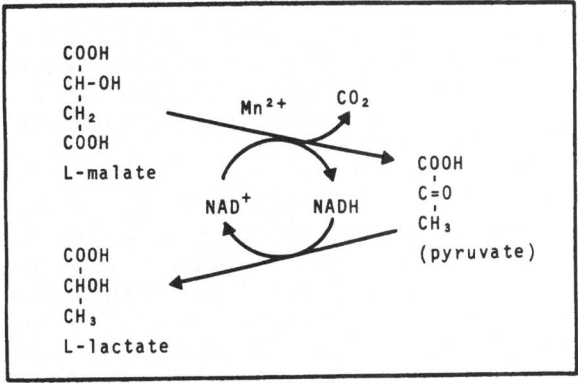

Fig. 1. Proposed reaction scheme for malolactic conversion by
         Lactobacilli (1).

in the wine the acid reduction is most important; while in areas
with less acidic wine the biological stabilization must be secured
in order not to get spontaneous growth of undesirable microorganisms. Several lactic acid bacteria have been found to carry
out malolactic fermentation, but only *Leuconostoc oenos* and a
few *Lactobacilli* strains have been judged to give the desirable
organoleptic improvement of treated wines.

In traditional wine making the malolactic fermentation occurs
spontaneously during storage of the new wines at slightly elevated
temperature for periods of months or years; and the bacteria origi-
nate from grape skins or the wooden vats. However, in the modern
wine industry with cleaner processing and much concern about minimized storage times, such spontaneous fermentation is often not
sufficiently rapid or reliable. *Leuconostocs* and *Lactobacilli*
grow extremely slowly, and sometimes not at all, in the harsh wine
milieu with pH 3-3.8, 10-14% ethanol, 10-100 ppm SO_2, and low contents of residual sugar. Inoculation with malolactic cultures is
sometimes successful, especially if a very large number of bacteria
is added; but a more safe approach is to separate the propagation
of the bacteria and their application as malolactic catalyst (2).

Immobilization of the viable bacteria after production
on an optimized growth medium may permit reuse of the cell preparation (3). In addition, performance of the malolactic conversion
by passing wine through a reactor containing immobilized cells
will have the advantages of speed, easy control of the degree of
conversion, and easy removal of the cells after conversion.

DESIGN OF AN IMMOBILIZED MALOLACTIC BIOCATALYST

Table 1 shows a number of important factors for the design
of an industrial malolactic catalyst. In an attempt to meet the
requirements for acceptance by wine consumers and health authori-
ties, a *Leuconostoc oenos* strain and Ca-alginate immobilization
were chosen for Novo's preparations.

The manufacture was performed essentially according to (4);
1 volume of wet cell sludge (about 15% dry weight) was mixed with
5-12 volumes 5% w/v sodium alginate solution (sterile) and the
mixture dripped into 2% $CaCl_2$ (sterile). After two hr the alginate sphere preparations were transferred to sterile filtered
grape juice/glycerol medium (reconstituted grape juice with 30 mM
L-malate, 10% ethanol and 35% glycerol, pH adjusted to 5.0), incubated overnight at 6°C, and finally stored at -20°C. The storage
half-life was about 3 months, as judged from the residual number
of viable bacteria by a plating technique. But, the residual
malolactic activity in the columns was hardly diminished even after
six months of storage, indicating that the preparations were over-

TABLE 1

IMPORTANT FACTORS FOR DESIGN OF MALOLACTIC CATALYST

Production:

       Selected microorganism
       Optimized growth conditions
       Aseptic cell separation and immobilization
       Good storage stability
       Easy shipping

Application:

       Quick processing = high volumetric capacity
       Good physical stability
       (fixed bed reactor with 0.3-0.5 atm overpressure)
       Useful at cellar temperature
       Useful in all types of wine
       Aseptic conditions
       Sufficient half-life for campaign period

Acceptance:

       Good organoleptic properties of treated wine
       Nontoxic immobilization reagents
       No leakage

---

loaded with cells and subjected to diffusion restrictions in the columns.

In this context it should be pointed out that it was proven necessary to use immobilized *intact* cells for carrying out malolactic conversion in wine with *Leuconostoc oenos* (5). All attempts to use ruptured cells (free or immobilized) in the presence of the cofactor, NAD, failed at the acid pH of wine. This presumably was due to the well-known instability of the cofactor molecule, especially in the reduced form, at low pH.

The pressure stability of the Ca-alginate spheres was tested according to the method of Norsker *et al.* (6) and was found surprisingly good. This permitted bed heights of 0.5-1 meter, in spite of the high contents of di- and tri-carboxylic acids in the wine and juice. Drying procedures as described by Klein *et al.* (7) could improve the physical stability further; but they were laborious to carry out without activity losses.

APPLICATION OF IMMOBILIZED MALOLACTIC BIOCATALYST

Temperature and pH optima for malolactic activity as well as operational stability of the Ca-alginate immobilized *L. oenos* preparations were investigated in laboratory size columns run with grape juice substrate (30 mM L-malate and 12% ethanol added). Under these conditions an optimum half-life of about 40 days was found at 20°C and pH 3.5-3.8.

One of the worst obstacles to malolactic fermentation is the presence of SO_2 in wine. SO_2 is added as an antioxidant and pre- servative during wine production. Although *L. oenos* grows ex- tremely slowly in SO_2-containing media, about 20% of the activity of immobilized *L. oenos* was maintained in columns during contin- uous malate conversion in grape juice containing 100 ppm SO_2 (total) at pH 3.5. Fig. 2 shows the performance of a laboratory column run with alternating juice (<15 ppm SO_2) and wine (<50 ppm SO_2). Although the activity was always higher in the grape juice medium, it was not possible to demonstrate any increase in cell

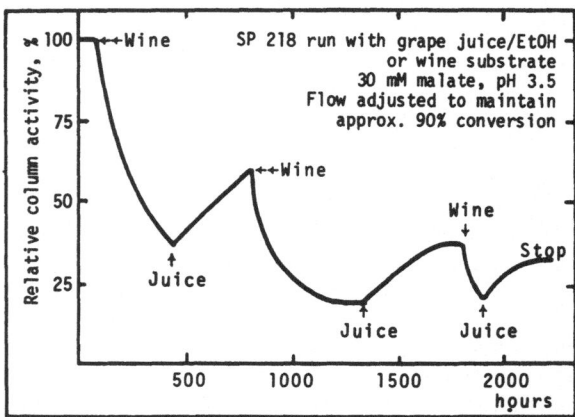

Fig. 2.   Malolactic activity in a packed bed column of Ca-alginate immobilized *Leuconostoc oenos* cells (preparation called SP 218) versus time.

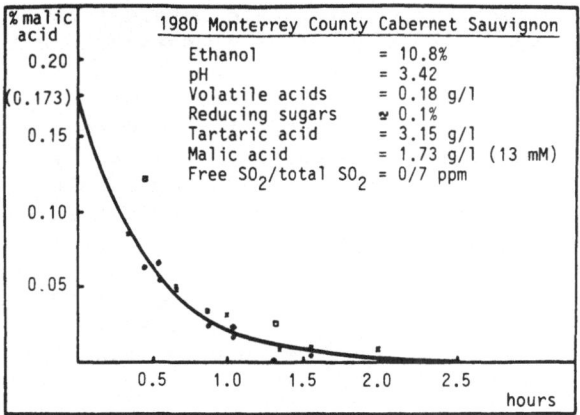

Fig. 3. Conversion of malic acid in a red wine as a function of
residence time in a pilot plant reactor system with Ca-
alginate immobilized *Leuconostoc oenos* cells.

number in the alginate spheres, neither with DNA- or ATP-analysis
not with culture plating techniques.

A pilot plant reactor arrangement consisting of three 16 L
packed bed reactors was designed to allow conversion of larger
quantities of new wine. A pressure of about 0.3 atm was maintained
in the reactor system in order to keep the produced carbon dioxide
in solution. As seen from Fig. 3 the total residence time was
about two hr when a red wine with low SO_2 was treated at normal
cellar temperature (16-17°C). Accurate flow regulation turned out
to be very important during long term runs. All three column loadings had to be provided with malate containing wine (i.e. substrate) in order to maintain activity; at the same time complete
malate conversion had to be secured.

The chemical analysis of column-treated wines showed no
anomalies in volatile acids or other compounds as compared to
spontaneously fermented or inoculated wines. Also, the organoleptic evaluations of a number of red wines from the 1979 and 1980
vintages were very satisfactory.

CONCLUSION

The laboratory and pilot plant experiments have demonstrated
the possibilities of performing malolactic fermentaiton continuously and quickly by passing new wine through reactors containing Ca-alginate cimobilized viable *Leuconostoc oenos* cells.

This type of immobilized malolactic catalyst is applicable in many types of wine at normal cellar temperatures without destroying the organoleptic properties of the wines. However, due to the need for maintaining the cells intact during application, the reactor equipment must be designed to allow continuous feed of malate containing wine to all parts of the catalyst. In addition, the SO_2 contents in wines to be treated should be minimized; and the pH should be held above 3.2 in order to secure a sufficient half-life for industrial application.

A continuous malolactic conversion process fits well with the current efforts in modern wine industry to perform all unit operations except yeast fermentation continuously (e.g. clarification and tartrate stabilization) (8).

ACKNOWLEDGMENTS

The valuable contributions of H. Møllgaard, L. H. Posorske, and J. C. Villettaz are gratefully acknowledged.

REFERENCES

1. MORENZONI, R., in "Chemistry of Winemaking," Adv. Chem. Series No. 137, American Chemical Soc., Washington (1974) p. 171.
2. LAFON-LAFOURCADE, S., in "Lactic Acid Bacteria in Beverages and Food," Academic Press, London (1975) p. 43.
3. DIVIES, C., British Patent Application 1,545,545 (1976).
4. KIERSTAN, M. and BUCKE, C. *Biotechnol. Bioeng. 19:* 387 (1977).
5. GESTRELIUS, S., unpublished observations.
6. NORSKER, O., GIBSON, K. & ZITTAN, L. *Starch/Starke 31:* 13 (1979).
7. KLEIN, J. & WAGNER, F. in "Biotechnology", Dechema Monographs No. 1693-1703, Vol. 82, Weinheim (1978) p. 142.
8. HAVIGHORTS, C. R. *Food Engineering*, 81 (1981).

PRODUCTION OF L-TRYPTOPHAN WITH IMMOBILIZED CELLS

F. Wagner, S. Lang, W.-G. Bang, K. D. Vorlop,* and
J. Klein*

Institute of Biochemistry and Biotechnology, and
Institute of Chemical Technology,* Technical University
Braunschweig, Federal Republic of Germany

One possibility for the production of L-tryptophan is a condensation reaction between indole and L-serine catalyzed by L-
tryptophan synthetase. In this work *Escherichia coli* B 10 cells
with high tryptophan synthetase were used. This strain is a tryptophan auxotrophic one, which is in addition tryptophanase and L-
serine desaminase negative. In a previous paper (1) several
methods for production of the amino acid by whole cells of *E.
coli*, grown in erlenmeyer flasks, were described. The best result was obtained using 5 g wet cells and 10 g nonionic detergent
Triton X - 100 (Merck). L-Tryptophan at 14.14 g/100 ml of phosphate buffer were produced at 37°C for 60 hr.

BATCH CULTIVATION OF *E. COLI* B 10 IN A 50 L BIOREACTOR

In order to ascertain in which growth phase of a discontinuous culture *Escherichia coli* B 10 cells have the highest specific
tryptophan productivity, investigations were made with a 50 L
batch culture. The cells were grown at 37°C in a mineral salt
medium (2) supplemented with 6 g/L $(NH_4)_2SO_4$, 0.06 g/L L-tryptophan, and 15 g/L glucose. During growth, the cells were harvested
periodically; and the tryptophan synthetase activity of the whole
cells was measured. In general this activity was determined as
complete production of L-tryptophan from indole and L-serine without regard to the initial velocity of the reaction. For the determination of L-tryptophan productivity, 0.3 g wet cell mass in
100 ml 0.1 M phosphate buffer, pH 8, were incubated in the presence
of 0.2 g indole, 0.2 g L-serine, and 1 mg pyridoxal-5'-phosphate
at 37°C for 3 hr and at 100 rpm agitator speed. After the

251

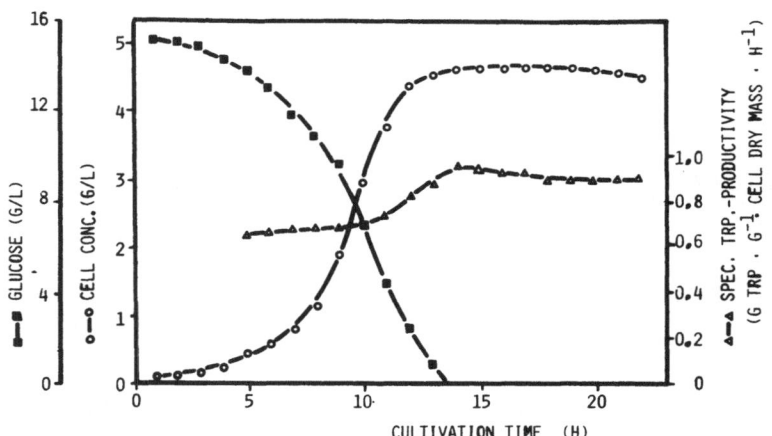

Fig. 1.   Specific L-tryptophan productivity of *E. coli* B 10 in
connection with growth and glucose consumption in a 50 L
bioreactor.   Conditions: intensor bioreactor type;
aeration rate 1.2 vvm; stirrer speed 1300 rpm; pH controlled at 7.0.

L-tryptophan was determined quantitatively, the specific tryptophan productivity, expressed as g L-tryptophan/g cell dry mass/hr,
was calculated.

Fig. 1 shows that the specific tryptophan productivity was
constant during the logarithmic growth phase and reached a maximum
of nearly 1.0 at the beginning of the stationary phase. This value
was unchanged within 8 hr after consumption of glucose. In the
case of *Hansenula anomala* the dependence of tryptophas synthetase
activity from growth is similar (3). *Escherichia coli* B 10
cells, grown for 14 to 22 hr in such 50 L batch cultures, were used
for several immobilization studies.

IMMOBILIZATION OF *E. COLI* CELLS

Because of a lack of long time stability the cells were entrapped in polyacrylamide (1), epoxy (4), eudragit (5), Ca-
alginate, and chitosan (6, 7, 8) to give different loadings and
remaining activity compared to the free cells (Table 1). The
polyacrylamide and chitosan preparations were investigated in detail. In comparison to chitosan one disadvantage of polyacrylamide
beads is their rather low value of cell loading. In order to
demonstrate the living state of the immobilized cells and to prove
that there were cavities within the polymeric matrix, the loading
of the beads was chosen to be low to allow for subsequent growth.
Only in the case of chitosan did the incubation in double concen-

TABLE 1

ACTIVITY OF IMMOBILIZED *E. COLI* B 10 CELLS WITH TRYPTOPHAN
SYNTHETASE

Immobilization Method	Dia. (mm)	Loading (%, w/v)	Relative Activity (%)	Productivity* (mg Trp/g cat./hr
Polyacrylamide	1	4.5	56	2.9
Epoxy	2	56	19	15
Eudragit[R]	3	15	60	9.1
Ca-alginate	3.5	10	25	2.5
Chitosan	2	11	57	11
Chitosan (dried)	1.3	43	32	24
Chitosan (dried)	0.7	52	41	35

*Productivity of free cells: 130 - 180 mg Trp/g cell mass/hr

trated growth medium (1), supplemented with 40 g/L K_2HPO_4, lead
to a high increase of activity. Alternatively, these experiments
were started a) with immobilization of the culture broth

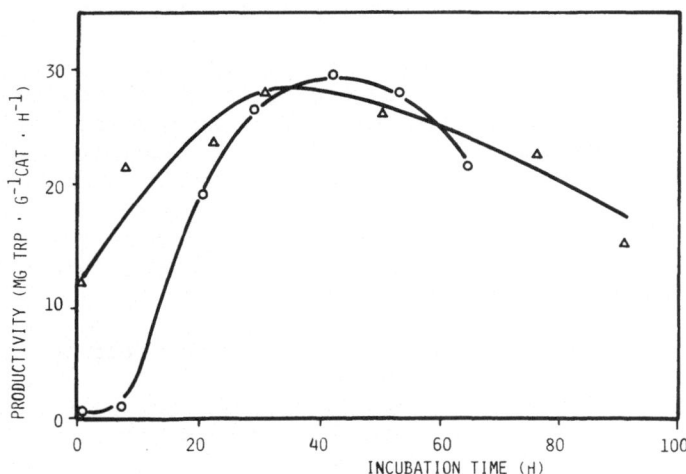

Fig. 2. Growth of *E. coli* B 10 in chitosan beads, expressed as
tryptophan productivity, as influenced by the starting
cell loading. Δ 11% (w/v) starting cell loading, O 0.02%
(w/v) starting cell loading.

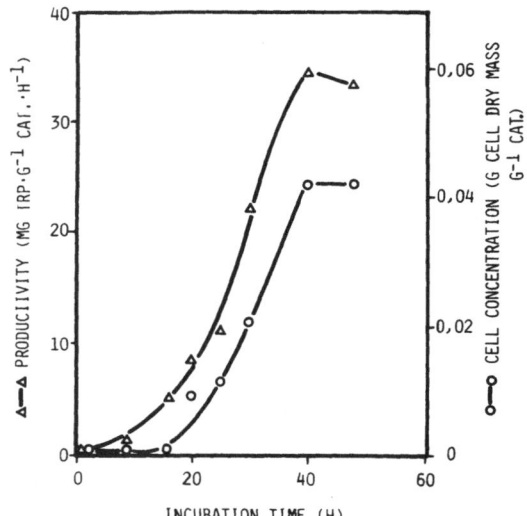

Fig. 3.   Cell growth inside of the chitosan beads and the corres-
          ponding productivity, as dependent on the incubation time.

and with a limited number of cells within the beads (0.02%, w/v)
or b) with entrapment of a large number of *E. coli* cells (11%,
w/v). It must be mentioned that the chitosan immobilization procedure generally gives a cell concentration of 3% (w/v) in the
chitosan-acetate solution that in turn leads to a starting cell
loading of about 11% after crosslinking and shrinking of the beads.

     Fig. 2 shows that the same level of specific productivity of
L-tryptophan can be obtained, depending on the incubation time.
This may be caused by the limited number and size of the cavities
within the polymeric matrix. In Fig. 3 the linear relationship
between the cell concentration within the beads and the L-tryptophan productivity under growth conditions is demonstrated. In this
experiment the initial cell concentration was 0.1 mg cell dry
mass/g wet catalyst. After 40 hr inculation in growth medium the
cell mass increased more than 400 fold.

     The influence of the initial indole concentration on the
stability of the tryptophan synthetase in *E. coli* B 10 cells
immobilized in chitosan is shown in Fig. 4. The half life of the
biocatalyst dropped from 8 batch runs (8 days) to 1 run (1 day)
by increasing the initial indole concentration from 50 to 200 mg
per 100 ml reaction volume and 200 mg biocatalyst·with a cell loading of 20%. In comparison the half life of *E. coli* B 10 cells
entrapped in polyacrylamide beads changed only two fold; in these
experiments 800 mg of biocatalyst were used since this matrix had
only 5% maximal cell loading.

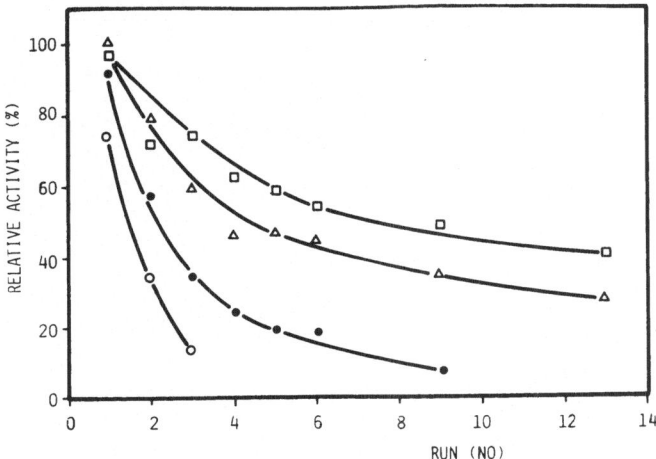

Fig. 4. Influence of the indole concentration on the stability
of the tryptophan synthetase in *E. coli* B 10 immobilized
in chitosan. Conditions: cell loading 20% (w/v); reaction time 24 hr for each run; reaction mixture 200 mg
catalyst in 100 ml phosphate buffer + 1 mg PLP; indole
+ L-serine respectively: □ 50 mg, △ 100 mg, ● 150 mg,
0 200 mg.

CONTINUOUS PRODUCTION OF L-TRYPTOPHAN

In addition to the batch runs continuous experiments have
been carried out with regard to the inhibitory effect of indole.
The production of L-tryptophan was studied in a continuous-flow
stirred tank reactor (CSTR). In the case of polyacrylamide under
substrate limiting conditions, 80% of the activity of the immobilized cells remained after continuous use for 50 days (9). Under
comparable cell loading, feed mixture, reaction volume, and dilution rate the chitosan catalyst showed a half life of only 30 days.

Additional experiments were carried out to determine the influence of flow rate of substrate solution on L-tryptophas production and the volumetric tryptophas productivity. Fig. 5 indicates
that up to a dilution rate of 0.75 (hr^{-1}), corresponding to a residence time of 1.33 hours, constant steady state values were obtained. At a dilution rate of 0.3 and a yield of 95%, the maximum
volumetric tryptophan productivity was reached with the chitosan
beads. With polyacrylamide beads a decrease in the volumetric
productivity by a factor of 3 was estimated for the same yield of
95% (9). This is in agreement with the lower maximal cell loading
in the polyacrylamide system.

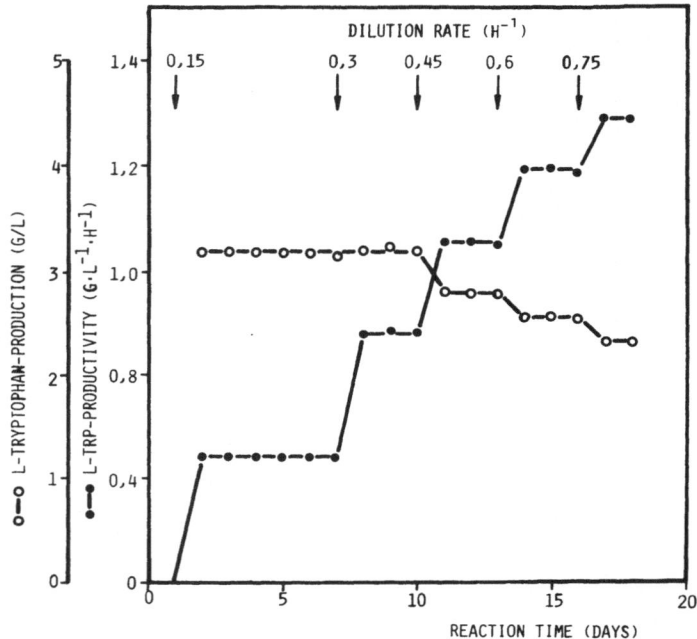

Fig. 5. Influence of dilution rate on the continuous production
of L-tryptophan using immobilized *E. coli* B 10 in
chitosan in a stirred reactor. Feed mixture: 2 g
indole, 2 g L-serine, 10 mg PLP in 1 L 0.1 M phosphate
buffer, pH 8; reaction volume 200 ml, containing 100 g
biocatalyst with 16% (w/v) cell loading.

CONCLUSION

Escherichia coli B 10, which has high tryptophan synthetase
activity, was grown in a 50 L batch culture in order to determine
in which growth phase the cells had the highest specific tryptophan productivity. Accordingly, whole cells of the stationary
phase were used for immobilization in various matrices. The production of L-tryptophan from indole and L-serine was studied in
detail in polyacrylamide and chitosan. The living state of the
cells, immobilized in chitosan beads, was demonstrated by means
of growth experiments. After investigation of the inhibitory effect of indole on the stability of the tryptophan synthetase of
entrapped *E. coli* cells, the production of the amino acid was
carried out in a continuous process. At a dilution rate of 0.3
hr^{-1} and a yield of 95%, the maximum volumetric tryptophan productivity was reached with the chitosan beads.

ACKNOWLEDGMENT

Support of this work by the Federal Government (BMFT) is gratefully acknowledged.

REFERENCES

1. BANG, W. G., LANG, S., SAHM, H. & WAGNER, F. *Abst. 1st Eur. Congr. Biotechnol.* 186 (1978).
2. VOGEL, H. T. & BONNER, D. M. *J. Biol. Chem. 218:* 97 (1956).
3. KIDA, T., EBIHARA, Y., ENATSU, T. & TERUI, G. *J. Ferment. Technol. 49:* 390 (1971).
4. KLEIN, J. & ENG., H. *Biotechnol. Lett. 1:* 171 (1979).
5. KLEIN, J., ENG., H., KLUGE, M., & VORLOP, K. D. *Abst. UIPAC Macro Mainz 3:* 1547 (1979).
6. VORLOP, K. D., KLEIN, J., & WAGNER, F. *Abst. VIth Intern. Ferment. Symp. London (Canada)* 122 (1980).
7. VORLOP, K. D. & KLEIN, J. *Biotechnol. Lett. 3:* (1981).
8. KLEIN, J. & MANECKE, G., this volume.
9. BANG, W. G., thesis, Techn. University Braunschweig (1979).

APPLICATIONS OF IMMOBILIZED TANNIN FOR PROTEIN AND METAL

ADSORPTION

I. Chibata, T. Tosa, T. Mori, T. Watanabe,
K. Yamashita and N. Sakata

Research Laboratory of Applied Biochemistry
Tanabe Seiyaku Co., Ltd., Yodogawa-ku, Osaka, Japan

Since the 1960's we have been studying techniques for immobilization of enzymes and cells; and we have succeeded in the
industrial application of immobilized enzymes for the continuous
optical resolution of DL-amino acids and immobilized microbial
cell systems for the productions of L-aspartic acid and L-malic
acid. On the basis of this background, we have expanded the
immobilization techniques to include separations, purifications,
and the recovery of proteins.

As protein adsorbents, there are many known inorganic and
organic compounds. However, they do not wholly satisfy the basic
requirements. For example, they are not specific for proteins
but adsorb various organic and inorganic compounds together with
proteins. Therefore if more specific adsorbents for proteins can
be prepared, they will be used advantageously in a variety of
fields. To design this adsorbent, we have extensively investigated the covalent binding of tannin to water insoluble matrices
and the characteristics and applications of the resulting immobilized tannin.

PREPARATION AND CHARACTERIZATION OF IMMOBILIZED TANNIN

To prepare immobilized tannin (1), cellulose is mercerized
with sodium hydroxide, activated with epichlorohydrin, and reacted
with hexamethylenediamine to give aminohexylated cellulose. This
is then activated with epichlorohydrin and reacted with chinese
gallotannin to give immobilized tannin of about 25% tannin (w/w).
The packed volume of one gram of dry weight in a column is about
20 ml.

259

TABLE 1

SPECIFICITY OF IMMOBILIZED TANNIN FOR VARIOUS PROTEINS

Protein		Amount Adsorbed*			
Type	Specific	pH 2	pH 4	pH 7	pH 10
Albumin	Ovalbumin	--	5.4	49.5	0
	Bovine serum albumin	--	8.7	58.8	0
Globulin	Bovine α-lactoglobulin	--	18.9	75.0	5.1
	Bovine serum β-globulin IV-1	--	24.0	83.4	27.5
	Concanavalin A	--	5.0	15.1	4.1
Glutelin	Wheat glutenin	21.3	12.3	--	--
Gliadin	Zein	--	13.8	8.4	4.8
Protamin	Salmon protamin	--	9.6	55.9	94.0
Scleroprotein	Gelatin	--	9.1	22.2	2.4
Phosphoprotein	Bovine milk casein	--	42.9	52.2	31.2
	Soybean casein		34.8	97.5	28.1
Chromoprotein	Bovine hemoglobin	--	0	44.6	16.4
Glycoprotein	Gastric mucin	--	27.6	58.8	25.4
Enzyme	α-Amylase	--	69.1	43.3	0
	Glucoamylase	--	26.4	51.2	0
	Lysozyme	--	0	38.4	80.2
	Pepsin	--	76.5	34.1	0
	Trypsin	--	0	11.4	11.7

*Amount of protein adsorbed/amount of protein used x 100.

To clarify the adsorption specificity of immobilized tannin, the adsorption rates of several proteins were compared in acid, neutral and alkaline conditions (2). As shown in Table 1, the proteins tested in this experiment were more or less adsorbed on immobilized tannin. On the other hand when various carbohydrates, amino acids, peptides, alkaloids, organic acids, and nucleic acid-related compounds were incubated with the adsorbent, none of them were adsorbed.

It is well known that tannin reacts with heavy metals. Thus, we investigated the adsorption of heavy metals on immobilized tannin (3). Iron, copper and lead were adsorbed well; but other metals were less adsorbed, as shown in Table 2. The adsorption specificity of immobilized tannin for various kinds of cationic and anionic ions also was investigated; and chlorine ion, fumaric ion, sodium ion, and calcium ion were little adsorbed.

Next, we investigated the factors influencing the adsorption and desorption of proteins and metals. The adsorption rate of protein and metal on immobilized tannin was affected by pH, temperature, and incubation time. Also, it was shown that the desorption of proteins and metals from immobilized tannin could be carried out with a weak acid solution (3,4).

APPLICATIONS OF IMMOBILIZED TANNIN

Since immobilized tannin specifically adsorbs proteins but not other organic and inorganic compounds except heavy metals and the adsorbed proteins are readily eluted, this adsorbent can be

TABLE 2

ADSORPTION SPECIFICITY OF IMMOBILIZED TANNIN
FOR VARIOUS METAL IONS

Kind of Metal Compound	Concentration of Metal Ion (ppm)				
	Charged Solution	Fraction of Column Effluent* (60 ml fraction)			
		1	2	3	4
$MnSO_4$	3.0	–	2.9	3.1	3.2
$Fe(NH_4)_2(SO_4)_2$	3.0	<0.05	0.08	0.08	0.08
$CoCl_2$	10.4	7.5	10.3	–	–
$Ni(CH_3COO)_2$	9.5	5.7	9.0	8.7	–
$CuSO_4$	3.0	–	<0.03	<0.03	<0.03
$Zn(CH_3COO)_2$	7.7	3.0	6.2	6.0	–
$HgCl_2$	1.9	0.81	2.4	÷	–
$Pb(CH_3COO)_2$	8.4	<0.1	<0.1	–	–

*Column volume 4 ml.

used for the recovery of proteins from aqueous solutions, such as culture broths, blood, and urine. For example, we tried to recover hesperiginase from culture broth of *Aspergillus niger*. When culture broth of pH 4.8 and 14 m mho was charged, both hesperiginase and undesired proteins were adsorbed to immobilized tannin. However, the results of batchwise experiments showed that at pH 6 and 10 m mho ionic strength, hesperiginase was rather selectively adsorbed. Thus, the culture broth was adjusted to pH 6 and 10 m mho ionic strength and passed through the immobilized tannin column; almost all the proteins except hesperiginase passed through the column. Hesperiginase was eluted with 0.02 N HCl; and the recovery of activity was 92%.

In order to confirm that immobilized tannin can be used for the separation of proteins, a model experiment for chromatographic separation of different kinds of protein was carried out (5). A mixture of trypsin, lysozyme, and ovalbumin dissolved in pH 7.5 buffer was applied to the immobilized tannin column. After washing the column with the same buffer the adsorbed proteins were eluted with 0.05 M carbonate buffer, pH 9, and subsequently with 0.05 M acetate buffer, pH 4. Trypsin was not adsorbed on the column at neutral pH, whereas the other two proteins were adsorbed and could be eluted with alkaline buffer and then with acid buffer without significant loss of protein. These results suggest that the separation, recovery, and removal of proteins using immobilized tannin may be carried out by selecting suitable conditions from adsorption and desorption.

The immobilization of enzymes on tannin was carried out very easily by adding the enzyme dissolved in an appropriate buffer to a suspension of immobilized tannin and shaking the mixture for $60 \sim 120$ min at $25°C$. The enzymes adsorbed on immobilized tannin displayed 25-70% activity, depending on the enzymes used (4,6).

Sake is a traditional alcoholic drink produced from rice by fermentation. Generally, fresh sake is pastuerized at $55 \sim 65°C$ for 15 min. This process causes turbidity during subsequent storage of sake due to heat denaturation of soluble proteins. As these denatured protein particles are too small to remove by ordinary filtration, soluble persimmon tannin has been used conventionally for coprecipitation. We tried to change this to a continuous process using immobilized tannin (7-10). Fresh sake was passed through a column packed with immobilized tannin; and the proteins causing turbidity in sake were adsorbed on the immobilized tannin column. The turbidity of the untreated sake increased during storage; but that of treated sake did not change. The capacity of the immobilized tannin was estimated to be around 5000 L sake/kg of adsorbent or 250 L sake/L of adsorbent.

TABLE 3

REMOVAL OF IRON ION FROM WATER FOR BREWING
AND TAP WATER BY IMMOBILIZED TANNIN

Volume of Charged Water (6 x 10^3 L/L of adsorbent)	Conc. of Iron Ion in Effluent (ppb)*	
	Water for Brewing	Tap Water
0.5	<10	<10
2.5	<10	<10
4.5	<10	<10
5.0	<10	10
5.5	<10	20
8.5	<10	-
10.0	<10	-

*Conc. of iron ion in charged water was 10-40 ppb in brewing water and 80 ppb in tap water.

It is important to remove iron completely from water used for brewing because trace amounts of iron lower the quality of sake, wine, and beer. However, as the water for brewing contains several forms of iron, it has been difficult to remove the trace amounts of various forms of iron by a single method. Immobilized tannin adsorbs almost all kinds of iron; and this adsorbent can be useful for the removal of iron from solutions (3). We carried out pilot scale runs with 60 L of immobilized tannin for removal of iron from water for brewing. This equipment was able to treat 6 x 10^3 L/hr for brewing; and 600 x 10^3 L water can be treated without regeneration in the case of water containing 50 ppb of iron. We removed trace amounts of iron from tap water and water from brewing in field tests using the pilot scale equipment. As shown in Table 3, the concentration of iron in the effluent was less than 10 ppb, even when 10 x 10^3 L of water for brewing or 5 x 10^3 L of tap water/L of adsorbent was passed through the immobilized tannin column.

As stated above, this new adsorbent holds promise for improving the quality of beverages, such as Japanese sake, beer, wine, and juice by removal of undesirable protein constituents from the solutions. Further, this adsorbent may be used as a tool for the recovery and purification of proteins from solutions containing various organic and inorganic compounds and for the immobilization

of enzymes with high activity. Also, this adsorbent shows promise
for the removal of heavy metals from solution.

REFERENCES

1. WATANABE, T., MATUO, Y., MORI, T., SANO, R., TOSA, T. &
    CHIBATA, I. *J. Solid-Phase Biochem.* *3:* 161 (1978)
2. WATANABE, T., MORI, T., TOSA, T. & CHIBATA, I. *Agric.
    Biol. Chem.* 45: 1001 (1981).
3. MORI, T., WATANABE, T., TOSA, T., CHIBATA, I., IWANO, K. &
    NUNOKAWA, Y. *J. Brew. Soc. Japan* 76: 111 (1981).
4. WATANABE, T., FUJIMURA, M., MORI, T., TOSA, T. & CHIBATA,
    I. *J. Appl. Biochem.* 1: 28 (1979).
5. WATANABE, T., MORI, T., TOSA, T. & CHIBATA, I. *J.
    Chromatogr.* 207: 13 (1981).
6. WATANABE, T., MORI, T., TOSA, T. & CHIBATA, I. *Biotechnol.
    Bioeng.* 21: 477 (1979).
7. NUNOKAWA, Y., MIKAMI, S., TOSA, T. & CHIBATA, I. *Hakoko-
    gakukaishi* 55: 343 (1977).
8. NUNOKAWA, Y., SHIINOKI, S. & WATANABE, T. *Hakkokogaku-
    kaishi* 56: 776 (1978).
9. NUKOKAWA, Y., SEKIGUCHI, S., WATANABE, T. & MORI, T. *J.
    Brew. Soc. Japan* 74: 399 (1979).
10. WATANABE, T., MORI, T., SAKATA, N., YAMASHITA, K., TOSA, T.,
    CHIBATA, I., NUNOKAWA, Y. & SHIINOKI, S. *Hakkokogaku-
    kaishi* 57: 141 (1973).

INDUSTRIAL APPLICATIONS OF IMMOBILIZED BIOMATERIALS IN CHINA

S. Z. Zhang

Institute of Microbiology, Academia Sinica,
Beijing, China

In China research on immobilized enzymes and microbial cells started at the beginning of the 1970's; and since then some useful results have been achieved. The following is a brief overview.

IONIC BINDING

Immobilization by ionic binding has been achieved with glucoamylase, aminoacylase, and glucose isomerase (1, 2). Glucoamylase from *Aspergillus niger* M85 was immobilized on DEAE-Sephadex A-50 by ionic binding with an activity of 1,000 U/g and 25% recovery (1 U = 1 μmol glucose/min). The culture filtrate of *Asp. niger* was passed through a packed column of DEAE-Sephadex A-50 (0.1 M, pH 6.0 phosphate buffer) or mixed with the resin and shook at 28-30°C for 2.5-5.0 hr. The optimal conditions were pH 4.0 - 4.5 and 55°C. DEAE-Sephadex A-50 as an adsorbent was much better than ion exchange resin 110, CM-cellulose, or DEAE-cellulose. The Km of the immobilized enzyme for starch was 2.3%; for the free enzyme it was only 0.57%.

Aminoacylase from rice koji *Asp. oryzae* was immobilized on home made DEAE-Sephadex A-25 with an activity of 700 - 800 μmol/hr/g and a recovery of 60 - 70%. The water extract of rice koji (60 ml) was adsorbed on 1 g wet DEAE-Sephadex gel at pH 7.0-7.5 during overnight stirring in the cold room. The optimal conditions were pH 7.0 and 70°C; Co^{2+} served as the activator. The enzyme was used to prepare optically active amino acids for reagents or for pharmaceutical purposes. Besides the naturally occurring amino acids p-methoxy phenylglycine and α-aminobutyric acid have been produced.

Glucose Isomerase from *Streptomyces roseoruber* 336 was adsorbed on a strong basic anion exchange resin 290 (made by Nankai University) with an activity of 2,000 U/g wet matrix and a recovery of 70-80%. The fermentation broth had an activity of 140 U/ml. After centrifugation, the clear enzyme solution (300-400 ml) was mixed with 2.5 ml resin and 50-67 ml phosphate buffer, pH 7.0 at $50^\circ C$ for 9-12 hr (optimum at pH 7.4, $80^\circ C$, Mg^{2+} or C^{2+} as activators). Pilot plant scale experiments have been carried out with 1.25-2.2 ton of syrup produced per kg of dry material.

COVALENT BINDING

A cheap bifunctional reagent, p-(β- sulfatoethylsulfonyl)-aniline(SESA), usually used in the dye industry to prepare active dyes, was used by C. L. Tsou in 1970 to prepare immobilized enzymes. The reaction scheme is shown elsewhere (22). Glucoamylase from the fermentation broth of *Asp. niger* M85 was treated with acid clay, precipitated with 75% sat. $(NH_4)_2SO_4$, and dialyzed. It was coupled to diazotized ABSE-cellulose (from bagasse) (3-5) and also to glass beads (12). The properties were similar to those with ionic binding, except for a higher optimum temperature of $65^\circ C$.

Several other enzymes have been coupled covalently to supports (Table 1). For 3'-RNase (6) the fermentation broth of *Rodotorula glutinis* was centrifuged; and (600 g/L) $(NH_4)_2SO_4$ was added to the supernatant. The enzyme precipitate was dissolved in water, coupled to diazotized ABSE-Sephadex G-200, and used for production of 3'-mononucleotides from RNA with a ten-fold increase in efficiency. The 3'-mononucleotides were used as reagents for research work and for pharmaceutical purposes. Crystalline beef pancreas trypsin (7) was coupled to ABSE-cellulose in the presence of a competitive inhibitor, n-butylamine, to protect the active site. It formed an inactive complex with mung bean trypsin inhibitor in a molar ratio of 2:1 with 80% of the active sites participating. Polynucleotide phosphorylase (PNPase) (8, 9) catalyzes the conversion of IDP and CDP to poly I and poly C, which combine to form poly I:C, an interferon inducer. The enzyme was purified from *E. coli* extract by streptomycin precipitation, $(NH_4)_2SO_4$ fractionation, and DEAE-cellulose column chromatography and coupled to diazotized ABSE-agarose.

Nuclease Pl was obtained from the culture filtrate of *Penicillium citrinum* by alcohol precipitation and coupled to ABSE-cellulose (from bagasse) (10). Pilot plant experiments carried out in 1976 were successful; and industrial production of 5'-nucleotides was achieved in 1977. It was the first industrial immobilized enzyme process in China. Alkaline phosphatase from *E. coli*, with activity of 1200-1900 U/ml was coupled to ABSE-agar gel beads, which had been crosslinked with epichlorohydrin, in the

TABLE 1

ENZYMES IMMOBILIZED ON ABSE-GLYCANS*

Property	3'RNase	Trypsin	PNPase	Nuclease P1	Alkaline Phosphatase
Activyty	--	--	8 U/g (wet)	2090 U/g (dry)	80 U/g
Relative activity (1%)	50	Casein:45-50 BANA:95-100	--	50	--
Recovery (%)	35	--	15	19	60-80
Opt. pH	4.1	--	10	4.8(5.1)	8.4
Opt. Temp. ($^{\circ}$C)	55	--	42-52	70-75	68
Operational Stability	Used for 30 times, a little decrease	More stable than free enzyme to heat and urea	No loss of activity in column reactor at pH 9, 37°, for 1.5-2 months	Ten times more stable than free enzyme	--

*See text for enzyme source and support.

presence of bovine serum albumin and then treated with β-Naphthol as a blocking agent (11).

IMMOBILIZED CELLS

E. coli strains AS 1.881 and AS 1.76, having high aspartase (13) and penicillin acylase (15-17) activities, respectively, were immobilized on agar gel. The high activity aspartate cells (1 x 10^5 units/g wet cells) were immobilized by entrapping the wet cells in 6% agar gel (Table 2). The efficiency was high since it took only 80g of the wet cells to convert 1000 L of 1 M fumarate to 110 kg aspartic acid in 20 days. The acylase cells were mixed with an equal volume of 8% agar gel and pored into an organic solvent with stirring to make gel beads. After washing, they were cross-linked with 1% glutaraldehyde solution and washed (Table 2). Cells of *E. coli* strain D816 entrapped in gelatin and crosslinked with glutaraldehyde also were very useful (Table 2). Both immobilized materials have been successfully used in the pharmaceutical indus-try to produce 6-APA from penicillin G and 7-ADCA from 7-phenyl-acetamido-deacetoxyl-cephalosporanic acid (23).

Candida rugosa C90 was used to produce fumaric acid from liquid paraffin. The cells harvested from the fermentation broth were able to convert fumaric acid to L-malic acid at alkaline pH (14). Cells were entrapped in 15% polyacrylamide gel (Table 2). A reactor, containing 6 g wet cells at 30°C, was operated for 60 days and converted 12 L of 1 M fumarate to L-malate with a yield of 82-85%. Pilot plant experiments and industrial production have been accomplished.

Cells of *Streptomyces roseoflulvus* Kcl3-5705 were entrapped in gelatin and crosslinked with glutaraldehyde (Table 2) (18) and tested on a pilot plant scale. The efficiency of the immobilized cells was 5-fold higher than that of the native cells. Microbial cells having 3-ketosteroid $-\Delta^1-$ dehydrogenase activity from *Arthrobacter simplex* (19) also were immobilized. The cells were entrapped in 10% polyacrylamide gel, or in a mixed gel composed of 5% calcium alginate and 5% gelatin in a 2:1 ratio. The immo-bilized cells were used to convert hydrocortisone to prednisolone or cortisone acetate to prednisone acetate (Table 2). One gram of dry material gave 791.7 mg product/day.

MISCELLANEOUS

Research on other immobilized enzymes or microbial cells and new carriers also has been undertaken. ABSE derivatives of poly-vinyl alcohol (in bead form) have been used to immobilize trypsin (20); and isothiocyano derivatives of ABSE-cellulose have been used to immobilize alkaline phosphatase. (21) Other works are in prog-ress.

TABLE 2

IMMOBILIZED MICROBIAL CELLS

Property	Aspartase	Fumarase	Penicillin Acylase		Isomerase	3-Ketosteroid-Δ^1-dehydro-genase
			In Agar	In Gelatin		
Activity	6×10^4U/g(wet)	7000U/g(wet)	--	--	--	55U/g(wet)
Recovery (%)	71	90	--	--	40	80
Opt. pH	9.0-9.5(9.0)	8.5	8.0	9.0	--	7.0-8.5
Opt. Temp. (^{o}C)	40	45	40	55	--	35
Operational Stability	37°C, 20 days 18% loss	Half life 95 days	Column reactor 115 days no loss	37°C, 103 days no loss	20 days rate decreased to 40%	--

REFERENCES

1. *Wei Sheng Wu Hsueh Pao* (Journal of Microbiology), *13* (1): 25 (1973).
2. YUAN ZHONGYI *et al. Abst. Third Mtg. Chinese Biochemical Soc.* 174 (1978).
3. LI, K. *et al. Wei Sheng Wu Hsueh Pao 13* (1): 31 (1973).
4. YANG, L., *et al. ibid 16* (4) 335 (1976).
5. LI, G. *et al. ibid 19* (2): 150 (1979).
6. *Sheng Wu Hua Hsueh Yu Sheng Wu Wu Li Hsueh Pao* (Journal of Biochemistry *9*(2): 187 (1977).
7. *ibid* (1): 63 (1977).
8. YANG, K., *et al. ibid 11*(1): 79 (1979).
9. YANG, K., *et al. ibid 11*(2): 97 (1979).
10. YUAN, Z., *et al. K'o Hsueh Tung Pao 25*(14): 654 (1980).
11. YUAN, Z., *et al. ibid*, in press.
12. LI, G., *et al.*, this volume.
13. MENG, G., *et al. Wei Sheng Wu Hsueh Pao 18*(1): 39 (1978).
14. YANG, L., *et al. ibid 20*(3): 296 (1980).
15. SUN, W., *et al. ibid 20*(4): 407 (1980).
16. SUN, W., *et al. Wei Sheng Wu Hsueh Tung Pao 8*(2): 63 (1981).
17. WANG, Q., *et al. Sheng Wu Hua Hsueh Yu Sheng Wu Wu Li Hsueh Pao 12*(4): 305 (1980).
18. *Sheng Wu Hua Hsueh Yu Sheng Wu Wu Li Chin Chan* No. 2, 65 (1980).
19. YANG, L., *et al. Wei Sheng Wu Hsueh Pao*, in press.
20. LI, F., *et al. K'o Hsueh T'ung Pao 25*(15): 695 (1980).
21. *Sheng Wu Hua Hsueh Yu Sheng Wu Wu Li Hsueh Pao 8*(3): 207 (1976).
22. LIU, S., *et al.*, this volume.
23. WANG, Z., *et al.*, this volume.

PRODUCTION OF L-ALANINE FROM AMMONIUM FUMARATE USING TWO TYPES OF IMMOBILIZED MICROBIAL CELLS

T. Sato, S. Takamatsu, K. Yamamoto, I. Umemura
T. Tosa, and I. Chibata

Research Laboratory of Applied Biochemistry
Tanabe Seiyaku Co., Ltd.,
Yodogawa-ku, Osaka, Japan

Since 1965 we have carried out the industrial production of L-alanine from L-aspartic acid by a batchwise enzymatic method, using L-aspartate β-decarboxylase from *Pseudomonas dacunhae*, To develop a more efficient process for production of L-alanine, we studied its continuous production from L-aspartic acid using *P. dacunhae* cells immobilized with κ-carrageenan (1). We also succeeded in the industrial production of L-aspartic acid from ammonium fumarate using *Escherichia coli* cells immobilized with polyacrylamide gel in 1973 (2); and this method was changed to the system using *E. coli* immobilized with κ-carrageenan in 1978 (3). So, we decided that if the above two immobilized microbial cells were employed in a single reactor, then L-alanine might be more efficiently produced from ammonium fumarate, according to the following equation:

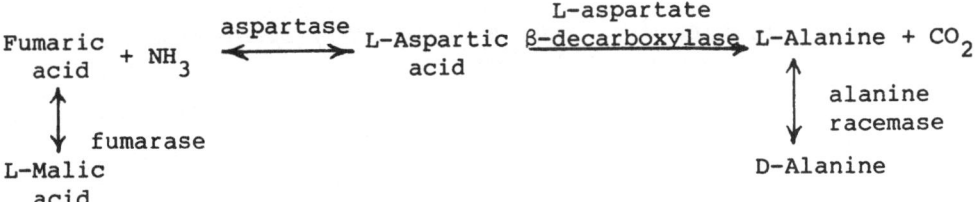

However, as these two microorganisms have alanine racemase and fumarase activities, D-alanine and L-malic acid were formed as by-products. This was disadvantageous for production of L-alanine, using the two immobilized microbial cells. Thus, for effective production of L-alanine, the actions of these undesirable enzymes had to be removed. Therefore, we tried various procedures for removal of these enzymes. It was found that when *E. coli* cells were treated at pH 5.0, 45°C for 1 hr and *P. dacunhae* cells at pH 4.75, 30°C for 1 hr, respectively, before immobilization with κ-carrageenan, the aspartase of *E. coli* and the L-aspartate β-decarboxylase of *P. dacunhae* were not decreased; but the alanine racemase and fumarase of the two microorganisms were almost completely inactivated. After these pH-treatments, both microorganisms were separately immobilized with κ-carrageenan. Production of L-alanine from ammonium fumarate was investigated by batchwise reaction using the two immobilized microbial cell lines. As a result, L-alanine was stoichiometrically produced; and the formations of D-alanine and L-malic acid were suppressed.

For further improvements of the L-alanine production, we co-immobilized *E. coli* and *P. dacunhae* cells together in κ-carrageenan. However, the efficiency of L-alanine production, using the mixture of the two immobilized cells, was higher than that of coimmobilized cells. The difference of the efficiency was thought to depend on the inhibition of L-aspartate β-decarboxylase by high concentrations of L-aspartic acid.

We are now studying the optimal conditions for continuous production of L-alanine from ammonium fumarate using the mixture of two immobilized cells; and we are going to industrialize this L-alanine production system in the near future.

REFERENCES

1. YAMAMOTO, K., TOSA, T. & CHIBATA, I. *Biotechnol. Bioeng.* 22: 2045 (1980).
2. TOSA, T., SATO, T., MORI, T. & CHIBATA, I. *Appl. Microbiol.* 27: 886 (1974).
3. SATO, T., NISHIDA, Y., TOSA, T. & CHIBATA, I. *Biochim. Biophys. Acta* 570: 179 (1979).

NEW PROCESS FOR PRODUCTION OF HIGH FRUCTOSE CORN SYRUP USING COMBINED ADSORPTION AND AN ENZYME REACTOR

K. Hashimoto, S. Adachi and H. Noujima

Department of Chemical Engineering, Kyoto University
Yoshida, Sakyo-ku, Kyoto, Japan

The most common high fructose syrup contains 42% fructose, 52% glucose, and 6% oligosaccharides on the basis of solid materials. Since fructose is sweeter and more soluble in water at low temperature than glucose, it is desirable to raise the content of fructose in the syrup to 55-90%. Isomerization of glucose to fructose by glucose isomerase is a reversible reaction with an equilibrium constant of 1.0. In order to get the fructose level above 50%, a separation process, such as adsorption with an ion exchanger or zeolite, is necessary. Chromatographic and simulated moving-bed adsorption processes have been in operation on a commercial scale. In this study, we present a new process for producing higher fructose syrup by combining immobilized-glucose isomerase reactors with a simulated moving-bed adsorber, which can reduce the amount of water used as the desorbent.

Although our proposed process does not include the real Movement of adsorbent, the basic idea behind the process is more easily understandable in terms of a hypothetical moving-bed adsorber. Fig. 1 shows schematic flows of solid and liquid in the hypothetical moving-bed reactive adsorber. The zone I oblique part represents an immobilized-enzyme reactor column, whereas the zone I white part bounded by the two dotted lines indicates an adsorption column. The feed, an equilibrated glucose-fructose mixture coming from the main reactor, enters at the boundary between zones I and II. The immobilized-enzyme reactors are stationary, while the adsorbent particles move counter to the liquid flow, skipping the enzyme reactors. The feed goes first to the last adsorber in zone I, where fructose is selectively adsorbed. This results in an increase in the fraction of glucose in the liquid stream. This liquid is next passed through the first zone I re-

actor, where most of the excess glucose over the equilibrated concentration is converted to fructose. The liquid from the above
reactor is now introduced to another zone I adsorber. By repeat-
ing the cycle of adsorption and reaction, the glucose originally
in the feed is almost completely converted to fructose, which in
time is adsorbed on the solid particles and transported to zones
II and III. Thus, the liquid stream leaving zone I is composed
mainly of desorbent (water), which is recycled to the make-up desorbent. The glucose adsorbed partially in zone I is desorbed in
Zone II and carried into zone I by the liquid stream. The fructose on the solid particles is mainly desorbed in zone III and
taken out as the product. The Ca^{2+} ion form of Y zeolite was used
as an adsorbent. The adsorption isotherms for glucose and fructose on the adsorbent were independent of each other and linear
over a wide range of concentrations. A mathematical model for
calculating the concentration profiles of glucose and fructose in
the adsorber has been proposed. The calculated concentration profiles coincided well with the experimental ones.

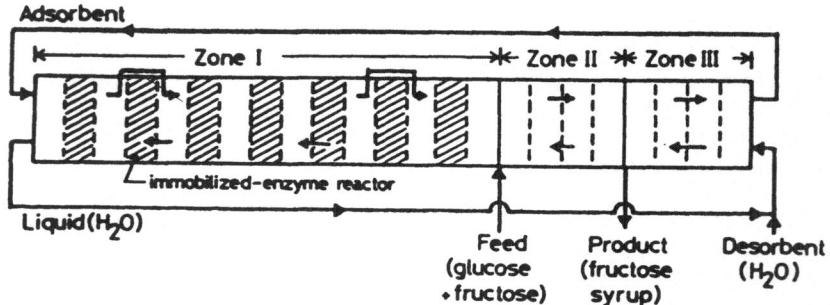

Fig. 1. Schematic flows of adsorbent and liquid in a hypotheti-
         cal moving-bed reactive adsorber.

IMMOBILIZED MULTIENZYME SYSTEMS FOR STARCH PROCESSING

A. Lindroos, Y. Y. Linko, and P. Linko

Helsinki University of Technology, Department
Of Chemistry, Espoo, Finland

Since the early immobilization of glucoamylase for starch
saccharification several authors have reported on different immobilization techniques and potential applications (1,2). α-Amylase,
β-amylase, and pullulanase have also been immobilized by various
methods (3-5). Furthermore, glucoamylase has been investigated
as an immobilized two-enzyme system together with glucose oxidase
(6), α-amylase (7), and glucose isomerase (2,8), and β-amylase with
pullulanase (9). In the present work, batch or continuous column
reactors either with immobilized glucoamylase or β-amylase alone,
or with immobilized glucoamylase, pullulanase and glucose isomerase
in various combinations were employed for processing of α-amylase
pretreated cereal starches.

Glucoamylase (*Aspergillus niger*, Novo 150L), β-amylase (barley, Wallerstein) and pullulanase *(Klebsiella aerogeneses,* ABM
Pulluzyme 2000K) were immobilized by adsorption on Duolite S-761
phenol-formaldehyde resin and subsequently cross-linked with 2.5%
glutaraldehyde. Immobilized glucose isomerase *(Bacillus coagu-
lans*, Novo Sweetzyme S) was a commercial preparation. Termamyl
60L *(Bacillus* sp., Novo) was employed for starch liquefaction.

Continuous saccharification of DE 28 (Bx 36-38) barley starch
syrup with immobilized glucoamylase resulted in glucose syrups of
DE ∿90-98, depending on the experimental conditions. Employing
a two-enzyme column reactor (50°C, pH 6.5) with equal weights of
immobilized glucoamylase (15 U/g) and glucose isomerase (169 U/g),
syrups of DE >90 and isomerization % of up to 47 could be obtained
in continuous operation. Using DE > 20 (Bx 23-25) barley starch
syrup as substrate for immobilized β-amylase (15 U/g, 40°C, pH 5.7),
maltose syrups of DE up to 47 were obtained. A typical syrup

275

contained nearly equal quantities of maltose and maltotriose, with little glucose and higher oligomers. When one unit of immobilized pullulanase was added per 20 U of immobilized glucoamylase and 10 g of starch dry matter, the rate of saccharification was considerably increased, with little effect on the final DE-value obtainable in a batch process (50°C, pH 5.0) (Fig. 1A). The inclusion of only 1 g of immobilized pullulanase (10 U/g) per 10 g each of immobilized glucoamylase (15 U/g) and glucose isomerase (169 U/g) clearly increased both the rate of hydrolysis and the final DE-value in continuous column operation (50°C, pH 6.5) (Fig. 1B). Consequently, enzymes having widely different temperature and pH-optima may be successfully used for continuous processing in a single reactor. The reactors investigated could be operated for several months with little decrease in activity, if substrate contamination was prevented with preservatives. The limiting factor in such multienzyme systems was the stability of immobilized glucose isomerase.

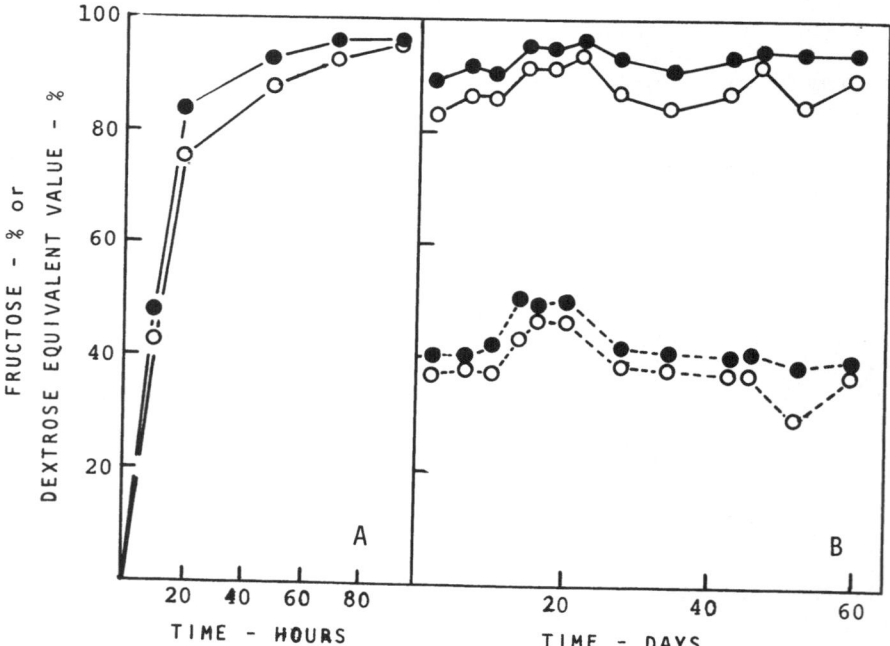

Fig. 1. Saccharification of barley starch syrup (DE 28, Bx 36): (A) in batch both with immobilized glucoamylase (o) and with immobilized glucoamylase and pullulanase (●), and (B) continuously with simultaneous isomerization both without (o) and with (●) immobilized pullulanase. (—— DE, ---- fructose).

REFERENCES

1. LINKO, Y. Y., LINDROIS, A., & LINKO, P. *Enzyme Microb. Technol. 1:* 273 (1979).

2. LINDROOS, A., LINKO, Y. Y. & LINKO, P. in "Enzyme Engineering in Food Processing" (P. Linko & J. Larinkari, eds), Applied Science Publ., London (1980) p. 92.

3. LINKO, Y. Y., SAARINEN, P., & LINKO, M. *Biotechnol. Bioeng. 17:* 153 (1975).

4. MÅRTENSON, K. & MOSBACH, K. *Biotechnol. Bioeng. 14:* 715 (1972).

5. OHBA, R., CHAEN, H., HAYASHI, S., & UEDA, S. *Biotechnol. Bioeng. 20:* 665 (1978).

6. GESTRELIUS, S., MATTIASSON, B., & MOSBACH, K. *Biochim, Biophys. Acta 276:* 339 (1972).

7. WALTON, H. M., EASTMAN, J. E., & STALEY, A. E. *Biotechnol. Bioeng. 15:* 951 (1973).

8. LEE, G. K. *Kansas State Univ. Inst. Syst. Des. Optim. 66:* 32 (1975).

9. OHBA, R. & UEDA, S. *Biotechnol. Bioeng. 22:* 2137 (1980).

GLUCOAMYLASE COVALENTLY COUPLED TO POROUS GLASS

G. X. Li, J. Y. Huang, X. F. Kou and S. Z. Zhang

Institute of Microbiology, Academia Sinica, Beijing
China

Glucoamylase from *Asp. niger* (A.S.3.4309) was obtained from the Wuxi Enzyme Factory in liquid form. It had an activity of about 1400 U/ml, where U represents 1 μmole of reducing sugar liberated from soluble starch/min at pH 4.5 and 55°C . Porous glass SB_3 (pore dia. 750-800 $\overset{\circ}{A}$; over 90% silica; pore volume 0.7-1.0 cm^3/g; surface area 65.5 m^2/g) was donated by the Institute of Silicate Chemistry, Shanghai. Glucoamylase was attached to the alkylamine porous glass with glutaraldehyde following a published procedure (1, 2). Treatment of the porous glass with 0.5 N KOH at 30°C for 1 hr greatly increased the surface area, the amount of enzyme coupled, and the activity of immobilized glucoamylase (IMG). The activity of IMG increased with increasing surface area of the porous glass with a maximum at 123 m^2/g. The efficiency of the Shanghai glasses SB 46-46, SB 43-43, and SB 43-48 were almost identical to those of the Corning glass and Bio-glass products. If the amounts of aminosilane and glutaraldehyde were decreased to one tenth of the original value, only a little decrease of IMG activity was found. The optimal temperature of the IMG and native enzymes was 65°C; and the optimal pH was 6 and 4.7, respectively. Both preparations were most stable at pH 4. The Km values for maltose with the IMG and the native enzyme were 0.384 and 0.33, respectively; those for starch were 0.83 and 0.196%.

Continuous conversion of 21-22% (w/w) enzyme liquified starch solution was carried out in a packed-bed reactor at a flow rate of 20.3 ml/hr at 45°C for 64 days. The product was glucose of DE 96. The hydrolysis of the starch remained nearly constant at about 95% during the 64 days. The apparent half-life was 104 days at 45°C (Table 1).

279

TABLE 1

ENZYME HALF-LIFE*

Operation Time (hr)	Initial Activity (U/g)	Remaining Activity (U/g) (%)	Decay Constant $(hr)^{-1}$	$t_{1/2}$ (day)
840	1704	1352 (79.3)	2.75×10^{-4}	104.8
1536	1704	1113 (65.3)	2.77×10^{-4}	104.2

*On SB 46-46 porous glass at 45°C

REFERENCES

1. WEETALL, H. H. & HAVAWALA, N. B. in "Enzyme Engineering"
    (L. B. Wingard Jr., ed.) Wiley, New York (1972) p. 241.
2. LEE, D. D., LEE, Y. Y., REILLY, P. J., COLLINS JR., E. V. &
    TSAO, G. T. *Biotechnol. Bioeng.* *18:* 253 (1976).

ROTARY MULTIDISC REACTOR OF COLLAGEN SUPPORTED IMMOBILIZED GLUCOAMYLASE

S. Gondo, H. Koya and M. Morishita

Fukuoka Institute of Technology
Shimowajiro, Higashi-ku, Fukuoka, Japan

A rotary multidisc reactor of immobilized glucoamylase was constructed; and the hydrolysis rates of maltose (Katayama Kagaku Co., Japan) and soluble starch (Katayama Kagaku Co., Japan) were measured. The parameters included in the kinetic equation were estimated for these reaction systems.

Glucoamylase (GA) from *Rhizopus niveus* of industrial grade, 7 units/mg (Toyobo Co., Japan), was immobilized in 55-60 μm thick wet solubilized collagen fibrils (0.15 g GA/0.5 g collagen/disc) by a method reported previously (1,2). The membraneous immobilized GA was fixed to the upper surface of a plastic rotary disc of 11 cm diameter to give an effective liquid contacting area of 60 cm.^{2} Five immobilized GA discs of the same diameter were fixed to a rotating shaft, with 2.2 cm between neighboring discs, to make the rotary multidisc configuration. An innert disc, i.e., a plastic disc of the same diameter without the immobilized enzyme membrane, was attached 2.2 cm above the top of the five active discs in order to have the same hydrodynamic turbulence on each upper surface of the five active discs. The rotary multidisc assembly was set vertically in a 15 cm diameter tank having four baffle plates each 1.5 cm wide. The volume of reacting solution was 3 L.

Assuming no radial distribution we obtained a kinetic equation where V = glucose production rate in g glucose/cm^3 of immobilized GA/sec, k_f = the liquid film mass transfer coefficient in cm/sec, L = membrane thickness in cm, C_b = bulk liquid concentration of substrate in g/cm^3, q_i = interfacial concentration of substrate in membrane in g/cm^3, m = distribution coefficient for

$$V = \frac{C_b}{\dfrac{L}{k_f} + \dfrac{K_m}{m\eta} + \dfrac{q_i}{V_{max}}}$$ (Eq. 1)

substrate, η = effectiveness factor, V_{max} = maximum rate of enzymatic reaction, and K_m = Michaelis constant. For soluble starch V was proportional to the stirring speed, N in rpm, of the multi-discs as N^a. As C_b increased, the expoent a decreased from 0.13 to 0.1 for maltose and from 0.11 to 0.07 for soluble starch at 35°C and pH 4.5.

The liquid film mass transfer coefficient, k_f, was estimated by separating the external mass transfer resistance, L/k_f, from the overall resistance of the reaction, i.e., the denominator of the right-hand side of Eq. 1. The other kinetic parameters also were estimated to give the following values. For maltose: $V_{max} = 1/7 \times 10^{-4}$ g/cm^3/sec; $mD = 1.3 \times 10^{-6}$ cm^2/sec; and $K_m/m = 5.9 \times 10^{-4}$ g/cm^3, for pH 4.5 and 35°C. For soluble starch: $V_{max} = 5.4 \times 10^{-5}$ g/cm^3/sec; $mD = 1.5 \times 10^{-7}$ cm^2/sec; and $K_m/m = 1.4 \times 10^{-3}$ g/cm^3, for pH 4.5 and 35°C. The details of the estimation procedures will be reported elsewhere.

The maximum glucose production rates obtained in this study were 500 and 150 g of glucose/day/meter^2 of immobilized GA$_3$ membrane of 60 μm thickness (wet) for C_b = 0.005 g maltose/cm^3 and for C_b = 0.01 g soluble starch/cm^3, respectively, at pH 4.5, 35°C, and 700 rpm.

REFERENCES

1. GONDO, S. & KOYA, H. *Biotechnol. Bioeng. 20:* 2007 (1978).
2. GONDO, S., KOYA, H. & OSAKI, T. *Enzyme Microb. Technol.*
    1: 125 (1979).

KINETICS FOR THE HYDROLYSIS OF SOLUBLE STARCH BY GLUCOAMYLASE

AND APPLICATION TO AN IMMOBILIZED ENZYME SYSTEM

K. Kusunoki and K. Kawakami

Department of Chemical Engineering
Kyushu University, Japan

The hydrolysis of starch by glucoamylase proceeds by step-wise removal of glucose units from the nonreducing ends of the starch chain; and the number of available substrate molecules may be unchanged in the course of the degradation. The rate of the hydrolysis generally increases with increase in MW of the substrate. In view of these aspects, a simple practical equation, consisting of modified Michaelis-Menten kinetics with product inhibition, is presented for the hydrolysis of soluble starch (1). It is assumed that the concentration of substrate does not change during the conversion, while the values of kinetic parameters V_m and K_m vary linearly with the reduction of the average MW of substrate from the values for starch toward those for maltose. The equation is expressed as follows:

$$-\frac{dS}{d\theta} = \frac{V_m\, S/1.1}{S + K_m(1 + G/K_i)} \qquad \text{(Eq. 1)}$$

$$V_m = V_{mo}\left[1 - a(0.7 - S/S_o)\right] \qquad S/S_o < 0.7 \qquad \text{(Eq. 2)}.$$

$$K_m = K_{mo}\left[1 + b(0.7 - S/S_o)\right] \qquad \text{(Eq. 3)}$$

$$a = (1 - V_{mM}/V_{mo})/0.7 \qquad \text{(Eq. 4)}$$

$$b = (K_{mM}/K_{mo} - 1)/0.7 \qquad\qquad\qquad \text{(Eq. 5)}$$

where S and G are the concentrations of substrate and glucose in
g/L, respectively, S_o is the initial value of S, and G is
$(1.1\ (S_o - S))$. V_{mo} and K_{mo} are for soluble starch and V_{mM} and K_{mM}
for maltose. It is assumed from the experimental results that when
$S/S_o \geq 0.7$, V_m and K_m equal V_{mo} and K_{mo}. K_i is the inhibitor constant. V_m is expressed as g/L/min.

The hydrolysis of two kinds of soluble starches and maltose
were performed in a 400 ml flask, using glucoamylase from *Rhizopus
delmar* in the free and immobilized forms. Immobilized glucoamylase was prepared by crosslinking with dialdehyde-starch on a
silanized ceramic monolith, composed of zircon and cordierite.
The monolith originally was 93 mm in diameter and 76 mm long with
parallel triangular channels ($21.7/cm^2$) separated by thin porous
walls of mean pore diameter 8 μm and 0.32 mm think. A fan-shaped
piece of the monolith was fixed to the stirring shaft in the reactor.

The kinetic parameters were determined from the analysis of
the initial reaction rates. Fig. 1 shows the time courses of hydrolysis of Wako starch (potato starch, DE 0.86, average MW
30,000) by free enzyme. The curves calculated from Eq. 1 to 3
are in good agreement with the experimental data. In this simulation, K_i was adjusted to fit all the data. The simulation for
the hydrolysis of Katayama starch (DE 0.17, average MW 3,000,000)
was also achieved successfully, using the same manner as above.
In the immobilized enzyme system, the reaction rates might be influenced by the diffusion of substrate through the porous walls
of the monolith because the plot of S_o/v versus S_o was nonlinear.
Therefore, V_m, K_m, and the effective diffusivity were determined

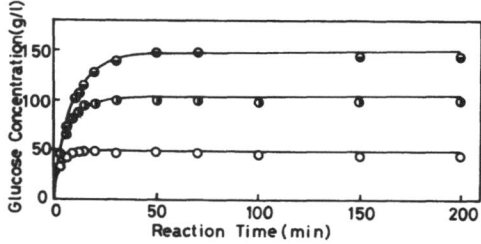

Fig. 1.   Time course of hydrolysis of Wako starch by free gluco-
          amylase; enzyme concentration 10 g/L; 50°C; pH 4.5;
          initial concentration of substrate: (●) 148 g/L; (◐)
          98.9 g/L; (O) 49.4 g/L; ——— calculated curves.

by the method of Kobayashi *et al.* (2), using the least square
estimation of nonlinear parameters. For the hydrolysis of maltose, no diffusion effect was detected. Fig. 2 shows the time
courses for the hydrolysis of Wako starch in the immobilized enzyme system. The simulation consisted of solving numerically the
steady state mass balance equations for starch and glucose over
a differential thickness within a wall of the immobilized enzyme
monolith and the equations expressing the change of concentrations
of starch and glucose with time in the reactor (3). The calculated curves in Fig. 2 are in good agreement with the experimental
data.

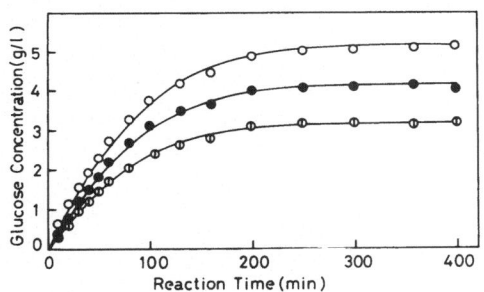

Fig. 2.   Time course of hydrolysis of soluble starch by immobi-
          lized glucoamylase; 50°C; pH 4.5; initial concentration
          of substrate: (O) 5 g/L; (●) 4 g/L; (Φ) 3 g/L; ——
          calculated curves.

REFERENCES

1.   KUSUNOKI, K., KAWAKAMI, K., SHIRAISHI, F., KATO, K. & KAI,
     M. *Biotechnol. Bioeng. 24:* 347 (1982).
2.   KOBAYASHI, T., OHMIYA, K. & SHIMIZU, S. *J. Ferment. Technol.*
     54: 260 (1976).
3.   SHIRAISHI, F., thesis, Kyushu University (1981).

REDUCED THERMOSTABILITY OF MODIFIED *MUCOR MIEHEI* RENNET

S. Branner-Jørgensen, P. Eigtved and P. Schneider

NOVO INDUSTRI A/S
Bagsvaerd, Denmark

During the last 10 years microbial rennets have found wide application in the dairy industry. The rennet derived from the fungus *Mucor miehei* has proved the most successful and can be used in most cheese making processes. One limitation is in the manufacture of cheese whey products, where residual milk-clotting activity after pasteurization may cause problems. Another limitation is the manufacture of some cheese types requiring high scalding temperatures, e.g. Emmental. In both cases the problems can be ascribed to the relatively high thermal stability of *Mucor miehei* rennet compared with calf rennet.

Due to the many favorable characteristics of the *Mucor miehei* rennet, we decided to modify the thermal stability of the enzyme by chemical means. Only two classes of agents were found to give significant destabilization: acylating agents and oxidizing agents. The acylations, exemplified by acetic acid anhydride, tend to result in a lower degree of destabilization and higher activity losses during the modification reaction than do the oxidations.

When heated in dilute solution at fixed pH, the unmodified rennet enzymes follow first order denaturation kinetics; and well-defined half lives for the milk-clotting activity can be estimated at a given temperature. Contrary to the unmodified enzymes, the oxidized enzymes deviate from first order denaturation kinetics. The half lives then depend on the degree of inactivation and are therefore only approximate. For each enzyme a temperature can be estimated to give a specified half life (Table 1). A physical method of analysis based on temperature scanning pH-stat titration can be used as an alternative stability measurement (1).

TABLE 1

STABILITY OF RENNET PREPARATIONS

Rennet Enzyme Source	Oxidizing Agent	Heating Temperature to Give Half Life of About 10 min (^{o}C)	
		pH 5.5	pH 6.0
Calf	None	59	53
M. miehei	None	68	65
M. miehei	Sodium hypochlorite	61	56
M. mienei	Hydrogen peroxide	61	56
M. miehei	Peracetic acid	61	56

The different oxidizing agents give products with very similar properties; but their specificity differs considerably when applied to crude *Mucor miehei* rennet. The following molar excess of oxidizing agent was used: 1,000-4,000 for hydrogen peroxide, 80 for hypochlorite, and 40 for peracetic acid, expressed as moles of reagent/mole of pure enzyme in the crude rennet. Methionine residues are the most obvious ones to be oxidized; but it remains to be proven that methionine oxidation is responsible for the destabilizations.

REFERENCE

1.   EIGTVED, P. *J. Biochem. Biophys. Meth. 5:* 37 (1981).

APPLICATION OF IMMOBILIZED ENZYMES TO MILK CURDLING

S. Shimizu, K. Ohmiya, S.-E. Yun and T. Kobayashi*

Department of Food Science and Technology, and
Department of Chemical Engineering,*
Nagoya University, Nagoya, Japan

Immobilized proteases with high proteolytic activity, other than the typical milk coagulants, were studied and used in clarification of the reaction mechanism of milk curdling.

Skim milk powder 10%(w/v) solution was used as the test milk in all trials. Thermolysin, alkaline-protease, trypsin, and *Mucor* rennin or rennet were examined as coagulants of milk. Each protease was immobilized on Dowex MWA-1 with glutaraldehyde (1). Phosphoprotein phosphatase from bovine spleen and chymosin purified from rennet were employed to modify β-casein. The extents of proteolysis and of dephosphorylation were determined by the Lowry-Folin method (2) and Martin and Doty method (3), respectively. Acrylamide slab-gel (3.75%) electrophoresis was performed in the presence of 8 M urea. The amounts of the proteins in the bands stained with Coomassie Blue were determined by a densitometric measurement. For milk curdling, each immobilized protease was stirred in the test milk with a propeller blade until the extent of hydrolysis of the milk proteins increased by about 2∿5% in the case of alkaline proteases. This procedure was performed at 10°C to retard the formation of milk clots during the enzymatic reactions of immobilized proteases. After removing the enzymes, the limit-hydrolyzed test milk was incubated at 35°C for 100 min; and the tension of the resulting curd was determined using the curd meter (4). For the fortification, the limit-hydrolyzed test milk (30 ml) was mixed well with α_s- or β-casein (5%, 2 ml) at 4°C for 10 min to disperse the added casein. Subsequent incubation at 35°C for 100 min was followed to determine the curd tension.

The curd tension of the coagulum treated by each acid protease in the soluble state increased linearly with the incubation time. The extent of proteolysis with the soluble proteases was constant (2∿3%) during 90 min incubation after clotting. On the contrary, the tension values of the curds prepared by alkaline proteases did not increase linearly with time and were about half those of rennet curd after 90 min incubation. The extent of proteolysis with the soluble alkaline proteases, except trypsin, was also 2∿3% at the initiation of clotting and increased gradually to 5∿7% at 90 min incubation with soluble trypsin, the value was 5.5% at the initiation of clotting and increased to 17% at 90 min incubation. By employing thermolysin, trypsin, or alkaline-protease in an immobilized form, the linear increase in curd tension was kept constant at around 2∿5%. From these data, it is essential that the proteolytic reaction in milk has to be stopped at the appropriate extent of proteolysis in order to prepare curd with desirable rigidity. Soft curd milk results from hydrolysis greater than 3%, and will be expected to be useful as an infant food. On the contrary, harder curd will be required by the cheese making industry; hydrolysis should be carried out to the extent required for initiation of clotting. Any demands for preparing milk curd with desired rigidity can be satisfied by using proteases with activity higher than typical coagulants.

The electrophoresis and densitometric measurements indicated that the amount of β-casein in the curds with lower tension was smaller than that in the curds with higher tension values. This suggested that the rigidity of the curd might depend on the content of β-casein. This was supported by the fact that the fortification of test milk with β-casein resulted in a remarkable increase of curd tension compared to α$_s$-casein. Further studies were performed to elucidate the role of any moiety of β-casein in curdling, by fortifying with modified β-caseins. The fortification with β-casein deleted the hydrophilic moiety, while dephosphorylated β-casein resulted in negligible increase of curd tension. The fortification with chymosin-modified β-casein devoid of hydrophobic moiety increased the tension of curd as did native β-casein. These results suggested that the phosphoryl group in β-casein contributes to milk curdling.

REFERENCES

1. OMIYA, K., TANIMURA, S., KOBAYASHI, T. & SHIMIZU, S. *Biotechnol. Bioeng. 20:* 1 (1978).
2. LOWRY, O. H., ROSEBROUGH, N. J., FARR, A. L. & RANDALL, R. J. *J. Biol. Chem. 193:* 265 (1951).
3. MARTIN, J. B. & DOTY, D. M. *Anal. Chem. 21:* 965 (1949).
4. IIO, N. *Sci. Cookery 2:* 54 (1969).

PRODUCTION OF 7-AMINODESACETOXYCEPHALOSPORANIC ACID BY IMMOBILIZED *E. COLI* CELLS

Z. Wang, H. Yuo, M. Wang, Q. Jiao,
W. Han, W. Sun and Q. Zhang

Institute of Microbiology, Academia Sinica,
Beijing, China

7-Aminodesacetoxycephalosporanic acid (7-ADCA) is the nucleus of cephalosporins and can be converted to various useful semisynthetic cephalosporins. The conversion of 7-phenyl-acetamidodesacetoxycephalosporanic acid (PA7-ADCA) to 7-ADCA may be carried out either by a biochemical route or by a chemical route. This paper reports the production of 7-ADCA from PA7-ADCA by immobilized *E. coli* cells.

Twenty *E. coli* strains having penicillin acylase activity were examined; all could produce 6-aminopenicillanic acid (6-APA) from penicillin G. and 7-ADCA from PA7-ADCA. Strain AS 1.76 had high activity for both substrates; the acylase activity of cells harvested by centrifugation reached 25-30 U/g wet cells when AS 1.76 was cultivated in a medium containing 1% pepton, 0.5% NaCl, 0.2% phenylacetic acid, and 0.3% corn steep liquor for 15 hr. *E. coli* AS 1.76 was immobilized by entrapment in 4% agar gel beads and crosslinked with glutaraldehyde to give 90-95% acylase activity entrapped and a recovery of activity of 20%. The apparent enzyme activity was 2.5-3.0 U/g immobilized cells (by NIPAB method). The optimal temperature and pH of the immobilized cells were 55°C and pH 8.0, while those of the intact cells were 50°C and pH 7.5. A conversion of 90% was achieved in 30, 60, 90, and 120 min for PA7-ADCA substrate concentrations of 2, 4, 6, and 8%, respectively. The flow diagram for production of 7-ADCA is shown in Fig. 1. Recycling of a 5% solution of PA7-ADCA through the immobilized cell column at 37°C for 2-2.5 hr resulted in a 91.5% conversion and 81.4% yield of 7-ADCA. The column was used 23 times during 30 days without a decrease of activity; a column was operated on an industrial scale for 72 times over 3 months.

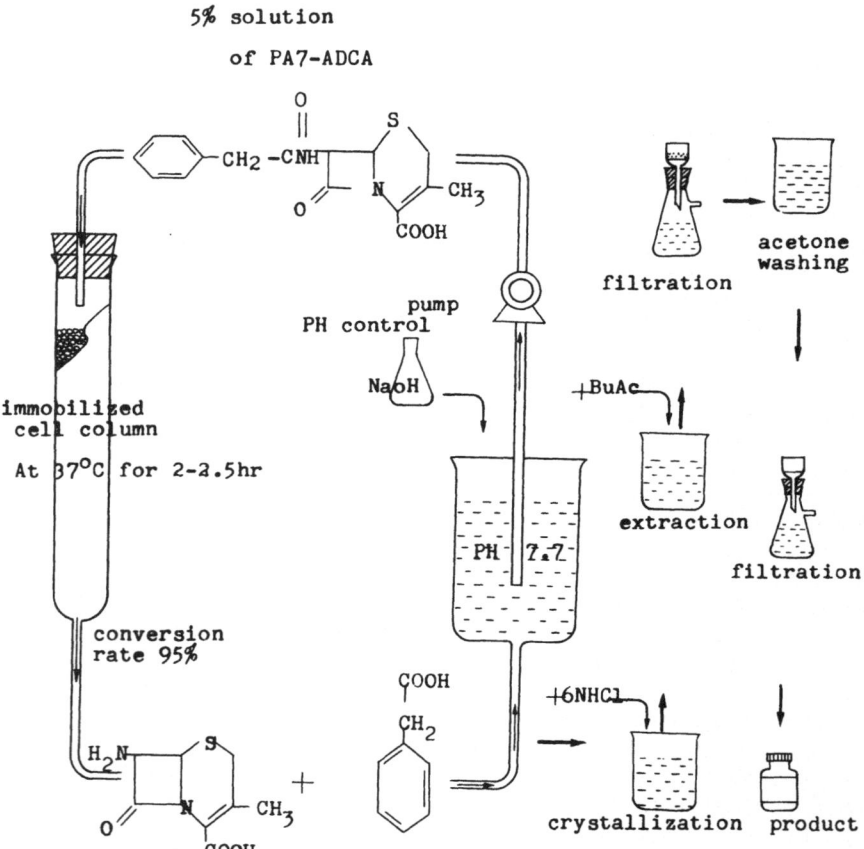

5% solution

of PA7-ADCA

Fig. 1. Flow diagram for production of 7-ADCA.

TECHNICAL APPLICATIONS OF LACTASE AND AMINOACID ACYLASE

IMMOBILIZED TO FORM PLEXAZYM

H. Plainer, B. G. Spörssler

Rohm GmbH, Darmstadt
Federal Republic of Germany

High activity and good flow properties ensure the successful performance of a packed bed reactor. We have achieved high activity per volume by using a carrier with a porosity of 3-4 mg/g. Enzymes are bound covalently by oxirane groups. Because of their rigid structure the 0.1-0.3 mm diameter beads are pressure stable and show good flow properties. The relationship between flow rate and pressure drop is linear with a slope of about 60 $(L/cm^3/hr)/$ (bar/cm bed length).

Asp. oryzae lactase was immobilized to form PLEXAZYM LA. On account of the pH-optimum of 4.5, acid whey is an ideal substrate for PLEXAZYM LA. At a space velocity of 50/hr lactose in filtered acid whey of pH 4.5 was hydrolyzed to glucose and galactose in a packed bed reactor at $35^{\circ}C$. The rate of hydrolysis decreased from an initial value of 100% to about 90% over 60 days of operation. The productivity of 70 tons of whey/kg PLEXAZYM LA was reached in 60 days. Cleaning with a 0.1% solution of a quaternary ammonium salt was necessary every 3 days.

Because of the optimal heat stability of PLEXAZYM LA at neutral pH, the reactor also could be run with milk at $55^{\circ}C$. At a space velocity of 20/hr a productivity of 13 tons of whole milk (3.5% fat, 4.6% lactose)/kg was reached in 30 days. The packed bed did not get plugged by the droplets of the fatty emulsion or by the casein micelles of milk because of the morphological properties of the support (Fig. 1).

Aminoacid acylase was immobilized to form PLEXAZYM AC. N-acetyl-D,L-aminoacids were converted to L-aminoacids in a packed bed reactor at $37^{\circ}C$ by PLEXAZYM AC. N-acetyl-D,L-methionine, 0.7M

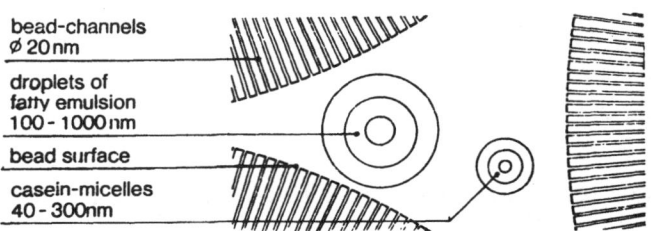

Fig. 1.   Passage of milk through PLEXAZYM LA.

at pH 8, was hydrolysed 80% at a space velocity of 6/hr.   The rate
of hydrolysis decreased to about 30% when the space velocity was
increased to 25/hr.   To demonstrate the high activity, 1 g PLEXAZYM
AC was as active as 0.4 g of the original purchased enzyme preparation.   A productivity of 500 kg L-methionine /kg was reached in
a 90 days reactor run.

CONTINUOUS HYDROLYSIS OF LACTOSE IN SKIM MILK AND ACID WHEY BY IMMOBILIZED LACTASE OF *ASPERGILLUS ORYZAE*

H. Hirohara, H. Yamamoto, E. Kawano and T. Nagase

Biological Science Laboratory, Sumitomo Chemical
Co., Ltd, Takatsuki, Osaka, Japan

We report here the development of a stable immobilized lactase to hydrolyze lactoses in both skim milk and acid whey. *Aspergillus oryzae* lactase was used. This fungal lactase has good thermal stability and effective activity in the pH ranges of both milk and acid whey (1,2).

The extracellular lactase, obtained from a solid-phase cultivation of a selected strain of *Aspergillus oryzae*, was purified by solvent precipitation and ultrafiltration techniques. The highly purified enzyme was adsorbed on a phenolic amphoteric ion-exchange resin, prepared by carboxylation of a macroporous anion-exchange resin. The carrier adsorbed enzyme was covalently immobilized by treatment with gluteraldehyde. This immobilization technique enabled the enzyme to hydrolyze lactase without difficulty, considerably lowering the product inhibition inherent in fungal lactase. The immobilized enzyme had 0.25 - 0.6 mm particle size, an apparent bulk density of 0.3 g/ml, and a standard activity of 1,000 ILU/g-IML and 30°C.

Table 1 summarizes the results for continuous operation at various temperatures. Water-jacked column reactors, containing 8 - 20 ml of immobilized lactase, were operated to maintain a constant conversion of 80% by decreasing the space velocity according to the activity decay. For whey powder solutions, insoluble substances were removed from the feed solution by centrifugation. It was remarkable that a space velocity (SV) of nearly 60 hr^{-1}, which corresponded to a residence time of 1 minute, gave 80% hydrolyzed whey at 45 - 50°C. The high estimated productivities, assuming the enzyme was used until the space velocity was 1 hr^{-1}, were due to high activity and stability. At lower temperature,

still higher productivities were obtained because of the long half-
life of 2,000 to 3,000 day; although the reaction rates were small.
When whey permeate instead of whey powder solution was used as a
feed, the productivity was much larger than that with whey powder
and close to that of pure lactose.

It should be noted that no changes except the conversion of
lactose to glucose and galactose were found in the skim milk passed
through the column reactors. Also the immobilized lactase had
outstanding resistance to various chemicals and disinfectants so
that sterilization was carried out successfully every day during
the experiments reported in Table 1.

It is conclused that this immobilized lactase is an excellent
industrial catalyst for hydrolyzing lactoses in skim milk and acid
whey with high productivities.

REFERENCES

1. TANAKA, Y., KAGAMIISHI, A., KIUCHI, A., & HORIUCHI, T. *J.
       Biochem.* 77: 241 (1975).
2. AKASAKI, M., SUZUKI, M., FUNAKOSHI, I., & YAMASHINA, I. *J.
       Biochem.* 80: 1195 (1976).

TABLE 1

HALF-LIVES AND PRODUCTIVITIES

Substrate	Initial Activity at 30°C (ILU/g-IML)	Initial SV^{-1} (hr^{-1})	Half-Life (hrs)	Productivity ($\frac{\text{L-solution}}{\text{ml IML}}$)	Days Run
10% Skim Milk	970[a]	9.3	2,920	35.1	74
	1,000[b]	1.1	2,300 ± 500 day	71.4	340
	970[c]	54	250; 680	34.5	96
7% Whey Powder[g]	970[d]	44	500; 1,780	69.2	94
	1,230[e]	8.0	3,000 ± 1,000 day	727	380
7% Lactose (USP)	990[f]	29	1,600 ± 100 day	1,550	650

a) 5.2% lactose, pH 6.65, 45°C., b) 4.8% lactose, pH 6.65, 4.5°C., c) 5.04% lactose; pH 4.5; 50°C, d) 5.04% lactose, pH 4.5, 45°, e) 5.04% lactose, pH 4.5, 4.5°, f) 7.0% lactose, pH 4.5, 40°, g) proteins 5.6 g/L, minerals 4.9 g/L.

LACTIC ACID FERMENTATION WITH IMMOBILIZED *LACTOBACILLUS* SP.

S. L. Stenroos, Y. Y. Linko, P. Linko, M. Harju*, and
M. Heikonen*

Helsinki University of Technology, Department of
Chemistry, Espoo and Valio Laboratory*, Kalevankatu
Helsinki, Finland

In addition to a number of food applications lactic acid is
an excellent feedstock for chemical synthesis, via lactonitrile,
and for lactide polymers. Homofermentative lactic acid bacteria
produce no carbon dioxide from sugars. Consequently, biotechnical
lactic acid production offers an interesting alternative to ethanol
fermentation. Lactic acid is produced industrially both by chemical synthesis and by batch fermentation; although continuous fermentation processes have been investigated on a laboratory scale
(1-4, 5). Lactic acid bacteria, grown on a solid support covered
by a gelatin-poly-electrolyte film in a packed-bed column reactor,
were able to increase the lactic acid content of acid whey from
1.4 to 2.1% in a single pass (6). We have shown recently that
lactic acid can be produced continuously in a column bioreactor
employing calcium alginate gel entrapped *Lactobacillus* sp. (7,
8). In more recent work, *L. delbrueckii* was used to produce
lactic acid from glucose, and *L. bulgaricus* to produce lactic
acid from lactose.

L. delbrueckii cell mass was grown on MRS-medium (9). For
L. bulgaricus glucose was replaced by lactose, if the lactose
was to be used as substrate in a subsequent fermentation. The cell
mass was suspended in 6% sodium alginate and extruded through 0.6
mm diameter hollow needles into 0.5 M calcium chloride to obtain
biocatalyst beads of about 2 mm diameter.

Up to 93% lactic acid yield could be obtained from 4.8% (w/v)
glucose with immobilized *L. delbrueckii* in a column reactor
operated continuously at 7.9 hr residence time. At 4.9 hr residence time the yield was still about 80%. Stabilization of the

299

process took about a week; after one month of continuous operation
the productivity was about 70% of the maximum obtainable (Fig. 1A).
In a recycle batch process 97% lactic acid yield could be obtained
in about 40 hr. The bioreactor column was stored at +4°C in between different experiments, and only about 10% of the original
activity was lost during a total of 150 day. With *L. delbrueckii*
>90% of lactic acid formed was L-lactic acid.

Up to 80% lactic acid yield could be obtained with immobilized
L. bulgaricus in a recycle batch process when 5% (w/v) glucose
was used as a carbon source, and about 60% when 5 – 9% (w/v) lactose was employed as substrate. With dairy industry by-products
of equivalent lactose concentration, such as whey, whey UF-permeate,
and demineralized whey, the yield varied from about 70 – 80%. In
continuous column operation with immobilized *L. bulgaricus* the
maximum lactic acid yield from lactose was about 40% at 12 hr residence time, with almost all of the lactic acid produced in the D-
form (Fig. 1B).

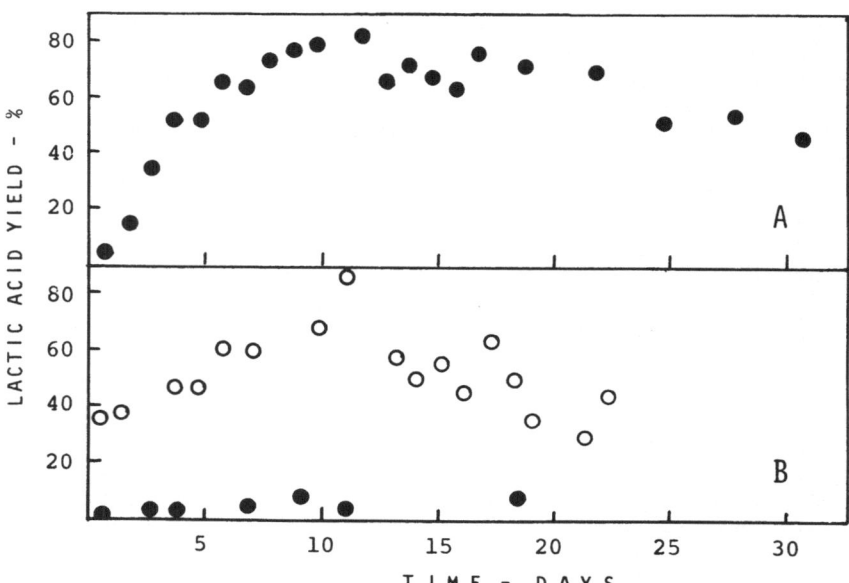

Fig. 1.   Continuous production of lactic acid: (A) from 4.8%
         (w/v) glucose (1% yeast extract, 4.8% CaCO$_3$, 43°C, pH
         5.7) with immobilized *L. delbrueckii*, and (B) from
         4.8% (w/v) lactose (1% yeast extract, 3% CaCO$_3$, 45°C,
         pH 6.0) with immobilized *L. bulgaricus* (● L-lactic
         acid, ○ total acid).

REFERENCES

1. WHITTIER, E. O. & ROGERS, L. A. *Ind. Eng. Chem. 23:* 532 (1931).
2. MARSHAL, K. R. *Proc. 18th Int'l Dairy Congr.,* Sydney 447 (1970).
3. KELLER, A. K. & GERHARDT, P. *Biotechnol. Bioeng. 17:* 997 (1975).
4. COX, G. C. & MACBEAN, R. D. *Aust. J. Dairy Technol. 32* (1): 13 (1977).
5. MILKO, E. S., SPERELUP, O. V., & ROBOTNOVA, I. L. *Z. Allgem. Mikrobiol. 6:* 297 (1966).
6. COMPERE, A. L. & GRIFFITH, W. L. *Dev. Ind. Microbiol. 17:* 247 (1975).
7. LINKO, P. in "Enzyme Engineering in Food Processing" (P. Linko & J. Larinkari, eds), Applied Science Publ., London (1980) p. 27.
8. LINKO, P. in "Advances in Biotechnology," vol. 1 (M. Moo-Young, C. W. Robinson, & C. Vezzina, eds.) Pergamon Press, New York (1981) p. 711.
9. MAN, J. C., ROGOSA, M., & SHARPE, M. E. *J. Appl. Bact. 23:* 23 (1960).

PRODUCTION OF l-MALIC ACID WITH IMMOBILIZED THERMOPHILIC

BACTERIUM, *THERMUS RUBENS* Nov. sp.

Y. Ado, T. Kawamoto, I. Masunaga, K. Takayama,
S. Takasawa, and K. Kimura*

Tokyo Research Laboratory, and Technical Research Lab
of Hofu Plant,* Kyowa Hakko Kogyo Co. Ltd., Tokyo and
Yamaguchi*, Japan

In order to establish an advantageous method for the production of L-malic acid, the conversion of fumaric acid to L-malic acid by thermostable fumarase was studied. A new thermophilic bacterium, *Thermus rubens* nov. sp., which has a high fumarase activity, was isolated from a hot spring in Japan. This bacterium could produce about twice the activity of fumarase in comparison with the mesophilic microorganisms, such as *Brevibacterium ammoniagenes* and *Candida lipolytica*. Past enzymatic production of L-malate from fumarate was done batchwise using Ca-fumarate as substrate (1); but the byproduct calcium sulfate was a disadvantage. Continuous production of L-malate with immobilized microorganism was developed (2, 3); with a maximum theoretical conversion of 81% (with Na-fumarate). The equilibilium constant (L-MA/FA) of L-malate and fumarate is 4.00 for Na, 4.75 for NH_4, 7.00 for Mg, and 12.3 for Ca, all at 60°C and pH 7.0. When Mg- or Ca-fumarate is used, the conversion can be increased to 87.5% or 92.5%, respectively; but these salts cannot be used on a practical basis because the solubilities are very low at room temperature.

To solve the problems described above, thermostable enzymes were screened for production of L-malate at high temperature (Table 1). With *Thermus rubens* nov. sp. the optimum temperature for growth was 60°C; and the optimum pH was 6.5. Fumarase activity was measured as the productivity of L-malate from fumarate for 30 min reaction at 60°C. The quantitative analysis of organic acids was carried out by HPLC using a Shodex ionpak C-811 column. Immobilization of *Thermus rubens* nov. sp. has been tested by adsorption of enzyme to a weakly basic anion exchange resin Duolite A-7 and by microcapsulation of cells with cellulose

TABLE 1

SCREENING OF FUMARASE FROM *THERMUS* SP.

Strains	Culture Temp. ($^{\circ}$C)	DCW (g/L)	Fumarase (U/g cell)
T. aquaticus KY-12261	70	1.39	150
T. lacteus nov. sp. KY-12264	70	0.52	60
T. rubens nov. sp. KY-12265	60	1.62	380
T. aquaticus ATCC-25104	70	1.10	120
T. aquaticus ATTCC-25105	65	1.25	130
T. thermophilus ATCC-27634	70	1.00	30

*Incubated in medium containing 0.8% polypepton and 0.4% yeast extract, pH 7.2 for 24 hr; DCW (dry cell weight) measured at 660 nm absorbance.

acetate butyrate (CAB). The retained activity was higher with Duolite A-7. Continuous production was carried out with the immobilized enzyme: Doulite A-7 resin and 1M Na-fumarate or Mg-fumarate as substrate at 60°C and pH 7.0. The substrate was passed through the reactor column at 0.32 hr^{-1} space velocity The conversion of the Na-salt was approximately 70%; for the Mg-salt it was about 80%.

REFERENCES

1. KITAHARA, K., et al. J. Gen. Appl. Microbiol. 6: 167 (1960).
2. YAMAMOTO, K., et al. Eur. J. Appl. Microbiol. 3: 169 (1976).
3. YAMAMOTO, K. et al. Biotechnol. Bioeng. 16: 1101 (1977).

CONTINUOUS HYDROLYSIS OF CONCENTRATED SUCROSE SOLUTIONS BY

IMMOBILIZED INVERTASE

D. Combes and P. Monsan

Departement de Genie Biochimique et Alimentaire
INSA, ERA CNRS, Toulouse, France

The aim of the present work is the pilot scale continuous invertase hydrolysis of high concentrated sucrose solutions to produce invert sugar, which does not show crystallization problems or colored byproducts.

Invertase (Rapidase) was covalently coupled to corn stover (EU-Grits, Eurama). The glucose units of the cellulosic fraction of the support were chemically modified by sodium meta-periodate oxidation to form aldehydes, amination by condensation with ethylene diamine, reduction of the resulting imino bonds to amino bonds with sodium cyanoborohydride, and activation of the amino groups with glutaraldehyde (1). Activities up to 83 mmoles reducing sugars/min/g of support were obtained. This immobilization method could be easily scaled up to prepare 10 kg of derivative without any important loss in efficiency.

The immobilized invertase, when used in a packed bed reactor for continuous sucrose inversion, showed very good operational stability, with a half-life of 365 day at $40 °C$ with 2 M sucrose at pH 4.5. Under similar conditions the half-life was 90 day at $45 - 55 °C$. The sucrose concentration had a very important effect on the stability of immobilized invertase. At $60 °C$, for instance, an important increase in immobilized invertase half-life was observed for sucrose concentrations higher than 1.5 M, resulting in a ten-fold increase for 2.5 M sucrose. As invertase is subject to both substrate and product inhibition, the efficiency was compared in continuous stirred tank (CSTR) and plug-flow (PFR) reactors. The PFR had a higher efficiency, corresponding to 30% - 40% higher maximal productivities for 0.4 M and 2.44 M initial sucrose concentration, respectively. A ten-fold scale-up in

packed bed volume from 0.1 to 1.0 L was achieved without any noticeable loss in efficiency.

Continuous sucrose hydrolysis was scaled up to a 17.6 L pilot
reactor. The reactor was fed with 65 - 71%, W/W, concentrated
sucrose solutions at 50 - 55°C, to produce invert sugar syrup with
the desired inversion degree. This reactor proved very efficient
as it allowed a productivity equal to 9.1 kg sucrose hydrolyzed
per hr in the case of a 69% (W/W) initial sucrose concentration,
with 72% hydrolysis. Inversion degrees up to 93% could be obtained with a 65% (W/W) initial sucrose concentration. The productivity then corresponded to 7.9 kg sucrose hydrolyzed/hr (2).

This pilot scale reactor allowed the processing of highly
concentrated sucrose solutions of close to 1 kg/L from a cane
sugar refinery. This direct treatment of the sugar solution
avoided any prior dilution and subsequent concentration of the
syrup. Furthermore, no colored byproducts were present in the
reactor effluent. The scale-up of this enzymatic sucrose inversion process to an industrial scale may thus be planned.

ACKNOWLEDGMENTS

We thank Beghin Say Co. for utilization of the pilot scale
reactor and placing it in a cane sugar refinery.

REFERENCES

1. MONSAN, P. French Patent 79-31387 (1979).
2. COMBES, D., thesis, INSA, Toulouse (1981).

A SERIES OF COVALENTLY BONDED ENZYMES AND THEIR APPLICATIONS

S. Liu, Z. Yuan, Q. Wang, J. Wang and Y. Zeng

Shanghai Institute of Biochemistry, Academia Sinica,
Shanghai, China

Since 1970, we have used a bifunctional reagent, p, β-, (sulphato ethylsulphonyl) aniline (SESA), to couple enzymes to different polysaccharides. The SESA was first reacted with various polysaccharides to obtain a series of p-aminobenzene sulphonylethyl derivatives (Fig. 1), which were then diazotized and coupled with enzymes. The immobilized enzymes (the support, the recovery of activity R on immobilization, and the relative activity A of the immobilized enzyme) are as follows: nuclease P_1 from *Penicillin citrinum* (1) (cellulase, 20-40% R, 50% A), 3'-RNase (Sephadex G-200, 25-35%R, 80%A) (2), penicillin acylase (crosslinked agar, 50-60% R) (4), trypsin (cellulose, 95-100% A with BANA as substrate) (3), subtilisin (cellulose, 32-35% R), snake venom phosphodiesterase (crosslinked agar, 98% R), and alkaline phosphatase (crosslinked agar, 60-80% R) (5, 6). In 1978 we established the first industrial application of immobilized en-

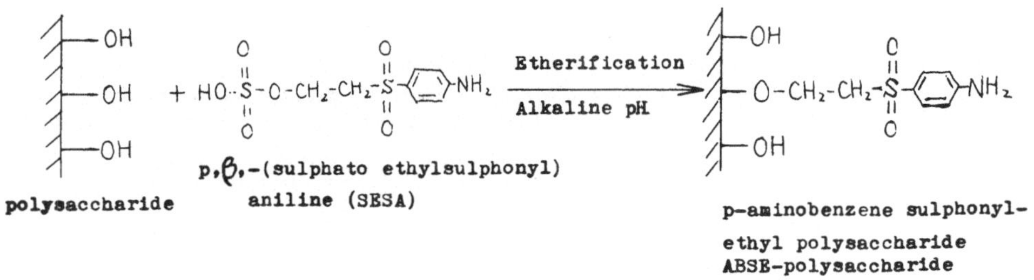

Fig. 1. SESA coupling chemistry.

zymes in China, i.e. the production of 5'-mononucleotides using immobilized nuclease P_1.

In the case of immobilized penicillin acylase one gram of ABSE-crosslinked agar (dry weight) could be coupled with about 750 mg of protein; the immobilized enzyme half life was 65 days at $37^{\circ}C$. The immobilized alkaline phosphatase showed no detectable loss of activity during 40 days of continuous operation. The heat stabilities of these enzymes were all enhanced by immobilization. For example, the inactivation velocity of immobilized nuclease P_1 was 10 times less than that of the native enzyme.

The ABSE-cellulose-nuclease P_1 industrial scale production of 5'-mononucleotides uses yeast RNA as the substrate. The enzyme can be reused for 23 cycles in a stirred tank reactor containing 2,000 L of 0.5% RNA solution. The practical efficiency of the enzyme was increased over 30 times. In addition immobilized 3'-RNase from *Rhodotorula glutinis* has been used to prepare 3'-mononucleotides from yeast RNA for use as reagents on a moderate scale with a practical efficiency 10 times over that of the broth process. Immobilized alkaline phosphatase has met success in the synthesis and analysis of oligonucleotides. The 3'-terminal phosphate of the tetranucleotide Cpm'IpΨpGp was successfully cleaved by the enzyme column with a dephosphorylation yield of 98%. This immobilized enzyme combined with immobilized 5'-phosphodiesterase was also utilized for sequencing oligonucleotides in the $[^3H]$ postlabeling method.

REFERENCES

1. YUAN, Z., LIU, S., WANG, J., MA, F., CHEN, Y. & QUAN, X. *Kexue Tongbao* (Engl. Ed.) *26*: 927 (1981).
2. Immobilized Enzymes Research Group of Shanghai Institute of Biochemistry, Academia Sinica, *Acta Biochimica Biophysica Sinica 9*: 187 (1977).
    Ibid. 9: 64 (1977).
4. WANG, Q., *et al.*, unpublished results.
5. YUAN, Z. & QIAN, X. *Kexue Tongbao* (Engl. Ed.), in press.
6. YUAN, Z., QIAN, X., YU, A., LIU, S. & MA, F. *Acta Biochimica Biophysica Sinica 13*: 291 (1981).

THE APPLICATION OF FLUIDIZED BEDS FOR IMPROVED ENZYME REACTOR PERFORMANCE

A. Renken, E. Flaschel and P. -F. Fauquex

Institute of Chemical Engineering
Swiss Federal Institute of Technology
Lausanne, Switzerland

Experiments to reduce fluidized bed channeling, circulation of biocatalyst, and flow instability (1) were done using a 40 mm dia by 2.96 m long column, with and without internal elements to reduce circulation of the solid biocatalyst and backmixing. As internals, two types of SULZER[R] static mixing elements SMX and SMV were used (2). The biocatalyst was *Aspergillus niger* lactase on chitosan coated silica gel particles of mean particle dia 0.14 mm and an apparent density of 1.47 g/cm^3. The residence time distribution of the liquid phase in the different reactors was determined using the stimulus response technique with pulse input of the tracer (aq. HCl). The concentration time function of the tracer was determined by conductivity at 0.725 m and 2.315 m from the column inlet. The measured input at the first position was convoluted with a residence time distribution function of reactor models for non-ideal flow. By fitting the calculated curve with the measured output signal at the second position, a model discrimination and model parameter were obtained: the number of tanks of the tanks-in-series model.

Due to the small particle size and the small density difference between biocatalyst and substrate, stable fluidized bed behavior in the empty tube was difficult to obtain. Great care had to be taken to align the tube exactly vertically. These problems were overcome using flow stabilizing internals. The ratio of the superficial velocity U to the terminal velocity U_t is related to the volume fraction of voids, ε, of the fluidized bed with and without internals as ε^n (3). Using this relation, the following results were obtained: a) without internals U_t = 4.5 mm/s , n = 5.0; b) with internals SMX U_t = 10.3 mm/s, n = 4.8; c) with internals SMV U_t = 16.0 mm/s, $\bar{n}$ = 5.3. As one can see, the catalyst

concentration for the same feed flow increased drastically by use
of static mixers in the fluidized beds. At the same time, large
dispersions (backmixing) of the fluid could be avoided and the behavior of the reactor approached that of an ideal tubular reactor;
and the productivity per unit volume of a constant diameter reactor
could be increased considerably (Fig. 1).

    The authors thank the board of the Swiss Federal Institutes
of Technology for financial support.

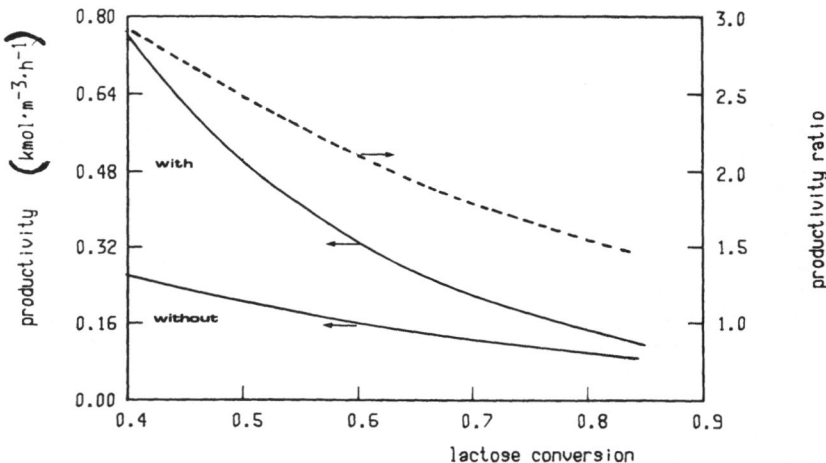

Fig. 1.   Hydrolysis of lactase in fluidized beds with and without
          internal static mixing elements.

REFERENCES

1. OESTERGAARD, J. & KNUDSEN, S. L. *Die Starke 28:* 350 (1976).
2. FLASCHEL, E. & RENKEN, A. Swiss Patent, 4707/81.
3. RICHARDSON, J. W. & ZAKI, W. N. *Trans. Inst. Chem. Engr. 32:*
        386 (1954).

CONTROL OF CONTINUOUS COENZYME REGENERATION

R. Wichmann and C. Wandrey

Institute of Biotechnology, Nuclear Research Centre
Juelich, Federal Republic of Germany

L-leucine can be produced from α-ketoisocaproic acid by a reductive amination, catalyzed by L-leucine dehydrogenase (LEUDH) (1). The coenzyme is regenerated simultaneously by reduction with formate, catalyzed by formate dehydrogenase (FDH). For continuous operation an enzyme membrane reactor (EMR) is used. NAD(H), co-valently bound to polyethylene glycol with a MW of 10,000 (PEG-10,000) is used. Thus, the coenzyme as well as the enzymes can be retained in the EMR by an ultrafiltration (UF) membrane. Using a mass balances for the reacting components, it is possible to calculate theoretically the optimal operating points for the EMR. Prerequisites are the kinetic models for the enzymes LEUDH and FDH.

At an optimal ratio of the enzyme activities most of the co-enzyme is oxidized. This was verified in a batch experiment (Fig. 1A). Up until almost complete substrate conversion, the L-leucine production rate was constant and most of the coenzyme was oxidized. After almost all of the substrate was consumed, all of the oxidized coenzyme was reduced in a short time because of a high excess of formate. Fig. 1B shows $d(NADH)/dt$ as a function of the fraction of NADH concentration for each of the times shown in Fig. 1A. The shape of the curve is influenced mainly by the total coenzyme concentration and the enzyme activities. As this curve can also be calculated from the mass balances, the results of a batch experiment can be used to determine the total coenzyme concentration and enzyme activities from an optimizing program.

The reaction mixture of a continuously operated EMR was pumped through a photometer cuvette and back to the reactor. The absorbance of the reaction mixture of the EMR was measured in comparison to the product solution to determine the NADH concentra-

tion in the EMR. The reactor arrangement then was modified, by
changing the setting of a 4-way valve so that the cuvette was no
longer part of the EMR and thus became an independent batch reactor. The formation of NADH was now observed; and the rate of NADH
formation as a function of NADH concentration was obtained. After
all active coenzyme was converted to NADH, the setting of the 4-
way valve was switched back again; and the content of the cuvette
was mixed with the content of the EMR. The cuvette again became
a part of the EMR.

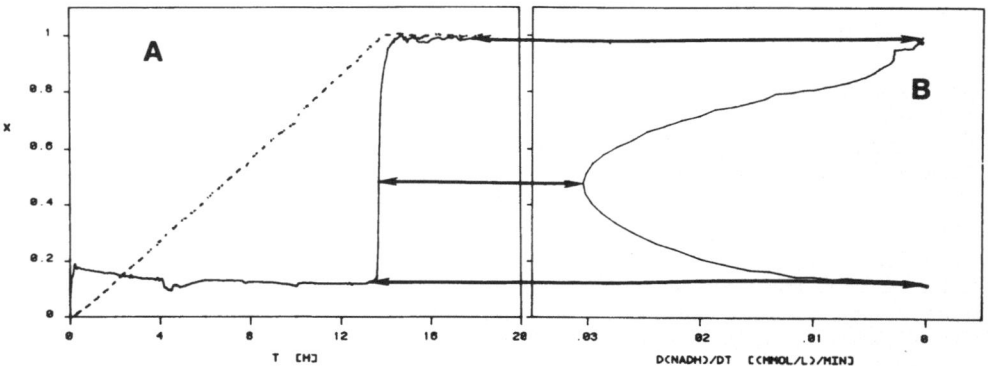

Fig. 1.   Enzymatic L-leucine production in a batch experiment.
          Initial substrate concentrations: α-ketoisocaproate
          100 mmol/L; NH_4-formate 400 mmol/L. (———) Fraction
          of reduced coenzyme, (---) fraction of L-leucine as a
          function of time, A, and rate of NADH formation, B.

REFERENCES

1. WANDREY, C., WICHMANN, R. & JANDEL, A.-S., this volume.

A NEW APPROACH TO MEMBRANE REACTOR DESIGN AND OPERATION

E. Flaschel, E. Raetz and A. Renken

Institute of Chemical Engineering
Swiss Federal Institute of Technology
Lausanne, Switzerland

Enzymes in their native state can be used continuously by means of physical immobilization in an ultrafiltration unit. The aim of this study is to design a tubular recycle membrane reactor (TRMR) that approximates plug flow operation and uses few movable parts and requires a small membrane area.

The TRMR is shown in Fig. 1. The sterilized substrate solution is passed by gravity through the tubular reactor and then pumped to the separation unit. A portion equivalent to the

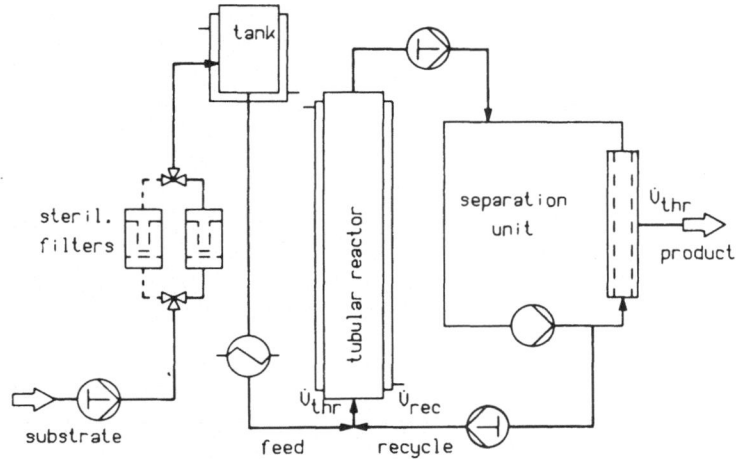

Fig. 1.   Tubular recycle membrane reactor.

313

throughput is forced across the membrane while another pump re-
cycles the concentrated enzyme solution back to the bottom of the
tubular reactor. Optimizing this concept means in the first approximation to reduce backmixing in the tubular reactor. This
can be achieved by introducing special motionless mixers (e.g.
SULZER type SMX or SMV) into the tube. The number of equivalent
stirred tanks is in the range of 35 to 40 per meter tube length
for a reactor of 4 cm inner diameter. The productivity of the TRMR
is very sensitive to changes in the recycle ratio (recycle rate/
throughput). If the recycle rate is kept very low or very high,
a simple continuous stirred tank reactor (CSTR) results. With intermediate recycle ratios, an optimum productivity is found, as
shown in Fig. 2, for lactose hydrolysis by lactase from *Asp. niger*
(Rapidase, 150 mmol/L lactose, pH 3.5, 50 $^{\circ}$C), the productivity
ratio CSTR/TRMR is given for 75% lactose conversion and an overall
enzyme concentration of 5 g/L. This ratio does not tend to unity
at very low recycle ratios due to the kinetics of lactase, which
are nonlinear with respect to enzyme concentration. Fig. 2 also
demonstrates the influence of the relative volume of the separation
unit (volume of sep. unit/total TRMR-volume). A low ratio is favorable, since most of the enzyme will then be present in the
tubular part of the reactor, where it reacts more efficiently.

    The authors thank the board of the Swiss Federal Institutes
of Technology for financial support.

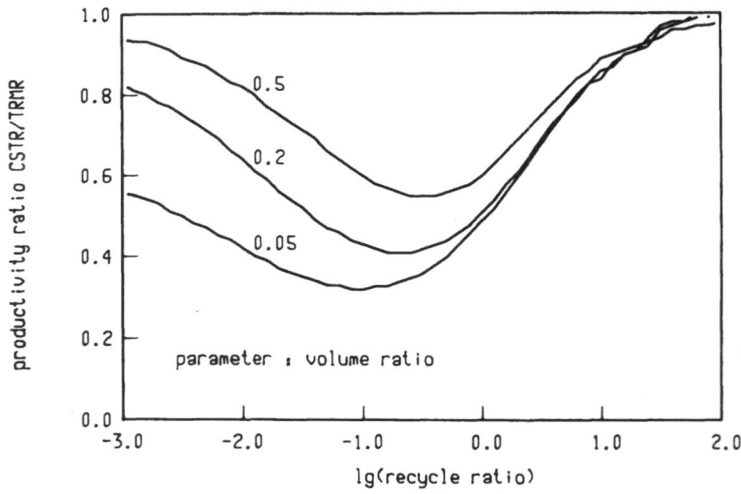

Fig. 2. Productivity ratio of the membrane reactor. See text for
         definitions.

INFLUENCE OF COMPACTION IN GEL-IMMOBILIZED ENZYME PACKED BED REACTORS

S. Furusaki, Y. Okamura and T. Miyauchi

Department of Chemical Engineering
University of Tokyo
Tokyo, Japan

Enzymes have been immobilized mostly on organic carriers, which may suffer from compaction when utilized in packed columns. Although packed columns are the most effective reactors, due to ideal flow characteristics, contact will be decreased significantly by compaction. Therefore, it is important to know the effect of compaction on enzymatic packed bed reactor design. The salient effect of compaction on conversion or on selectivity has been pointed out in practical use; but it has never been investigated from an engineering viewpoint.

The reaction investigated was the hydrolysis of sucrose by immobilized invertase. The enzyme was immobilized in polyacrylamide gel (1,2) of 1 mm particle diameter and narrow size distribution. Compaction was considered to be caused by shear stress due to the flow of liquid. Variation of the void fraction was obtained as a function of the liquid velocity (2). Hysteresis was observed with increasing or decreasing velocity. The reaction was carried out using the same packed column for investigating the variation of void fraction. Details of the experimental apparatus and procedures are shown elsewhere. The apparent reaction rate decreased with decrease in void fraction to less than half in extreme cases. Thus, compaction was shown to decrease the apparent reation rate.

The theoretical specific surface area of the particles was calculated as per the referenced theory (3) for spherical particles for Michaelis-Menton reaction kinetics. The results (2) showed that the specific surface decreased significantly with the decrease of void faction.

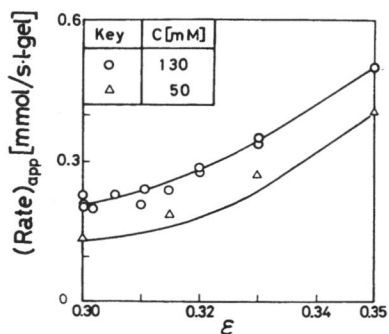

Fig. 1.   Change in apparent reaction rate with void fraction.

REFERENCES

1. NILSSON, H., MOSBACH, R. & MOSBACH, K. *Biochim. Biophys.*
    Acta 268:   253 (1972).
2. FURUSAKI, S., OKAMURA, Y. & MIYAUCHI, T. *J. Chem. Eng. Japan*,
    15:   148 (1982).
3. MOO-YOUNG, M. & KOBAYASHI, T. *Can. J. Chem. Eng. 50:*   162,
    (1972).

Session V
BIOMASS CONVERSION WITH ENERGY PRODUCTION
Chairmen: W. Vieth and S. Suzuki

ENZYMATIC REMOVAL OF HAZARDOUS POLLUTANTS FROM INDUSTRIAL AQUEOUS EFFLUENTS

A. M. Klibanov

Department of Nutrition and Food Science
Massachusetts Institute of Technology
Cambridge, Massachusetts, USA

This work deals with two different approaches to enzymatic waste water purification: a) peroxidase assisted removal of toxic phenols, aromatic amines, and some other organics from waste waters, and b) immobilized hydrogenase-catalyzed detritiation of water.

REMOVAL OF TOXIC POLLUTANTS FROM INDUSTRIAL WASTE WATERS

Various phenols and aromatic amines are inherent in waste waters of a number of industries, such as coal conversion, petroleum refining, resins and plastics, textiles, dyes and organic chemicals, timber, soaps and detergents, paving and roofing, iron and steel, and ore mining and dressing. Nearly all phenols and aromatic amines are toxic. Furthermore, some of them have been determined to be human carcinogens and mutagens (e.g. benzidine and its derivatives, naphthylamines, and aminoazobenzines); and many others are in the OSHA list of suspected carcinogens (e.g. aniline and its derivatives, 2,4,6-trichlorophenol, and 4-amino-2-nitrophenol). Numerous phenols and aromatic amines have been declared "consent decree priority pollutants" (e.g. phenol, 2,4-dimethylphenol, 2-chlorophenol, and nitrophenols). Therefore, removal of phenols and aromatic amines from industrial aqueous effluents is of great practical significance.

Existing methods for removal of these chemicals from industrial waste waters include microbial and chemical oxidation, adsorption on activated carbon, extraction, incineration, electrochemical techniques, and irradiation. These methods, although certainly feasible and useful, suffer from the serious shortcom-

ings of high cost, incompleteness of purification, formation of
hazardous by-products, and relatively low efficiency. Hence
alternative methods for advanced water treatment are highly desirable.

We have developed a novel, enzymatic approach to the removal
of phenols and aromatic amines from industrial waste waters (1-3).
It includes the treatment of aqueous solutions containing the
pollutants with horseradish peroxidase and hydrogen peroxide.
Horseradish peroxidase catalyzes the oxidation of a number of
phenols and aromatic amines by H_2O_2. During this oxidation, the
corresponding phenolic and aromatic amine free radicals are generated, which then polymerize to polyaromatic products. In contrast to their monomeric precursors, these polymers are nearly
insoluble in water and can be easily separated by filtration or
sedimentation. Therefore, peroxidase-catalyzed oxidation in fact
precipitates the pollutants from aqueous solution due to their
enzymatic crosslinking.

This method has been used by us to remove over thirty different phenols and aromatic amines from water (Tables 1 and 2).
The dependence of the removal efficiency on the concentrations of
enzyme, H_2O_2, and pollutant and also on the Ph and duration of
treatment have been investigated. For some pollutants the efficiency of enzymatic removal after a 3 hr treatment exceeded 99%,
whereas for others it was significantly lower. It should be
pointed out that treatment with either peroxidase or H_2O_2 alone
resulted in no appreciable pollutant removal.

We discovered that easily removed phenols and aromatic amines
greatly enhanced the enzymatic precipitation of those that had
low removal efficiency. For example, at pH 5.5 the efficiency of
the enzymatic removal of 100 mg/L phenol was 74.6% (3 hr treatment, 2.5 mM H_2O_2, 1000 units/L peroxidase). However, in the
presence of 100 mg/L 3,3'-dimethoxybenzidine, benzidine, or 8-
hydroxyquinoline, the removal efficiency for phenol was 99.7%,
99.5%, and 99.8%, respectively (under the same conditions).
Apparently, free radicals enzymatically produced from easily removed substrates attacked and precipitated the phenols and ani-
lines that could not be easily removed. The same mechanism
appeared to work with non-phenolic and non-amine compounds. Thus
we found that naphthalene and azobenzene (both hazardous pollutants which do not react with peroxidase at all) could be precipitated in the presence of 2,3-dimethylphenol or 8-hydroxyquinoline.
The aforementioned findings are of great practical importance for
waste water detoxification, since real industrial aqueous effluents always contain many different pollutants and therefore
even if only a few of them are easily precipitated by peroxidase,
those few will facilitate the removal of others.

TABLE 1

REMOVAL OF AROMATIC AMINES FROM WATER BY PEROXIDASE

Pollutant	Removal Efficiency* (%)	Optimal pH
Benzidine	99.9	5.5
3,3'-Dimethoxybenzidine	99.9	5.5
3,3'-Dichlorobenzidine	>99.9	5.5
3,3'-Diaminobenzidine	99.6	5.5
3,3'-Dimethylbenzidine	99.6	5.5
1-Naphthylamine	99.7	5.5
2-Naphthylamine	98.3	7.0
5-Nitro-1-naphthylamine	99.6	4.5
4-Aminobiphenyl	95.4	7.0
Aniline	72.9	7.0
4-Chloroaniline	62.5	5.5
4-Bromoaniline	84.5	5.5
4-Fluoroaniline	86.4	7.0
4-Bromo-2-methylaniline	84.5	7.0
4,4'-Methylenedianiline	88.9	7.0
p-Phenylazoaniline	98.5	4.0
4'-Amino-2,3'-dimethylazobenzene	95.5	7.0
m-Phenylenediamine	98.6	7.0

*Conditions: 100 ppm aqueous solution of pollutant, 3 hr treatment at room temperature, 1 mM H_2O_2, 100 units/L peroxidase (1,2).

We have successfully tested our enzymatic purification method using a real industrial waste water sample obtained from a chemical plant. Treatment of this sample (containing over a hundred different phenols) at $4^{\circ}C$ (bacteria cannot be employed at such a low temperature) for 24 hr (2.5 mM H_2O_2, 100 units/L peroxidase) resulted in 96% efficiency of the total phenol removed. The treatment could be carried out equally efficiently with either purified or crude preparations of the enzyme. Preliminary assessments indicate that the enzymatic detoxification process described above can be economically feasible.

TABLE 2

REMOVAL OF PHENOLS FROM WATER BY PEROXIDASE

Pollutant	Removal Efficiency* (%)	Optimal pH
Phenol	85.3	3.5
o-Methoxyphenol	98.0	5.5
M-Methoxyphenol	98.6	5.5
p-Methoxyphenol	89.1	7.0
o-Cresol	86.2	4.0
m-Cresol	95.3	4.0
p-Cresol	85.0	5.5
o-Chlorophenol	99.8	7.0
m-Chlorophenol	66.9	7.0
p-Chlorophenol	98.7	5.5
Resorcinol	84.1	3.5
5-Methylresorcinol	90.8	3.5
1-Naphthol	99.6	4.0
2-Nitroso-1-naphthol	98.9	4.0
p-Phenylphenol	99.9	4.0
8-Hydroxyquinoline	99.8	7.0

*Conditions: 100 ppm aqueous solution of pollutant, 3 hr treatment at room temperature, 1 mM H_2O_2, 100 units/L peroxidase (1,2).

IMMOBILIZED HYDROGENASE FOR THE DETRITIATION OF WATER

Tritiated water is produced in nuclear power plants on neutron capture by deuterium and on uranium fission. Since tritium is radioactive, its isolation from these sources is necessary to prevent the release of tritium into the environment and for nuclear reactor personnel safety. The concentrated tritium produced as a result of the detritiation of contaminated water can be used for synthesis of labeled biochemicals for various biomedical applications. Tritium isolation on a much greater scale will be needed in the future when fusion reactors are developed.

The most promising methods for detritiation of water are based on catalyzed hydrogen isotope exchange reactions. The first step in such methods is electrolysis of contaminated water. Inas-

much as the rate of electrolysis of THO is considerably lower than
tat of H_2O, the water in the electrolysis cell gradually becomes
enriched with tritium. However, since the electrolytic separation
is by no means complete, a step is required in which the escaped
tritium in the hydrogen gas stream can be recovered. This is
accomplished in an exchanger which catalyzes the reaction HT +
H_2O H_2 + HTO. Thetritiated water produced here is then returned
to the electrolysis cell.

Nobel metals, in particular platinum, have been traditionally employed as catalysts in the aforementioned hydrogen-water
isotope exchange. Unfortunately, such catalysts suffer from a)
very high cost, b) limited availability of platinum, c) relatively
low catalytic activity in liquid water, and d) significant sus-
ceptibility to poisons. Therefore alternative catalysts for the
exchange reaction are highly desirable.

In this work (4) we replaced the platinum catalysts with im-
mcbilized hydrogenase. Whole bacterial cells of *Alcaligenes
eutrophus* immobilized in calcium alginate or κ-carrageenan gels
were found to be efficient catalysts for the hydrogen-tritium exchange in both batch tank and column reactors. The dependence of
the reaction rate on the amount of immobilized cells and on the
concentration of cells in the matrix indicated that enzymatic H-T
exchange was not controlled by diffusion. Immobilized *A.
eutrophus* cells were enzymatically active over a wide range of
pH (with a broad maximum from pH 6.0 to 8.0) and temperatures;
they were quite resistant to inhibitors of hydrogenases, such as
O_2 and CO. From the standpoint of catalytic efficiency, 1 g of
PtO_2 was approximately equivalent to 10 g of cells (wet weight).
In contrast to platinum-based catalysts, bacterial hydrogenases
a) are potentially inexpensive, b) can be readily available in
bulk quantities and c) are maximally active in liquid water.

ACKNOWLEDGMENT

This work was supported in part by grants from the U.S.
Environmental Protection Agency and from the National Oceanic and
Atmospheric Administration.

REFERENCES

1. KLIBANOV, A. M. & MORRIS, E. D. *Enzyme Microb. Technol.*
      119 (1981).
2. KLIBANOV, A. M., ALBERTI, B. N., MORRIS, E. D. & FELSHIN,
      L. M. *J. Appl. Biochem. 2:* 414 (1980).
3. ALBERTI, B. N. & KLIBANOV, A. M. *Biotechnol. Bioeng. Symp.*,
      11: 373 (1981).

4. KLIBANOV, A. M. & HUBER, J. *Biotechnol. Bioeng.* *23:* 1537
    (1981).

ENZYMATIC HYDROLYSIS OF CELLULOSE: EFFECTS OF STRUCTURAL PROPERTIES OF CELLULOSE ON HYDROLYSIS KINETICS

D. D. Y. Ryu and S. B. Lee

The Korea Advanced Institute of Science
Seoul, Korea (See Appendix for current address)

The effects of changes in structural properties of cellulose during enzymatic hydrolysis have been studied in relation to cellulose reaction kinetics. The crystallinity and specific surface

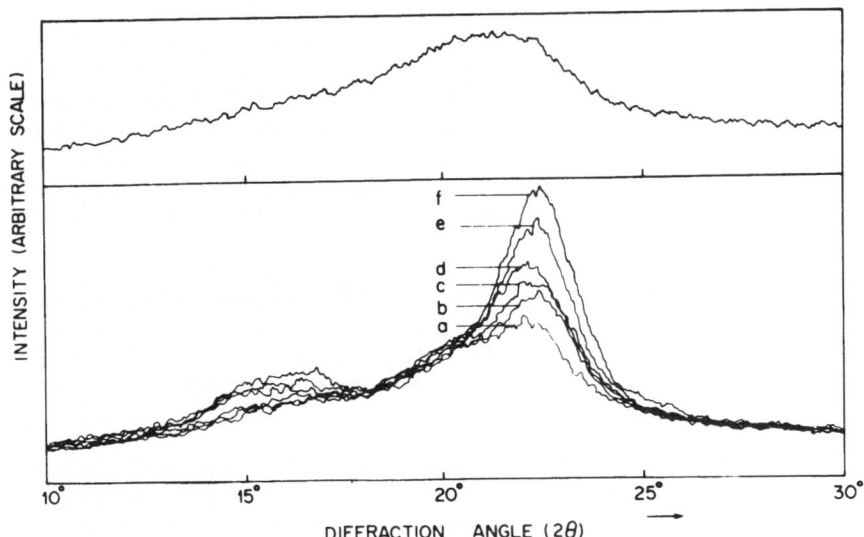

Fig. 1. X-ray diffractograms of Sweco 270 during hydrolysis reaction. Upper diffractogram shows vacuum dried cellulose as a control. a) diffractogram of a sample taken at reaction time t = 0; b) sample taken after 3 hr hydrolysis; c) t = 6 hr; d) t = 12 hr; e) t = 24 hr; f) t = 36 hr.

area of cellulose, the adsorption of cellulase on the solid cellulose substrate, and the degree of polymerization were found to be
important parameters that govern the enzymatic hydrolysis of cellulose. These parameters and their changing patterns during hydrolysis vary also with the source of cellulose and the pretreatment
methods. A predictive kinetic model is proposed, based on experimental evidence and a simplified reaction scheme for the enzymatic
hydrolysis of a well defined cellulose substrate system. The simulations show reasonably good agreement with experimental results
(1, 2).

For the purpose of developing a practical technology for the
enzymatic hydrolysis of cellulose and effective utilization of this
renewable carbon source, a further understanding of the basic reaction mechanisms and the cellulose-cellulase interactions involved
in the hydrolysis reaction is of primary importance (3).

The crystallinity of partially crystalline cellulose increases
as the hydrolysis reaction proceeds; and a significant slowing down
of the reaction rate during the enzymatic hydrolysis is, in large
part, attributable to this change of relative crystallinity of the
cellulose substrate (Fig. 1 and 2). X-ray diffractograms of cellulose samples taken at given time intervals during enzymatic hydrolysis were superimposed to determine whether there was any
preferential attack on certain lattice planes by the cellulase
(Fig. 1). It was found that the changes in crystallinity during
hydrolysis were closely related with the initial state of the
cellulose structure (pretreatment) and that cellulase preferentially attacks the 002 (for cellulose I) and 10$\bar{1}$ (for cellulose
II) crystal planes. These results indicate that the crystallite
width decreases while the crystallite length remains practically
constant during enzymatic hydrolysis.

The crystallinity index of Sweco 270 cellulose increased from
34.9 to 71.6 during a 48 hr hydrolysis (Fig. 2). The increase in
the crystallinity index is good evidence that the amorphous portion
of the cellulose was more readily and quickly hydrolyzed than was
the crystalline portion (4,5).

Fig. 2 also shows the changes in specific surface area and
cellulase adsorption during hydrolysis of Sweco 270 cellulose.
During the initial 6 hr, the specific surface area decreased
rapidly from 72.1 to 19.1 m^2/g cellulose. A high specific surface
area at time zero indicates that the amorphous cellulose was swol-
len greatly by water action. Rapid hydrolysis and removal of the
amorphous portion of Sweco 270 cellulose, which was confirmed by
the increase in the crystallinity index and rapid sugar production,
caused a rapid decrease in the specific surface area. As the rate
of hydrolysis slowed down further, the changes in specific surface
area became insignificant. This result reveals that the specific

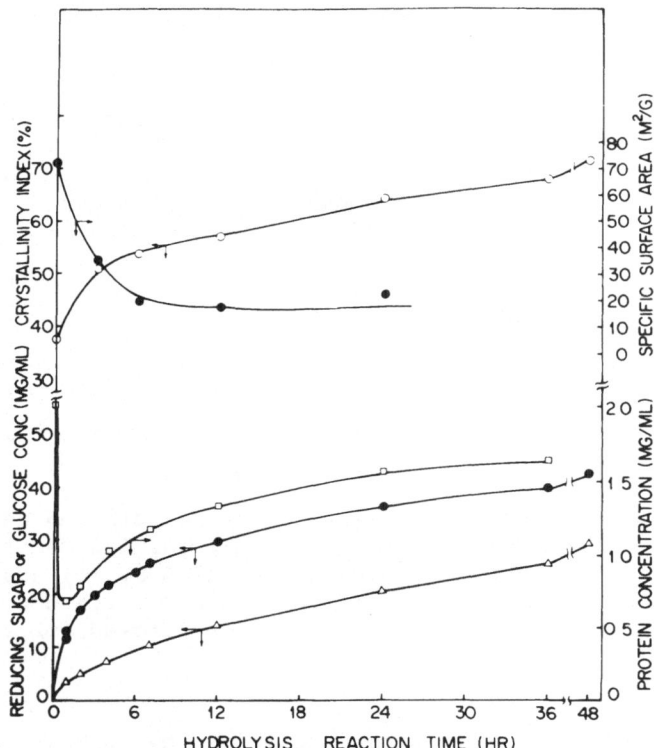

Fig. 2. Changes in structural parameters during hydrolysis of
         Sweco 270. For upper figure: (o) crystallinity index,
         (●) specific surface area. Lower figure: (□) concen-
         tration of cellulase protein in reaction solution, (●)
         reducing sugar concentration, (Δ) glucose concentration.

surface area did not change significantly as the hydrolysis reaction proceeded, when the crystallinity of the residual cellulose
substrate remained high.

     The adsorption of cellulase usually parallels the rate of
cellulose hydrolysis (Fig. 3). In Fig. 2, it was also shown that
initially the cellulase adsorption onto Sweco 270 cellulose increased sharply; but the cellulase protein was gradually released
as the hydrolysis reaction proceeded. The specific surface area
decreased due to a significant lowering in the amorphous fraction
of cellulose.

     The pattern of the adsorption profile of cellulase during the
hydrolysis reaction varied with the source of cellulose and with
the pretreatment methods. As shown in Fig. 4, two distinctly dif-

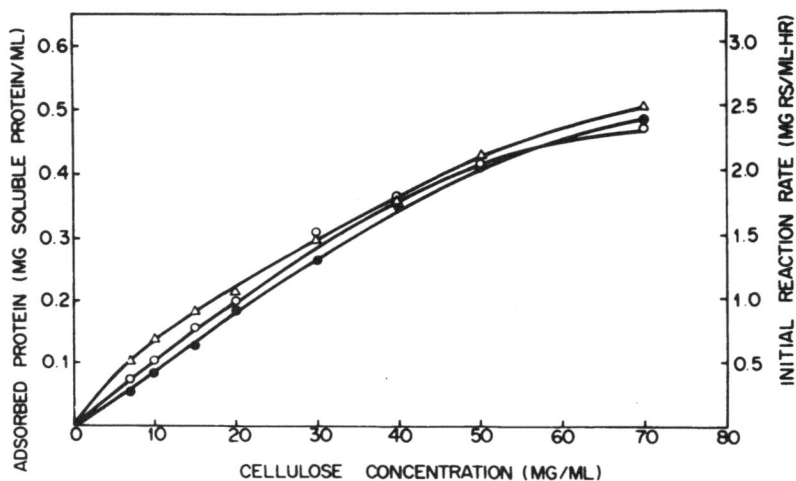

Fig. 3. Michaelis curves of cellulose hydrolysis and adsprption
        of cellulase enzyme protein. (Δ) initial reaction rate,
        (●) adsorbed cellulase protein at 50°C, (o) adsorbed
        cellulase at 4°C. Total cellulase protein concentration
        used was 0.67 mg/ml; and the time allowed for adsorption
        was 30 min.

ferent adsorption patterns during cellulose hydrolysis could be
observed. One was a continuous increase in adsorption of cellulase
protein (cotton and NEP 40); and the other was a gradual release
of enzyme protein from cellulose (pretreated Avicel and BW-200).
It was found that the cellulase adsorption continued throughout
the hydrolysis period if the initial adsorption was restricted
or hindered by inaccessibility of substrates. When the cellulase
was adsorbed to its maximum at the beginning of hydrolysis for pretreated cellulose, the cellulase protein was gradually released
as the crystalline and inaccessible fractions of the cellulose increased.

     The changes in the degree of polymerization (DP) during hy-
drolysis were examined for two cellulose samples: untreated Solka
Floc SW-40 and 85% phosphoric acid treated Sokla Floc SW-40 (Table
1). For untreated cellulose, little change in DP was observed;
although the conversion of cellulose to reducing sugar was as high
as 46.1% in 60 hr. On the other hand, the DP of the regenerated
cellulose decreased dramatically during the hydrolysis, while
achieving 59.1% conversion in 1.5 hr. Both cellulose samples had
practically the same initial DP. These results suggest that the
enzymatic hydrolysis reaction of relatively crystalline cellulose
takes place mainly on the surface of crystallites, and results in
only a small change in the DP of the substrate cellulose. When

TABLE 1

CHANGES IN DEGREE OF POLYMERIZATION DURING HYDROLYSIS

Hydrolysis Time (hr)	DP	Conversion* (%)

A. Untreated Solka Floc SW-40 (native cellulose)

0	1210	0
1	1180	7.5
4	1250	15.2
25	1200	35.6
36	1100	39.3
48	1190	43.1
60	1030	46.1

B. 85% H_3PO_4 treated Solka Floc SW-40 (regenerated cellulose)

0**	1080	0
15**	291	25.8
30**	178	39.6
60**	142	52.4
90**	104	59.1

*Initial cellulose concentration, 5% (wt) for A and 2% (wt) for
 B; E_o/S_o = 20 FPU/g cellulose for A and B.
**Time in min.

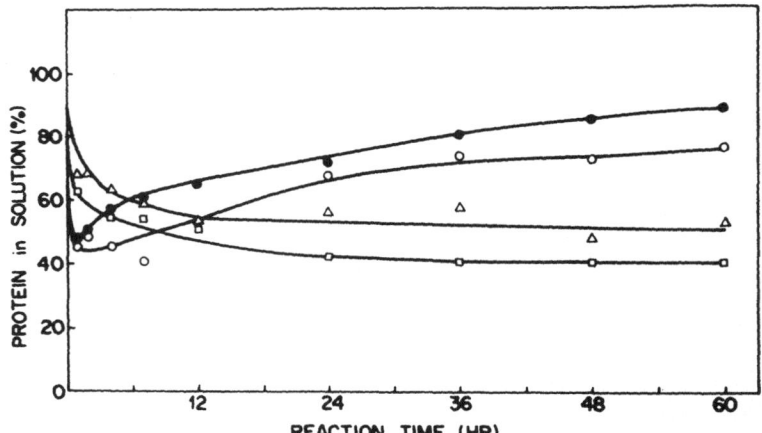

Fig. 4.   Adsorption pattern during hydrolysis reaction: (●) avicel, (o) Solka Floc BW-200, (Δ) NEP-40, (□) adsorbent
cotton.   The cellulose concentration (S_o) and amount of
enzyme used (E_o/S_o) were 20 mg/ml and 20 filter paper
unit/g cellulose, respectively.

the penetration of cellulase molecules into the interior of cellulose fibers becomes easy, by means of pretreatment or swelling,
the DP is lowered significantly for an equivalent level of sugar
yield.

Based on experimental evidence and current views on the
structural properties of cellulosic materials, a kinetic model for
the enzymatic hydrolysis of cellulose is proposed as follows:

$$E \underset{k_d}{\overset{k_{ad}}{\rightleftharpoons}} E^* \qquad\qquad (Eq. 1)$$

$$E^* + S_a \underset{k_{2a}}{\overset{k_{1a}}{\rightleftharpoons}} E^* S_a \xrightarrow{k_{3a}} E^* + P \qquad (Eq. 2)$$

$$E^* + S_c \underset{k_{2c}}{\overset{k_{1c}}{\rightleftharpoons}} E^* S_c \xrightarrow{k_{3c}} E^* + P \qquad (Eq. 3)$$

$$E^* + S_x \underset{k_{2x}}{\overset{k_{1x}}{\rightleftharpoons}} E^* S_x \tag{Eq. 4}$$

$$E + P \underset{k'_{ip}}{\overset{k_{ip}}{\rightleftharpoons}} E P \tag{Eq. 5}$$

Usually, most cellulosic materials (S_o) contain both amorphous (S_a) and crystalline (S_c) fractions as well as non-hydrolyzable inert materials (S_x), such as lignin. The first reaction step describes adsorption of cellulase on the cellulose surface (Eq. 1). E and E* denotes the cellulase in the free state and in the adsorbed state, respectively. The adsorbed cellulase complex simultaneously hydrolyzes both the amorphous and crystalline celluloses at different rates (Eqs. 2 and 3). The inert fraction forms an inactive complex when combined with cellulase (Eq. 4). The hydrolysis product inhibits cellulase competitively (Eq. 5).

From the proposed reaction scheme a rate expression (Eq. 6) can be derived in terms of the kinetic parameters shown below.

$$-\frac{dS}{dt} = \frac{\left[V_{max,a} (1-\phi)(1-\gamma) + V_{max,c} \phi(1-\gamma)(K_a/K_c) \right] S}{\left[K_a(1 + P/K_{ip}) + S \left[(1-\phi)(1-\gamma) + (K_a/K_c)\phi(1-\gamma) + \overline{(K_a/K_x)\gamma} \right] \right]} \tag{Eq. 6}$$

where, $V_{max,a} = k_{3a} E_o$; $V_{max,c} = K_{3c} E_o$; $K_D = (E^*/E)$

$$K_a = K_a^* (1 + K_D)/K_D \; ; \qquad K_a^* = (k_{2a} + k_{3a})/k_{1a}$$

$$K_c = K_c^* (1 + K_D)/K_D \; ; \qquad K_c^* = (k_{2c} + k_{3c})/K_{1c}$$

$$K_x = K_x^* (1 + K_D)/K_D \; ; \qquad K_x^* = k_{2x}/k_{1x} \; ; \quad \phi = S_c/S_H$$

$$K_{ip} = K_{ip}^* (1 + K_D)/K_D \; ; \qquad K_{ip}^* = k'_{ip}/k_{ip} \; ; \quad \gamma = S_x/S$$

$$S = S_H + S_x = S_a + S_c + S_x$$

Also, ϕ represents the fraction of crystalline phase, while γ represents the fraction of non-hydrolyzables. α_1 is assumed to be 1.05.

   Although the proposed reaction scheme appears to be very simplified, it is a reasonably good predictive kinetic model for a well defined substrate system where the most important structural properties are understood and known.

   The following two simultaneous differential equations (Eq. 7 and 8) can be derived by rearrangement of Eq. 6; and they can be used to predict the fractional conversion of well defined cellulose substrate as a function of hydrolysis reaction time.

$$\frac{dX_a}{dt} = \frac{V_{max,a}\,(1 - X_a)}{K_a(1+\alpha_1 S_o X/K_{ip}) + S_o\left[(1 - \phi)(1 - X_a) + (K_a/K_c)\,\phi\,(1 - X_c)\right]} \tag{Eq. 7}$$

$$\frac{dX_c}{dt} = \frac{V_{max,c}\,(1 - X_c)\,(K_a/K_c)}{K_a(1+\alpha_1 S_o X/K_{ip}) + S_o\left[(1 - \phi)(1 - X_a) + (K_a/K_c)\,\phi\,(1 - X_c)\right]} \tag{Eq. 8}$$

where, $X = (1 - \phi)\,X_a + \phi X_c$

The simulation showed good agreement with the experimental results (Fig. 5).

   The results and discussion presented here represent only a small part of the overall problems and complexities involved in the enzymatic hydrolysis of cellulose. Indeed, there is a great deal more to learn before we are ready to find a practival biotechnology for effective utilization of this renewable carbon source (6). However, an attempt is made here to show a small example of how we might systematically study the complex problem and begin to understand the fundamentals of the cellulose-cellulase system.

REFERENCES

1. RYU, D. D. Y., LEE, S. B., TASSINARI, T., & MACY, C. *Biotechnol. Bioeng.*, in press.

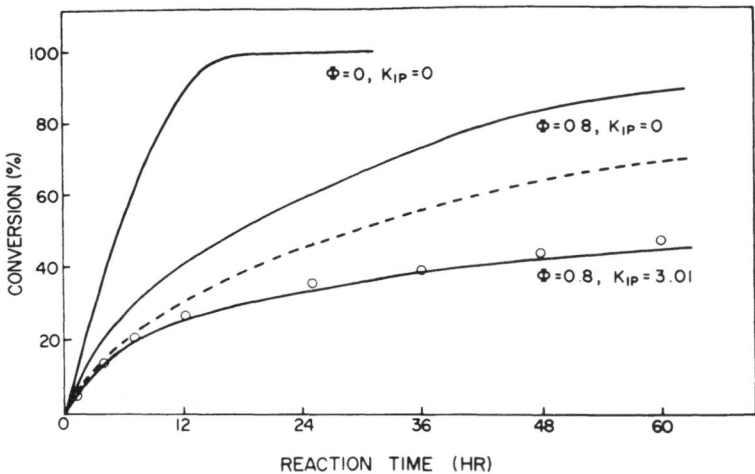

Fig. 5. Conversion profile based on the proposed kinetic model. Parameters used in simulation: S_o = 50 g/L; $V_{max,a}$ = 5 g/L/hr; K_a = 7 g/L; K_c = 70 g/L. When ϕ = 0.8 and K_{ip} = 3 mg/ml, the theoretical curve (solid line) agrees very well with the experimental data (o) for hydrolysis of Solka Floc SW-40. Ratio of enzyme/substrate, E_o/S_o = 20 filter paper unit/g cellulose. Dashed line shows the theoretical curve for simple Michaelis-Menten kinetics when ϕ = 0.8 and K_{ip} = 3 mg/ml. Other solid lines show theoretical curves for ϕ and K_{ip} as indicated.

2. RYU, D. D. Y., ANDREOTTI, R., MEDEIROS, J., & MANDELS, M in "Enzyme Engineering", vol. 5 (H. Weetall and G. Royer, eds.) Plenum, New York (1980) p. 33.

3. MANDELS, M., ANDREOTTI, R., & ROCHE, C. *Biotechnol. Bioeng. Symp. 6:* 21 (1976).

4. COWLING, E. B. & KIRK, T. K. *Biotechnol. Bioeng. Symp. 6:* 95 (1976).

5. CAULFIELD, D. F. & MOORE, W. E. *Wood Sci. 6:* 375 (1974).

6. REESE, E. T. *Recent Adv. Phytochem. 11:* 311 (1977).

CONTINUOUS ETHANOL FERMENTATION BY IMMOBILIZED BIOCATALYSTS

P. Linko and Y. Y. Linko

Helsinki University of Technology, Department
of Chemistry, Espoo, Finland

Various renewable carbohydrate sources recently have been
looked at as alternate feedstocks for the chemical industry and
as a source both of energy and liquid fuel. Immobilized biocatalyst engineering, in particular the development of continuous
heterogeneous biocatalysis techniques involving immobilized living
microbial cells, has opened up entirely new possibilities towards
more economic biotechnical conversions.

Living yeast cells able to carry out ethanol fermentation
while attached to an insoluble carrier were first suggested for
alcoholic beverage manufacture in a patent assigned to Intermag-
Getranke Industri A.G. in 1969 (1,2). Similar techniques developed
later both for brewing and for ethanol production have been reviewed recently (3,4); and the patents granted reflect the developments. Divies (5) showed that yeast cells entrapped in polyacrylamide gel could ferment glucose to ethanol. In a patent assigned
to the United States Department of Energy, Griffith and Compere
(6) described a biocatalyst column reactor based on yeast grown
on conventional distillation column packing in the presence of a
polyelectrolyte and a crosslinking agent. In another patent
assigned to Sanraku-Ocean Co., Ltd Hino *et al*. (7) described the
entrapment of yeast in natural or synthetic hydrophilic gels,
preferably polyvinyl alcohol. Pioneering work towards large scale
applications of κ-carrageenan entrapped living yeast cell reactors has been done by Chibata and coworkers (8,9). It has also
been reported that higher production rates and productivities may
be obtained employing immobilized *Zymomonas mobilis* bacterium
cells (10-13), although problems involving biocatalyst reactor
stability and scale-up still remain to be solved.

We have employed calcium alginate gel bead entrapped *Saccharomyces cerevisiae* (3,4,14,15) and *Kluyveromyces fragilis* (16) yeast, and κ-carrageenan locust bean gum blend gel bead entrapped *Z. mobilis* (13) bacterium cells for ethanol fermentation of fermentable sugars from a number of sources. The present paper describes some of our more recent experiences.

MATERIALS AND METHODS

Commercial *Saccharomyces cerevisiae* baker's and/or distiller's yeast obtained from Oy Alko Ab and selected strains of *Kluyveromyces fragilis* and *Zymomonas mobilis* were used in ethanol fermentations.

Sodium alginate, κ-carrageenan, locust bean gum, and gelatin from various sources were employed as carriers for living microbial cells; but most experiments were carried out with BDH sodium alginate (BDH Chemicals Ltd, Poole, England). Glucose, cane molasses, starch hydrolysates, electrodialytically demineralized spray-dried whey powder (DEMI), ultrafiltered whey permeate powder (UF-P), spruce sulfite spent liquor, barley malt wort, and grape juice were used as substrates. Thermomechanical pretreatment of starchy and cellulosic materials by high temperature short time (HTST) extrusion cooking for subsequent enzymatic hydrolysis also was used. Typically barley starch conditioned to 25% moisture was extruded at 150°C mass temperature, feed rate 175 g (d.s.)/min, screw revolution rate 50 rpm, and $\tau < 60$ sec. The product was hammermilled through 0.8 mm screen, and hydrolyzed as about 20% suspension for 24 hr with Novo 150L *Aspergillus niger* gluocoamylase at 60°C, pH 4.5 to yield 17% (w/v) DE 96 glucose syrup.

Yeast cells were entrapped in calcium alginate gel according to Linko *et al.* (17), with the exception of omitting crosslinking with glutaraldehyde. In most cases cells were suspended in 6-8% sodium alginate and extruded under pressure through hollow needles of 0.8 mm inner diameter into 0.5 M $CaCl_2$ to yield biocatalyst beads of about 2 mm diameter. For some experiments beads prepared with 2% alginate were hardened by partial drying. Yeast was able to grow and multiply in the gel matrix; the typical relatively stable biocatalyst cell count was 2 to 5 x 10^9 cells/g. Alginate bead entrapped *Z. mobilis* required the adition of $CaCl_2$ to substrate. Consequently, *Z. mobilis* cells were entrapped in a gel of a blend of κ-carrageenan and either locust bean gum or gelatin. Typically, 4 g of κ-carrageenan and 1 g of locust bean gum were dry-mixed and dissolved in 90 ml of 0.9% NaCl, cells suspended in the saline solution, and biocatalyst beads formed by extrusion of the 45°C suspension into 23°C 2% KCl.

Continuous ethanol fermentations were carried out in jacketed packed-bed glass column reactors varying in bed volume from 40-500 ml. Stainless steel wire nets on top and bottom prevented bed expansion. Substrate was fed from the bottom; and the rate was regulated by a peristaltic pump. Most fermentations were carried out at 25°C. The biocatalyst was stable for several weeks of continuous operation, with no breakage of beads due to CO_2 evolution. Ethanol was determined enzymatically by the Boehringer Mannheim GmbH Alcohol Test Combination.

RESULTS AND DISCUSSION

Most failures in early scale-up experiments of continuous ethanol fermentation with immobilized living cells have been associated with the rapid carbon dioxide evolution on one hand and the mechanical properties of the carrier gel on the other. Such problems could be largely eliminated by employing as high as 6-8% sodium alginate solution. The high viscosity caused no problems with the techniques used for biocatalyst bead preparation. It is well known that the chemical composition affects alginate gel characteristics. On the other hand, it could be shown that commercially available sodium alginate of widely varying price could be successfully used with comparable results.

Dilute glucose solutions, such as would be obtained from the hydrolysis of certain cellulosic waste materials, could be continuously processed with nearly quantitative ethanol yields at residence times as low as 1 hr or less. Fig. 1 illustrates the effect of residence time on product ethanol concentration during continuous fermentation of 4.5% (w/v) glucose with sufficient added nutrients to maintain stable production. When pure 10% (w/v) glucose was used as substrate, nearly quantitative conversion at 10

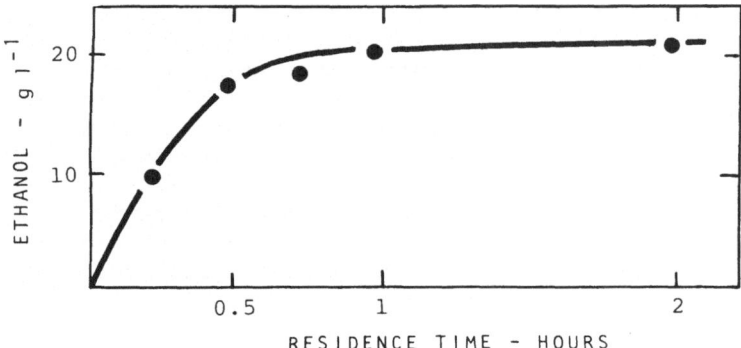

Fig. 1. The effect of residence time on continuous ethanol production from 4.5% (w/v) glucose with nutrients.

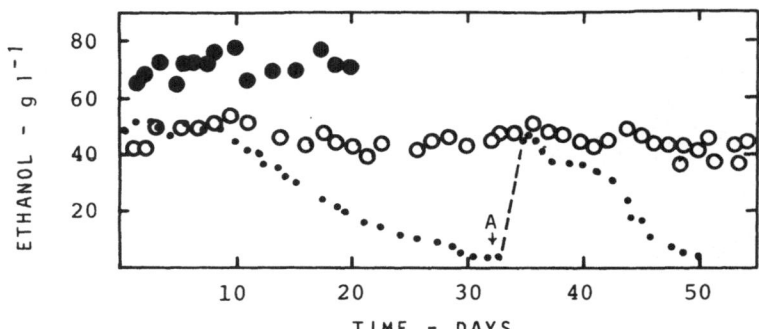

Fig. 2. Continuous ethanol production from grape juice (●, 14.8%
w/v sugar, τ ∿ 5 hr), barley malt wort (O, τ ∿ 2 hr), and
glucose (...., 10% w/v, τ ∿ 10 hr with nutrients added
at point A during 24 hr under aeration).

hr residence time could be maintained for about a week, followed
by a rapid decrease in bioreactor activity (Fig. 2). However,
nearly original biocatalyst activity could be regained by supply-
ing nutrients under aeration for about 24 hr. Fig. 2 also shows
that nutritious materials of similar glucose content, such as barley malt wort or grape juice, allowed stable continuous production
for several weeks at 2 to 5 hr residence times. Somewhat lower
ethanol yields were obtained with spruce sulfite spent liquor of
about 2% (w/v) sugar content (Fig. 3); but nevertheless nearly
10 g/L product ethanol level could be maintained at 1 hr residence
time. Under similar conditions birch sulfite spent liquor yielded
only 4 g/L of ethanol even at 7 hr residence time, owing to the
high pentose content. Corn glucose syrup fortified with necessary
nutrients is an excellent substrate, allowing stable continuous
high rate production (Fig. 3). Fig. 3 also shows results obtained
with barley starch glucose syrup of 17% (w/v) glucose, produced
from HTST-extrution pretreated starch with subsequent enzymatic
hydrolysis. This illustrates the possibilities in combining
efficient continuous pretreatment with continuous ethanol production. Starchy and cellulosic materials have also been extrusion
cooked under mild acid or alkaline conditions. Alternately, thermostable α-amylase could be added before extrusion of about 55-70%
moisture starch or ground whole grain at about 120-140°C mass temperature to obtain less viscous suspensions for rapid saccharification. Such slurries may also be employed in simultaneous saccharification and ethanol fermentation. Results from some of these experiments will be reported elsewhere.

Fig. 4 shows results from batch [4A] and continuous [4B]
ethanol fermentation experiments with immobilized *K. fragilis*
[4A,B]. Results also are shown for a bioreactor consisting of a

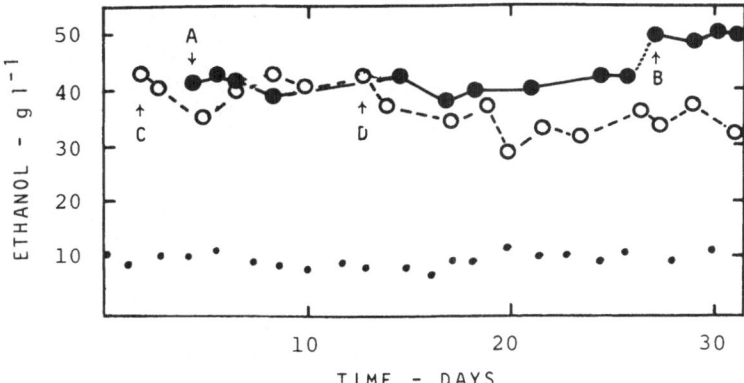

Fig. 3.   Continuous ethanol production from corn glucose syrup (●,
          from A 17% w/v glucose, τ ∿ 4.4 hr, from B 22% w/v
          glucose, τ ∿ 8 hr), extrusion cooking pretreated barley
          starch hydrolysate (O, 17% w/v glucose, from C τ ∿ 4.2
          hr, from D τ ∿ 3.2 hr), and spruce sulfite spent liquor
          (...., τ ∿ 1 hr).

50/50 mixture of calcium alginate gel entrapped *S. cerevisiae*
cells and Duolite S-761 phenol-formaldehyde resin adsorbed glutaraldehyde crosslinked *A. niger* β-*galactosidase* [4B]. Demineralized

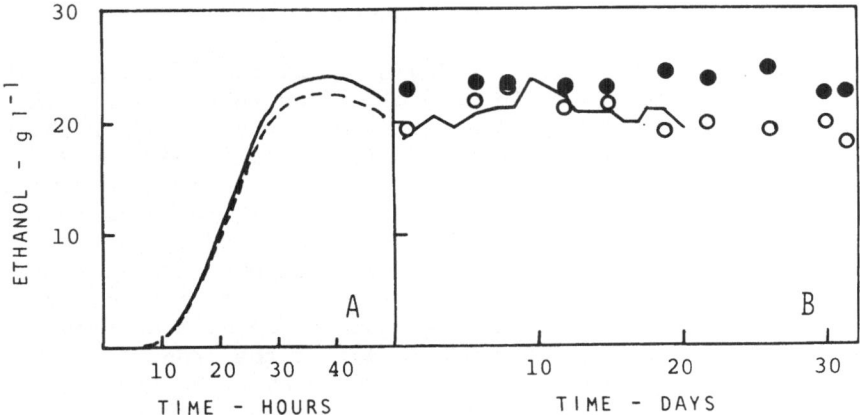

Fig. 4.   Batch (A) and continuous (B) ethanol production from
          whey ultrafiltration permeate (A, ----; B, ●) and de-
          mineralized whey (A, ———; B, O and ———) with immobilized
          K. fragilis, except (B, ———) with immobilized *S.
          cerevisiae* and β-galactosidase in same reactor.
          Lactose 5% w/v, τ 4 - 5 hr in continuous operation.

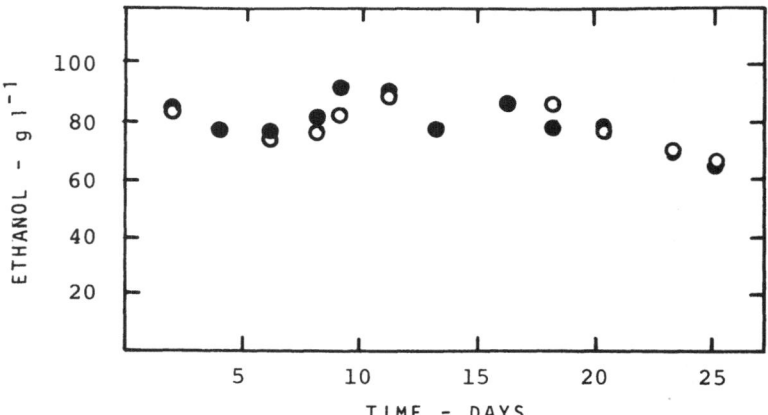

Fig. 5. Continuous ethanol production from cane molasses
         (17.5% w/v sugar), using fresh (o, τ ∿ 5 hr) or
         partially dried (●, τ ∿ 1.5 hr) calcium alginate gel
         bead entrapped *S. cerevisiae*.

whey (DEMI) and whey ultrafiltration permeate (UF-P) were used as
substrates. Whey is an abundantly available dairy industry by-
product well suited for alcoholic fermentations (15,16). Liquid
whey containing about 5% (w/v) of lactose yielded about 20 g/L of
ethanol. Nearly quantitative conversions were obtained at 4 - 5
hr residence times with both whey based materials employed in continuous fermentation for at least a month.

     Whey contains about 0.3 - 0.4 g/L of calcium. Consequently,
UF-P at 5% (w/v) lactose could be employed as substrate for the
calcium alginate gel entrapped yeast biocatalyst reactor without
problems. Substrate demineralization is normally too costly for
ethanol fermentations; but in other cases the biocatalyst may be
stabilized by the addition of 2 - 3 g/L of $CaCl_2$ to the substrate.
Alternately, the biocatalyst beads may be hardened by partial drying. The shrinking of beads during drying allows increased cell
densities and, thus, higher productivities. According to several
authors, typical mean productivities obtainable with continuous
packed-bed immobilized yeast reactors are of the order of 20 ± 5
g/L/hr, with the ethanol yield varying from about 70 to >90% of
theoretical. Fig. 5 illustrates laboratory scale trials carried
out with 17.5% (w/v) reducing sugar molasses as substrate; partial
drying of the biocatalyst beads allowed the reduction of the residence time (total bed volume basis) from 5 hr to about 1.5 hr with
an increase in productivity from about 20 to >40 g/L/hr at about
80% yield. This is of the same order of magnitude as reported by
Wada *et al.* (9) with κ-carrageenan entrapped yeast (τ = 2.6
hr, 95% yield), and Grote *et al.* (10 with calcium alginate or κ-

carrageenan entrapped Z. mobilis (τ = 1.18 - 1.25, 75 - 85%
yield). In our experiments with κ-carrageenan locust bean gum
entrapped Z. mobilis, the mean productivity from 10% (w/v) glucose was 40 g/L/hr (τ = 1 hr based on bed volume, 82% yield).
When the residence time was reduced to 20 min, productivity increased to 78 g/L/hr, with the ethanol yield decreasing to about
57%. Typically the relatively stable biocatalyst cell count was
$\sim 10^{10}$ cells/g; and the effluent cell count was 10^7 -10^8 cells/g,
sufficient to allow relatively rapid conversion after fermentation
of the remaining sugar.

Unfortunately the data reported in the literature are not always sufficient to allow exact comparisons of results. The varying
practice to calculate the residence time further complicates the
picture, as has been recently pointed out (12,18). In some cases,
the differences may be small, but in others productivity values
reported may differ considerably. Furthermore, reported high productivities are often obtained at high substrate flow rates, resulting in low conversions. In calcium alginate and κ-carrageenan
gel systems the total reactor void volume (including biocatalyst
porosity) is about 80%; and the reactor void volume (excluding biocatalyst beads) is about 40% of the total bed volume. Furthermore,
in ethanol fermentations, carbon dioxide fills a considerable part
of the void space, leaving less than 40% of the total bed volume
as efficient liquid volume. If such solid supports as glass beads
or wood shavings are employed for the microbial film, the effective
liquid volume may differ considerably from the reactor bed volume.
Margaritis et al. (12) have emphasized that the liquid mean
residence time is kinetically important in determining the conversion. Clearly then the total reactor void volume should be considered. It should be noted here that in most reported calculations
the space taken up by the carbon dioxide has been neglected. On
the other hand, the productivity per unit of installed cost should
not be overlooked in feasibility studies (18). Consequently, if
42 g/L/hr productivity with Z. mobilis at 1 hr residence time
(based on total biocatalyst bed volume) is given, the figure would
be approximately 52 g/L/hr if based on total reactor void volume
(liquid phase), and about 105 g/L/hr if based on reactor void
volume excluding the gel entrapped biocatalyst. If space taken
up by the carbon dioxide would also be considered, the apparent
productivity figure would be still higher.

Nevertheless, recent claims for increased ethanol productivities with immobilized Z. mobilis appear justified. However,
considerable technical problems remain to be solved before scale-
up to economic commercial production can be realized. The operational biocatalyst stability, in comparison to immobilized yeast,
at high production rates should be improved (10). The rapid carbon
dioxide evolution necessitates specific reactor designs. We have

partially been able to eliminate such problems by employing a
specially designed conical reactor; while Margaritis *et al.* (12)
have suggested a slightly inclined horizontal design. It should
also be noted that *Z. mobilis* only ferments sucrose, glucose,
and fructose, but not starch or maltose, and thus is not applicable
to all types of substrates.

REFERENCES

1. INTERMAG GETRANKE INDUSTRI A. G., Ger. Offen. 1,517,814 (1969).
2. BERDELLN-HILGE, P., U.S. Patent 3,769,175 (1973).
3. LINKO, P. in "Food Process Engineering," vol. 2 (P. Linko and
    J. Larinkari, eds) Applied Science Publ., London. (1980)
    p. 27.
4. LINKO, P. & LINKO, Y. Y. in "Applied Biochemistry and Bio-
    engineering," vol. 4 (I. Chibata and L. B. Wingard Jr.,
    eds), Academic Press, New York (in press).
5. DIVIES, C., French Patent 844,786 (1977).
6. GRIFFITH, W. L. & COMPERE, A. L., U.S. Patent 4,127,447 (1978).
7. HINO, T., YAMADA, H. & OKAMURA, S., U.S. Patent 4,148,689
    (1979).
8. CHIBATA, I. in "Food Process Engineering," vol. 2 (P. Linko
    and J. Larinkari, eds) Applied Science Publ., London (1980)
    p. 1.
9. WADA, M., KATO, J. & CHIBATA, I. *Eur. J. Appl. Microbiol.
    Biotechnol. 11:* 67 (1981).
10. GROTE, W., LEE, K. J. & ROGERS, P. L. *Biotechnol. Lett. 2:*
    481 (1980).
11. ARCURI, E. J., WARDEN, R. M. & SHUMATE, S. E. II *Biotechnol.
    Lett. 2:* 499 (1980).
12. MARGARITIS, A., BAJPAI, P. K. & WALLACE, J. B. *Biotechnol.
    Lett. 3:* 613 (1981).
13. LINKO, P. & LINKO, Y. Y. *Abst. 74 Ann. Am. Inst. Chem. Eng.
    Mtg.* New Orleans Louisiana, November (1981).
14. LINKO, Y. Y. & LINKO, P. *Biotechnol. Lett. 3:* 21 (1981).
15. LINKO, P. in "Advances in Biotechnology," vol. 1 (M. Moo-
    Young, C. W. Robinson, and C. Vezina, eds), Pergamon Press,
    New York, (1981) 711.
16. LINKO, Y.-Y., JALANKA, H. & LINKO, P. *Biotechnol. Lett. 3:*
    263 (1981).
17. LINKO, Y.-Y., WECKSTROM, L. & LINKO, P. in "Food Process
    Engineering," vol. 2 (P. Linko & J. Larinkari, eds) Applied
    Science Publ., London (1980) p. 81.
18. ANON. *Biotechnol. Lett. 3:* 600 (1981).

AN IMMOBILIZED YEAST CELL COLUMN FOR THE FERMENTATION OF

MOSASSES

D. F. Day and D. Sarkar

Louisiana State University, Baton Rouge
Louisiana, USA

Immobilized whole yeast cells have been shown to be efficient converters of sugar to ethanol under laboratory conditions.
However, this technology has not been extensively tested either
on the pilot scale or with commercial feedstocks. We have employed a packed bed reactor containing *Saccharomyces cerevisiae*
adsorbed on agar beads to establish design (1) and operating
parameters for the optimum production of ethanol from cane blackstrap molasses.

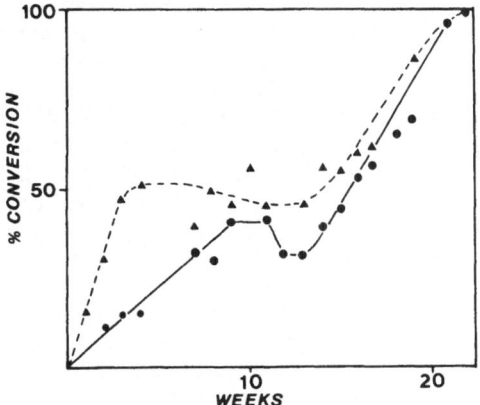

Fig. 1. Ethanol production as a function of time, from an immo-
         bilized yeast column fed clarified (▲) and unclarified
         (●) molasses.

343

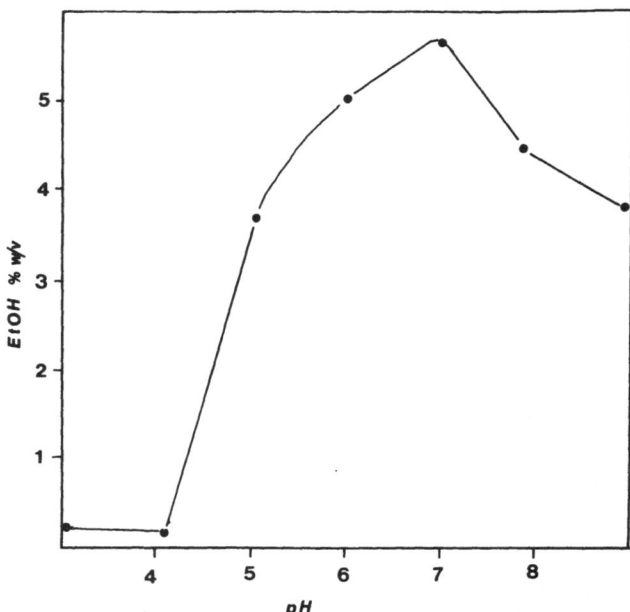

Fig. 2.   The effect of molasses pH on ethanol production from an
          immobilized yeast cell column.

        Beads containing *S. cerevisiae* were packed into a 150 ml
bed volume column. The column was fed from the bottom with
diluted blackstrap molasses. We found that it was not necessary
to maintain aseptic conditions once the column had been operating
for 24 hr. Two columns were run in parallel, one with untreated
molasses the other with molasses which had been clarified by
centrifugation (Fig. 1). The clarified molasses was found to
reach a 57% conversion plateau at a flow of 0.25 flow volumes/bed
volume/hr, more quickly than non-clarified molasses. The actual
conversion rates were a linear function of the throughput. After
15 weeks of operation the % conversion started to rise and reached
100% within 22 weeks of operation. This plateau was maintained
for the rest of the experiment (30 weeks). It made no difference
in efficiency of operation between the treated and untreated
molasses once the initial plateau had been reached. The temperature optimum for conversion was found to be 33°C; but unexpect-
edly, there was a shift in the pH optimum from 5.5, which was
optimum for batch fermentations, to 7.0 in this system (Fig. 2).
The other major optimization factor for maximum alcohol production was found to be sugar concentration (Fig. 3). Complete
conversion was achieved with substrate levels below 10%; between
10% and 18% sugar the conversion was 80-85%; both the conversion
and ethanol yield rapidly decreased when the sugar levels were

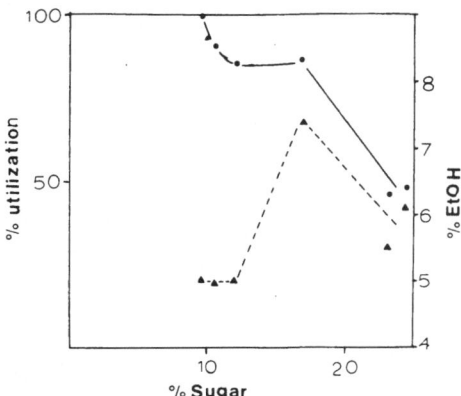

Fig. 3. The effect of fermentable sugar concentration of molasses on utilization of an immobilized yeast cell system. Sugar % utilized (●) and ethanol produced, % w/w (▲), are shown as a function of the percent fermentable sugar in the feedstock.

above 18%. This figure probably represents the maximum practical sugar level which can be achieved with molasses in this system.

We are currently constructing a small pilot plant designed for fermenting molasses. It will have an output of 500 ml/min. of fermented molasses. The results from this plant will be used to establish a design for a commercial operation utilizing cane molasses and other sugar juices for fuel ethanol production.

ACKNOWLEDGMENT

The work was supported by the Bioenergy Council.

REFERENCES

1. POLACK, J. A., KUU, W. Y., CHO, Y. K. & DAY, D. F. *Abst. 74th Ann. Meet. Am. Inst. Chem. Eng.* (1981).

PILOT OPERATION FOR CONTINUOUS ALCOHOL FERMENTATION

OF MOLASSES IN AN IMMOBILIZED BIOREACTOR

S. Fukushima and S. Hanai*

Department of Chemical Engineering
Kansai University, Suita, and
Takara Shuzo Co. Ltd.,* Kyoto, Japan

The industrial alcohol fermentation of blackstrap cane molasses in a batch system under aerobic conditions, with intact yeast cells and malt, gives 12-13 % (v/v) alcohol. Semi-batch fermentation of cane juice produces 8% (v/v) alcohol by reuse of intact cells, as widely used in Brazil. Continuous processing of beet molasses with recycling cells is also operated to produce 7-8% (v/v) alcohol (1). Since Kierstan and Bucke (2) presented a process for continuous alcohol production from glucose by yeast cells entrapped in Ca alginate, many researchers have studied continuous fermentation with entrapped living cells. The following criteria should be met in order to produce a high concentration of alcohol from blackstrap cane molasses on a large scale by rapid continuous fermentation over a long time: a) the biocatalyst gel particles must not become mechanically cracked or crushed in a bioreactor and b) the bioreactor must not develop local accumulations of CO_2 gas bubbles, alcohol, or byproducts.

The yeast strain used was *S. formosensis* M-111, which was the mutant given by Hanai *et al.* (3). It was observed that the Al alginate entrapped cells had strong mechanical properties and gave rapid alcohol fermentation in a fluidized bed, even at pH 2.85 and 30°C. A multi-stage three-phase fluidized bed was designed as a new type of bioreactor. The unit was constructed by connecting the rhombus-shaped units vertically. The working volumes for the laboratory and pilot plant units were 0.5 and 150 L, respectively, packed with 1-2 mm particles. The molasses solution was fed to the lowest unit at pH 2.8-3.6 and 30°C; and the inert gas, namely CO_2 or N_2, was also introduced into the bed at 0.002 to 0.05 v/v/min through a sintered or perforated plate located at the bottom of the lowest unit.

The alcohol concentration in the effluent P reached 78 g/L from a feed sugar content of 162 g/L at a reciprocal residence time D_s of 0.36 hr^{-1} for both the laboratory and pilot plant units. At 198 g/L feed sugar content, P became 96 g/L or 12% (v/v) at a D_s of 0.33 hr^{-1} for the laboratory unit. This value is equal to that given by industrial batch fermentations. The pilot operation to produce 78 g/L alcohol was continued for 90 days under anaerobic conditions. These results show that this new immobilized bioreactor satisfies the items mentioned above.

Simulations agreed with the observed data for D_s of 0.33-1.2 hr^{-1} on the assumption that the resistance to chemical reaction in the biocatalyst was the predominant factor and that the bioreactor was equivalent to a two-stage ideal mixing tank. A process was designed for producing 100,000 L of 99.5% (v/v) alcohol per day, where the immobilized bioreactors gave 12% (v/v) alcohol. A cost estimation for investment of equipment and operation of the process, excluding the cost of blackstrap cane molasses showed 28 and 30% lower costs, respectively, for this process than for those for the Melle-Boinot process. Thus, this process is expected to be useful for ethanol production on a large scale.

REFERENCES

1. ROSEN, K. *Process Biochem.* 25 (1978).
2. KIERSTAN, M. & BUCKE, C. *Biotechnol. Bioeng. 19:* 387 (1977).
3. HANAI, S., HIRAMATSU, J., TERAMOTO, M. & YANO, T. *Abst. Soc. Ferment. Technol. Meet.* 119 (1980).

ENZYMATIC HYDROLYSIS OF CELLULOSE

Y. Harano, H. Ooshima, K. Ohmine and M. Sakata

Faculty of Engineering Osaka City University
Osaka, Japan

Despite many kinetic studies on the enzymatic hydrolysis of cellulose, the attempts to deduce the rate expression for various kinds of cellulosic materials have not been successful. At present, it would be difficult to obtain a rate equation based on the reaction mechanism because of too many physical and chemical rate controlling factors, such as the crystallinity of the cellulose or the inhibition of enzymes by substrate and products. In this work, an attempt has been made to develop an empirical rate expression for the enzymatic hydrolysis of cellulose.

Avicel (Asahi Chem. Co., Japan) and a tissue paper were reacted batchwise at 50°C and pH 4.8 with *Trichoderma* cellulase (Meicelase, Meiji Conf. Co., Japan). Enzyme at 0.33 mg/ml and cellulose initially at 30.8 - 308 mM (cellobiose base 1 - 10 wt%) were used. The concentration of total reducing sugar (TRS = cellobiose + glucose) produced were determined by the DNS and glucose oxidase method.

In the enzymatic hydrolysis of cellulose, the most characteristic phenomenon is that the hydrolysis rate decreases rapidly during the initial stage of the reaction. On the basis of our data, an empirical rate expression was developed in terms of k as the overall rate retarding factor.

$$- \frac{dv}{dX} = k \, v \qquad \text{(Eq. 1)}$$

or

$$S_0 X = \frac{S_0}{k} \ln\left(1 + \frac{v_0}{S_0} k t\right) \qquad\qquad \text{(Eq. 2)}$$

where t is the reaction time in hr; v and v_0 are the hydrolysis rates in mM/hr at $t = t$ and $t = 0$, respectively; X is the fractional conversion of cellulose to sugar; and S_0 is the initial cellulose concentration in mM.

The experimental data obtained with different initial concentrations of Avicel and tissue paper showed good agreement with Eq. 1 and 2. Eqs. 1 and 2 also were applied successfully to the hydrolysis of other kinds of cellulosic material, such as dewaxing cotton, wood chips, and rice straw (1).

REFERENCES

1. OOSHIMA, H., OHMINE, K., SAKATA, M. & HARANO, Y. unpublished data.

ATP REGENERATION BY ENZYMES OF ALCOHOL FERMENTATION AND KINASES OF YEAST AND ITS COMPUTER SIMULATION

R. Matsuno, M. Asada, K. Nakanishi and T. Kamikubo

Department of Food Science and Technology
Faculty of Agriculture, Kyoto University
Kyoto, Japan

Bioreactors which synthesize valuable substances using ATP as an energy source must be combined with ATP regeneration systems. We have been studying continuous ATP regeneration from adenosine catalyzed by enzymes from yeast alcohol fermentation, especially adenosine kinase and adenylate kinase from yeast (1-3). A crude preparation of the enzymes was extracted by incubation of 150 mg of acetone-dried baker's yeast in 1 ml of medium containing 1000 mM glucose. When necessary, gel filtration with Sephadex G-50 was included to remove intermediates formed during the alcohol fermentation (FBP). ATP was continuously regenerated from adenosine in a plug flow reactor, equipped with a semi-permeable collodion membrane with the enzymes as catalysts and glucose as the energy source. The reactor consisted of 8 chambers, each separated into 2 compartments by the 40-50 μm thick collodion membrane. The height, width and thickness of each reaction compartment was 80, 20 and 2 mm, respectively. The total volume of the compartments was 25.6 ml. The crude enzyme solution was injected into the compartment on the left side of each chamber. Substrate solution, containing 600 mM glucose, 600 mM potassium phosphate pH 7.3, 20 mM $MgSO_4$, 20 mM adenosine, 1 mM NAD and 1 mM ATP, was pumped into the compartment on the right side of the first chamber and flowed through each chamber in turn to contact the crude enzyme preparation through the collodion membranes. The flow rate gave a residence time of 15 hr. Steady state operation at an ATP yield of 75% continued for 42 hr. The addition of 2 mM dithiothreitol(DTT) to the substrate solution increased the duration of the steady state to 100 hr. The crude enzyme preparation was heated for 3 min to 100°C to inactivaae enzymes; the large amount of precipitate formed was removed by centrifugation at 3,000 x g for 10 min. The resulting supernatant contained no

protein precipitable by 5% trichloroacetic acid; but it did contain
a considerable amount of alcohol fermentation intermediates called
the yeast extract. The addition of 10% yeast extract extended the
duration of the steady state to 2 weeks. At steady state, the remaining glucose and accumulated FBP were 120 mM and 140 mM, respectively. The substrate adenosine was not detected. Accumulated adenosine nucleotides, ATP, ADP and AMP were 15 mM, 2.5 mM
and 1 mM, respectively. The significant effect of the yeast extract was ascribable to the protection of the enzymes against
inactivation by the reaction intermediates. For a practical reactor, the additions of DTT and yeast extract could be replaced
by degasification of the substrate solution and by recirculation
of one tenth of the product solution into the feed stream. For
a better understanding of the characteristics of the present reaction system, batchwise ATP regeneration was performed using
gel-filtered enzymes. More than 7 mM FBP or 10 mM ATP in the reaction mixture was necessary for initiation of the alcohol fermentation. These were called threshold concentrations. The changes
in nucleotide concentrations with reaction time were not mono-
tonical; ATP especially accumulated suddenly after 6 hr of reaction. These nonlinear phenomena might be due to the action of
FBPase and/or ATPase as well as a complex multi-enzyme system.
A mathematical model, consisting of 17 simultaneous differential
equations for the concentrations of 17 reactants, was constructed
with respect to reaction time, based on 11 enzyme reactions (4).
For the sake of simplicity, the enzymes were classified into two
groups: those catalyzing reversible reactions and those key enzymes catalyzing irreversible reactions. The calculated time-
dependent concentrations of glucose, FBP, adenosine, and nucleotides were in agreement with experimental values. Moreover, the
existence of threshold concentrations of FBP and ATP and abrupt
changes of nucleotide concentrations were also recognized. The
threshold concentrations were 7 mM and 8 mM, respectively, for FBP
and ATP.

REFERENCES

1. ASADA, M., NAKANISHI, K., MATSUNO, R., KARIYA, Y., KIMURA, A.
    & KAMIKUBO, T. *Agric. Biol. Chem. 42:* 1533 (1978).
2. ASADA, M., MORIMOTO, K., NAKANISHI, K., MATSUNO, R., TANAKA,
    A., KIMURA, A. & KAMIKUBO, T. *Agric. Biol. Chem. 43:*
    1773 (1979).
3. ASADA, M., YANAMOTO, K., NAKANISHI, K., MATSUNO, R., KIMURA,
    A. & KAMIKUBO, T. *Eur. J. Appl. Microbiol. Biotechnol. 12:*
    198 (1981).
4. ASADA, M., SHIRAI, Y., NAKANISHI, K., MATSUNO, R., KIMURA, A.
    & KAMIKUBO, T. *J. Ferment. Technol. 59:* 239 (1981).

EXTRACELLULAR CELLULASES PRODUCED BY A YEAST-LIKE FUNGUS

G. Larios, A. Gilbon, Y. Lara, and C. Huitron

Department of Biotechnology, Institute of
Biomedical Research, National University of
Mexico, UNAM. Mexico, D.F.

Cellulose occurs in abundance in nature and constitutes a
third to one-half of the approximately 150 billion tons of organic
matter that is photosynthesized annually (1). The possibility of
using preparations of cellulases to produce glucose, alcohol, or
single cell protein from cellulose has been under intensive study
during the last years (2-6). One of the major problems encountered
is obtaining sufficiently active cellulolytic enzymes so that high
concentrations of glucose can be obtained from crystalline cellulose in a reasonable period of time. Our group has been working
on the isolation of new cellulolytic microorganisms, which can be
used to produce extracellular cellulases active in the degradation
of crystalline cellulose (7). From several microorganisms we have
selected a strain with morphological characteristics of a yeast-
like fungus, which was identified as *Aureobasidium* sp.

The maximal cellulase activity produced by this strain when
it is grown on microcrystalline cellulose as the sole carbon
source is reached after about 72 hr of fermentation. This is
shorter time than that required by other microorganisms. The strain
grows well on different carbon sources but only produces filter
paper activity in the presence of cellulose. The filter paper assay
used was that developed at Natick (8). In order to measure the
extracellular activities on different celluloses and related substrates, cell free filtrates were obtained after growing for 72
hr on four different carbon sources: microcrystalline cellulose
(0.75%), carboxymethyl cellulose (0.6%), cellobiose (0.6%), and
xylan (0.6%). The filtrate from the first carbon source contained
the following activities: filter paper activity (2.25 mg/ml reducing sugars), CMC activity (4.40 mg/ml reducing sugars), β-
glucosidase measured on p-nitrophenyl-β-D-glucoside (3.96 mg/ml

reducing sugars), and xylanase (4.15 mg/ml reducing sugars). The
filtrate from the second carbon source contained the same activities but in lower amount: 23%, 30%, 25%, and 60%, respectively.

On cellobiose the filtrate obtained showed activity on p-
nitrophenyl-β-D-glucoside but only 9% compared to that obtained
when microcrystalline cellulose was utilized at the carbon source.
The filtrate from the carbon source xylan only contained xylanase
activity (5.15 mg/ml reducing sugars).   It is interesting that xylanase activity was produced either on xylan or on cellulose; however, when the microorganism was grown on xylan alone, xylanase
but not cellulase was produced.   At the moment it is not known if
there are one or more xylanases.

The only two hydrolysis products obtained from filter paper
were identified as cellobiose and glucose by paper chromatography.

The proteins of the cell-free filtrate from *Aureobasidium*
sp., grown on microcrystalline cellulose, were separated on DEAE-
Sephadex A-50 into three peaks.   Peak 1 showed filter paper activity, β-glucosidase, and xylanase activities; while peaks II and
III had filter paper activity only.   The protein recuperation into
the three peaks was 99%; and the filter paper activity was 64% of
the original.   These low activities seem to indicate a lack of expression of synergistic effect of the different enzymes forming
the cellulase system.   When we combined the three fractions from
the DEAE-Sephadex A-50 chromatogram, we found that the mixture had
more activity than the sum of the filter paper activities of each
fraction.   The results make this newly discovered microorganism
an interesting subject for further study of cellulase production.

REFERENCES

1.   GHOSE, T. K. & PATHAK, A. N. *Process Biochem.*   (May.) 20 (1973).
2.   MANDELS, M. & STERNBERG, D. *J. Ferment. Technol. 54:*   267
        (1976).
3.   NYSTROM, J. M. & ANDREN, R. K. *Process Biochem.* (Dec.) 26
        (1976).
4.   GHOSE, T. K. & GHOSE, P. *Process Biochem.* (Nov.) 20 (1979).
5.   HUMPHREY, A. E. *et al.*, in "Continuous Culture, Applications
        and new Fields" (A.C.R. Dean, ed.) Soc. of Chemical Insustry,
        London (1976) p. 85.
6.   MAIORELLA, B., WILKE, C. R. & BLANCH, H. W. *Adv. Biochemical
        Eng. 20:*   43 (1981).
7.   GILBON, A., LARIOS, G. & HUITRON, C. *Rev. Technol. Aliment.*
        (Mex) *16:*   24 (1981).
8.   MANDELS, M., ANDREOTTI, R. & ROCHE, C. *Biotechnol. Bioeng.
        Symp. No. 6:*   21 (1976).

RELATIONSHIP BETWEEN EXTRACELLULAR PROTEASES AND THE CELLULASE

COMPLEX OF *TRICHODERMA REESEI*

C. P. Dunne

U.S. Army Natick Research & Development Laboratories
Natick, Massachusetts, USA

Trichoderma reesei is a fungus best known for production
of an effective extracellular cellulase. When grown on cellulose,
the cellulase enzyme complex normally comprises up to 90% of the
extracellular proteins; minor protein components include several
carbohydrases and proteases. There are several reasons for study
of the *Trichoderma* proteases: a) proteases may be involved with
the control of release of extracellular enzymes, e.g., cellulase;
b) protease may have important effects on the stability and activity of the cellulase components; and c) proteases may play a role
in the production of multiple forms of cellulase components.

An assay was developed for the extracellular proteases using
the insoluble substrate Azocoll; the pH optimum for Azocoll
hydrolysis by *Trichoderma* protease was 4.2-4.8. The assay was
used to survey cellulalse preparations from different strains of
Trichoderma for the presence of acid proteases. Two commercial
cellulases had high levels of protease; little protease was found
in the active cellulase preparations from Rutgers University
mutant strains C30 and NG14. The role of growth conditions in
the production of extracellular proteases was explored in shake
flask cultures of different strains of *T. reesei*. The Natick
strains, QM9414 and MCG77, produced high levels of protease when
grown on either glucose plus NH_4 or casein as a carbon/nitrogen
source. Rutgers strains NG14 and C30 had low levels of protease
on glucose; but they showed a 15-fold or greater increase in protease level when 0.3% casein was the nitrogen source.

Two simple strategies were devised for purification of
proteases from *T. reesei* strain QM9414. In the first the

fungus was grown on cellulose; the majority of the extracellular proteins (cellulase components) were removed by adsorption on cellulose giving a 6-fold purification of proteases, which were then separated into two fractions with pI = 4.42 (90 units protease/mg) and pI = 4.55 (124 units protease/mg) by free boundary isoelectric focusing. In the second strategy casein (0.3%) was used to induce the protease when grown in the absence of cellulose. The initial culture filtrate had a specific activity of 110 protease units/mg, which yielded a peak fraction of MW 35,500 and 190 units protease/mg by gel filtration on Bio Gel P-60. Isoelectric focusing yielded three acid proteases with pI values of 4.3, 4.65, and 4.85. The protease was inhibited 95% by equimolar pepstatin.

The acid proteases from *T. reesei* 9414 hydrolyzed denatured hemoglobin with a pH optimum of 3.0 at 40°C; and the enzymes also hydrolyzed casein. Cellulase from the C30 strain was a substrate for hydrolysis by the acid protease when denatured; the pH optimum was temperature dependent; there was no hydrolysis of cellulase above pH 4 in 30 min at 40°C.

Two findings suggest that the acid proteases from *T. reesei* are not solely responsible for the multiple forms of cellulase: a) similar HPLC patterns for cellulases from high protease strains (QM9414) and low protease (C30) were observed; and b) cellulase in its native conformation at pH greater than 4 was quite resistant to action of acid protease. The extracellular acid proteases of *T. reesei* will act upon cellulase in fermentations if growth conditions (pH and temperature) are not carefully controlled. The role of acid proteases in the secretion of cellulase and other extracellular enzymes is still unclear.

Trichoderma and related fungi may be useful sources of other enzymes besides cellulases. The extracellular nature of inducible enzymes like the acid proteases offers advantages in purification and utilization of the enzymes.

ACKNOWLEDGMENT

This research was done on an Intergovernmental Personnel Act Assignment from California State University, Long Beach. The author would like to thank M. Mandels, F. Robbins, E. Reese, and R. Andreotti for their help and advice.

Session VI

ANALYTICAL APPLICATIONS OF
IMMOBILIZED BIOMATERIALS

Chairmen: L. Wingard Jr. and H. Okada

BINDING ASSAYS INVOLVING SEPARATION IN AQUEOUS TWO-PHASE SYSTEMS: PARTITION AFFINITY LIGAND ASSAY (PALA)

B. Mattiasson, T. G. I. Ling and M. Ramstorp

Pure and Applied Biochemistry, Chemical Center
University of Lund
Lund, Sweden

Many of the routine analytical procedures that use specific binding between biological molecules involve various methods for separating the free from bound labelled ligands after the binding has taken place. The most commonly used separation methods are repeated centrifugation and washing steps, making the overall analytical procedure laborious and time consuming. We have studied the possibility of using differences in the partition of the binding entities in aqueous phase systems to diminish the need for conventional separation methods. When aqueous solutions of some water soluble polymers, e.g. poly-(ethylene glycol) (PEG) and dextran are mixed in sufficient concentrations, the mixture separates into two phases, with the top phase rich in PEG and the bottom phase rich in dextran.

Aqueous phase polymer systems have been used for separation and purification of cells, cell organelles, proteins, and nucleic acids (1). The partition separation exploits differences in the surface properties of the molecules to be partitioned. The selectivity in the distribution of material between the two phases is characterized by the partition coefficient, K_{part}, which is defined as C_{top}/C_{bottom}, where C_{top} and C_{bottom} denote the concentration of the material in the two phases. The partition coefficient is dependent on the characteristics of the phase systems, e.g. the type of polymers and MW of the polymers, as well as the characteristics of the material to be partitioned, e.g. size, charge, and hydrophobicity/hydrophilicity. The overall partition coefficient can be regarded as composed of factors due to electrical, hydrophobic, hydrophilic and conformational effects.

The PALA method can be used in assays for low MW antigens (haptens) as well as macromolecules and cells using the same basic protocol. The only requirement is that the antigen and the corresponding antibody (or other pair of reactants) partition to different phases. If both reactants partition to the same phase, it may be possible to modify the surface properties of one of the reactants in order to secure the desired partition. The modification can, in principle, be carried out to affect any of the factors included in the partition coefficient. We have used covalent attachment of the same polymers, as used in phase systems, to the reactant and thereby slightly changed its hydrophobicity. In order to get proteins or cells to partition to the PEG-rich top phase in a PEG/salt or PEG/dextran system, the modification was done with PEG molecules. To make proteins partition to the bottom phase, they were coupled to Sephadex particles, which are made of crosslinked dextran. As for cells and other particles, there may be a significant adsorption to the interface. This is no disadvantate since this distribution in the phase system does not *per se* prevent efficient separation.

The analytical procedure involves incubation of the reactants followed by separation of bound from free reactants, by addition

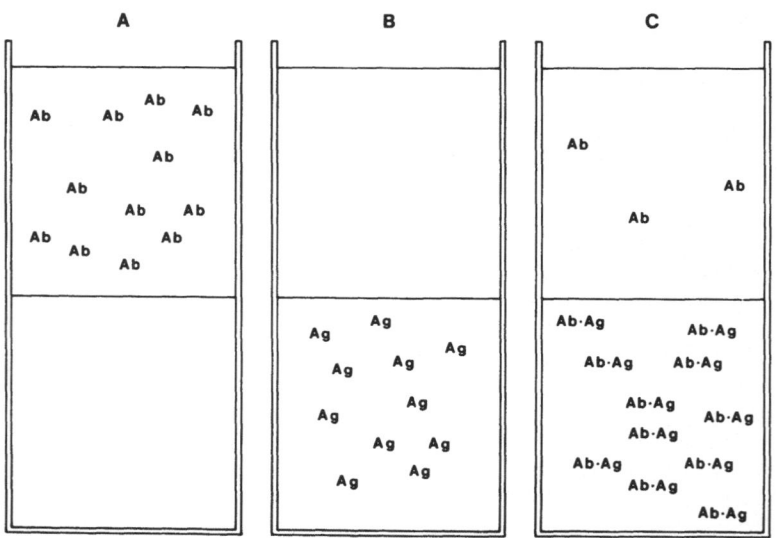

Fig. 1.   Schematic representation of direct binding assay. Labelled antibodies partition mainly to the top phase and antigens to the bottom phase. The antibody-antigen complex, formed during incubation, partitions to the bottom phase, resulting in a decrease in activity found in the top phase.

of a well-mixed phase system. The PALA-method requires no separate
washing or centrifugation and can therefore be performed in one
and the same test tube. Fig. 1 shows a schematical representation
of a direct binding assay in which the binding of a labelled molecule is utilized to quantify an unlabelled molecule. The reactants
are incubated in a test tube. After binding has taken place, the
bound and free labelled reactants are separated by partition in
a well-mixed system. The amount of label can be determined in
aliquots taken from the phases after phase separation and can be
correlated to the amount of unlabelled molecule present in the
incubation mixture. In a typical run the reactants are mixed and
incubated for a fixed time. The phase system is mixed, added to
the incubation mixture, and vortexed vigorously. After phase
separation an aliqout is removed from one or both phases for determination of the amount of label.

SYSTEMS STUDIED

     Assays for Haptens: Triiodothyronine, denoted T_3, and digoxin
(2) were anlayzed in competitive assays where ^{125}I-labelled antigen
and antigen from the sample were allowed to compete for binding

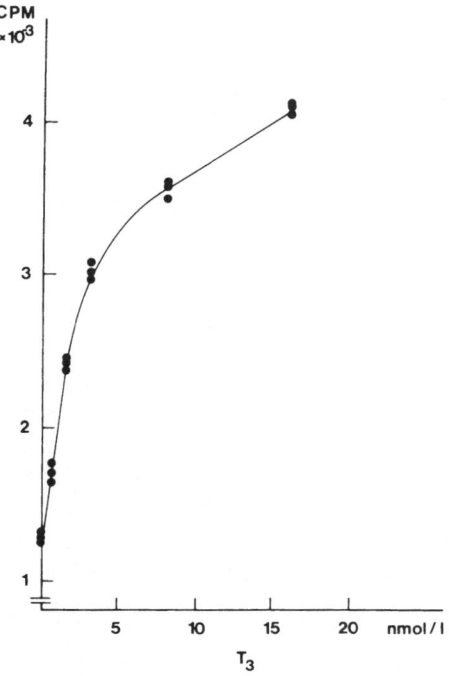

Fig. 2.   Radio-immunoassay of T_3 obtained from 12 hr incubation
          at $4°C$.. (see text).

to a limiting amount of specific antibodies. It was found that
in a phase system consisting of PEG-4000 and magnesium sulfate the
haptens partitioned to the PEG-rich top phase and the antibodies
to the salt-rich bottom phase. The T_3 had a K_{part} of about 10 and
the antibody about 0.2; so that after separation, free antigen was
found in the top phase and bound antigen in the bottom phase.
With an incubation time of 30 min the operational concentration
range in the analysis was 1.5 - 16 nM. When the incubation time
was prolonged to 12 hr, the sensitivity limit was 0.5 nM; and the
reproducibility was considerably improved. Fig. 2 shows a standard curve for the quantitation of T_3. The correlation between
the PALA method and a conventional RIA was 0.981 for the T_3 and
0.979 for the digoxin assay.

Assays for Macromolecules: The interaction between the lectin
conA and various carbohydrates was used in a model system for quan-
tifying macromolecules (3). In this system the glucoenzyme horseradish peroxidase was used as an enzyme labelled hapten. In a
phase system consisting of PEG and salt, conA has a partitioning

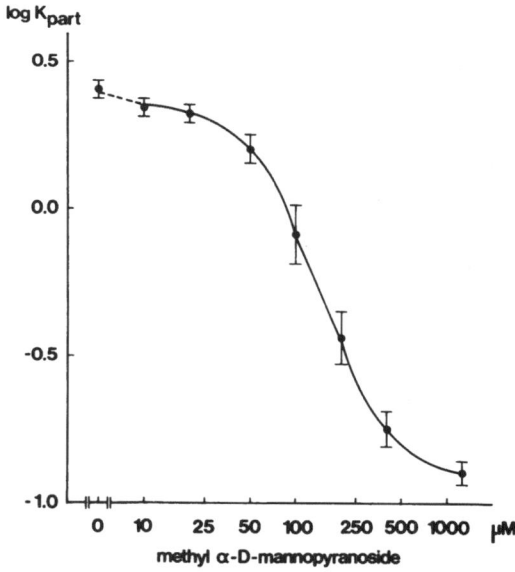

Fig. 3. Competitive assay of a carbohydrate. Methyl-α-D-manno-
         pyranoside, horseradish peroxidase, and PEG-modified
         conA were incubated for 10 min in a total volume 100 μl.
         The phase components, PEG and $MgSO_4$, were added to a
         final volume in 1 ml; after being mixed the phases were
         allowed to separate for 10 min. Then 200 μl was taken
         from each phase for determination of enzyme activity.
         (Reproduced from (4) with permission.)

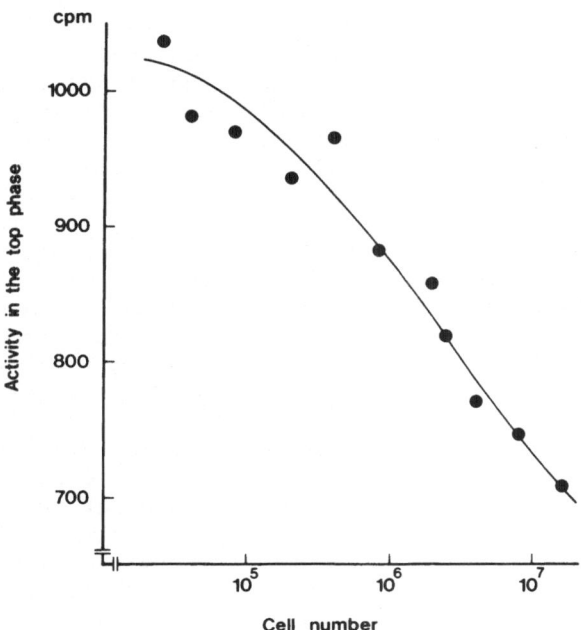

Fig. 4. *Staphylococcus aureus* in a competitive binding assay using a constant amount of ^{125}I-labelled IgG. Activity is shown as a function of the number of native *Staphylococcus aureus* cells mixed with 6.3×10^5 M-PEG-modified cells. The mixture was incubated with 27×10^{-15} mol of ^{125}I-labelled IgG for 30 min and then allowed to separate in an aqueous phase system consisting of PEG 6000 and dextran T250. A 200 μl sample was taken from the top phase for measuring of activity. (Reproduced from (4) with permission.)

coefficient of 0.031 and peroxidase 0.063. It is thus necessary to alter the partitioning of one of the reactants before setting up an assay. It was found that conA could be made to partition to the top phase after covalent attachment of PEG. The modification did not cause any detrimental loss of binding capacity. Activated monomethoxy-PEG (M-PEG), prepared according to Abuchowski, was allowed to react with amino groups on the molecule. The K_{part} for the M-PEG-conA obtained was about 80. When the enzyme was bound to the modified lectin, the complex partitioned to the top phase; so that the degree of binding could be determined by measuring the distribution of the enzyme activity. Using this basic set-up, methyl-α-D-mannopyranoside was quantified in a competitive assay (Fig. 3). It was also possible to quantify dextran T-40 with a detection limit of 90 nM (MW 40,000).

Assays for Cells: When phase systems are used for separating cells and particles, the possibility of adsorption to the interface must be considered. One may use either a phase system which minimizes the adsorption, or the analytical procedure may be designed in such a way that the adsorption could be ignored. For example, if the cells to be quantified distribute between the bottom phase and the interface, and the specific labelled antibodies partition mainly to the top phase, then the activity in the top phase after separation can be used to monitor the outcome of the binding step. In this way the distribution of the cells will not affect the analysis, as long as they do not partition significantly to the top phase. The interaction between *Staphylococcus aureus* Cowan I and labelled human immunoglobulin G (IgG) was used as a model system (4). On the surface of these cells there is a protein (protein A) that binds IgG specifically. It was found that both the cells and the antibodies partitioned to the bottom phase in a phase system consisting of PEG and magnesium sulfate. It was therefore necessary to modify one of the reactants to get a asymmetric distribution in the phase system. One alternative is to use a direct binding assay and modify the antibodies in order to create an analytical situation similar to that illustrated in Fig. 1. This direct binding assay proved useful in only a very narrow concentration range. A competitive assay was based on competition between native and PEG-modified *Staphylococcus aureus* for the binding of a limiting amount of labelled IgG (Fig. 4). When the number of native cells in the incubation mixture was increased, a decrease was found in the amount of labelled IgG in the top phase. The operational concentration range was 10^5 - 10^7 cells.

Yeast cells, *Saccharomyces cerevisiae*, were quantified with the use of ^{125}I labelled concanavalin A (conA) (5). In a competitive assay 300,000 M-PEG-modified yeast cells were mixed with a varying number of native yeast cells and incubated with labelled conA. An analytical range of 10^4 - 10^5 was obtained. In order to improve the sensitivity a two-step competitive assay was designed. The cells to be assayed were first incubated with a fixed amount of labelled conA; after a predetermined period an excess of M-PEG-modified cells were added in order to interact with unbound labelled ligand. After a short time a well-mixed phase system was added. This two-step analytical procedure made it possible to improve the sensitivity even further. Fig. 5 shows a typical calibration curve obtained in the two-step competitive assay of yeast cells.

Another immunological quantification was performed on streptococci. In this study antibodies against *Streptococcus* Bl cells were used either labelled with ^{125}I or with the enzyme horseradish peroxidase. When studying the distribution of each individual entity in a phase system consisting of PEG and dextran, both the cells and the I-labelled antibodies partitioned to the bottom

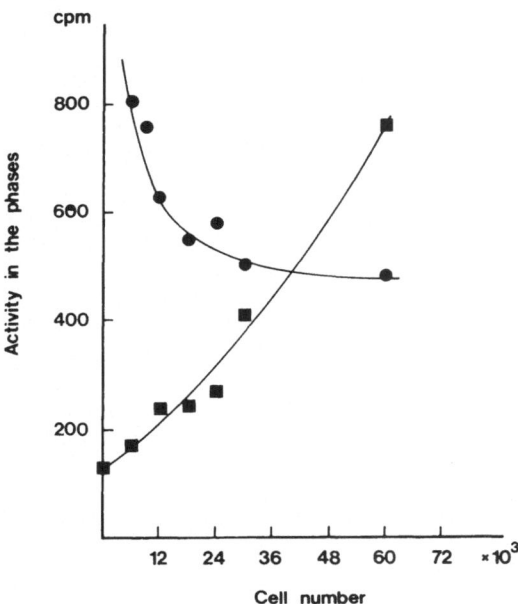

Fig. 5. Two-step competitive binding assay of *Saccharomyces*
cerevisiae. Native yeast cells were incubated with a
predetermined amount of ^{125}I-labelled conA. After 30
min 225,000 M-PEG-modified yeast cells were added
followed, after 1 min, by addition of a PEG 4000 and
$MgSO_4$ phase system. Activity in top (●) and bottom
(□) phases is a function of number of native yeast
cells present in the incubation mixture. (Reproduced
from (5) with permission.)

phase. This made it necessary to modify either the cells or the
antibodies. The enzyme-labelled antibodies, however, had an increased partition coefficient, thereby enabling the use of a direct
binding assay. Fig. 6 shows a calibration curve obtained for
various numbers of unmodified streptococci in a direct binding
assay.

CONCLUSION

The data presented here clearly demonstrate that separation
within aqueous two-phase systems can be a good alternative when
designing binding assays for haptens, macromolecules, and cells.
Furthermore, the fact that binding, separation, and measurement
can take place in one and the same vessel in both enzyme- (3) and

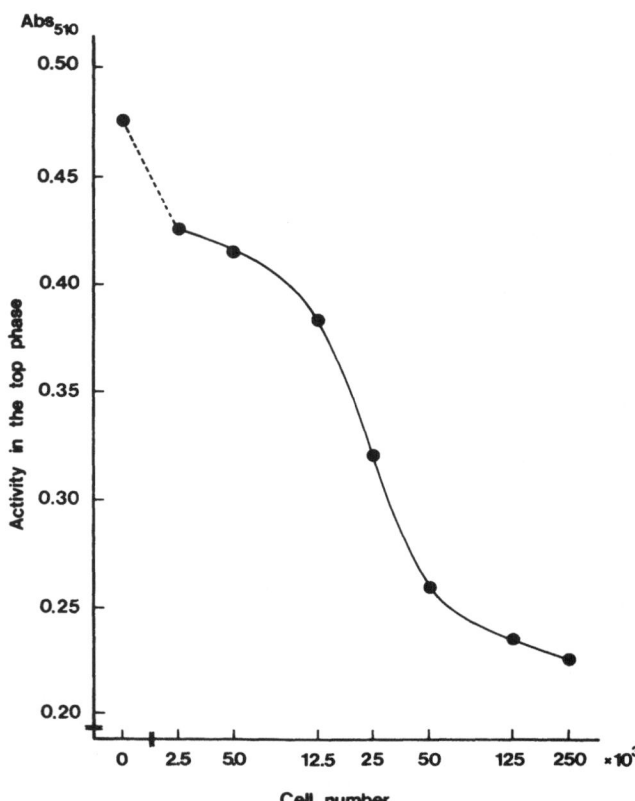

Fig. 6.   Direct binding assay of streptococci. Varying numbers
          of cells were incubated with a fixed amount of anti-
          bodies labelled with peroxidase in a total volume of
          100 µl. A well mixed PEG 6000 and dextran T250 phase
          system (900 µl) was added and mixed. A 200 µl aliquot
          was taken from the top phase, after phase separation, to
          which 800 µl reagent solution was added. The enzyme
          activity was measured in a spectrophotometer at 510 nm.

radio-immuno-assays makes the partition affinity ligand assay a
promising alternative for automated anlayses.

REFERENCES

1.   ALBERTSSON, P-Å. "Partition of Cell Particles and Macromole-
     cules", 2nd Edit., Almqvist and Wiksell, Uppsala, Sweden
     (1971).
2.   MATTIASSON, B. *J. Immunol. Meth. 35:* 137 (1980).

3. MATTIASSON, B. & LING, T. G. I. *J. Immunol. Meth. 38:* 217 (1980).
4. MATTIASSON, B., LING, T. G. I., & RAMSTORP, M. *J. Immunol. Meth. 41:* 105 (1981).
5. MATTIASSON, B., LING, T. G. I., NILSSON, J., & DURHOLT, M. "Lectins: Biology, Biochemistry, Clinical Biochemistry", vol. 2, Waltelr de Gruyter, Berlin, in press.

USE OF IMMOBILIZED ENZYME REACTORS IN AUTOMATED CLINICAL ANLAYSES

T. Murachi

Department of Clinical Science
Kyoto University Faculty of Medicine
Sakyo-ku, Kyoto, Japan

The diagnosis of diseases is heavily dependent on clinical laboratory data. Biochemical analyses of blood and urine samples have increased nearly exponentially both in number and in variety during the past two decades. The introduction of automated instruments and the use of enzymes as analytical reagents were the two major innovations experienced in this period.

We have studied various applications of immobilized enzymes for automated clinical analyses and have found that a mini-column type reactor, packed with immobilized enzyme-bearing glass beads, is very useful (1,2,3). Our enzyme column measures only 1.5 - 2.0 mm in inner diameter and 5 - 40 mm in length and can be inserted into a flow-type automated analyzer. The whole system can be operated routinely without change of reactor for several thousands samples or for 1 - 3 months.

Table 1 summarizes the successful application of immobilized enzyme reactors developed in our laboratory. We used alkylamine derivatives of porous glass beads as the solid support, to which enzymes were immobilized either by glutaraldehyde or by coupling through Schiff-base formation after periodate oxidation of the carbohydrate moiety of the enzyme protein. In each case, the results of analyses on serum or urine correlated satisfactorily with those obtained by other well-established chemical and enzymatic methods. The immobilized enzyme method always gave reliable and reproducible data for a long period of use, resulting in a considerable reduction of the reagent cost per assay.

TABLE 1

USE OF IMMOBILIZED ENZYMES IN CLINICAL ANALYSES

Assay	Enzyme (Source)	Method of Immobilization	References
Uric acid	Uricase (*Candida utilis*)	carbohydrate	(1,8)
Cholesterol	Cholesterol oxidase (*Nocardia erythropolis*)	glutaraldehyde	(9)
Cholesterol ester	Cholesterol ester hydrolase (*Pseudomonas* sp.)	glutaraldehyde	(9)
Glutamate	Glutamate dehydrogenase (Beef liver)	glutaraldehyde	(11)
Lactate	Lactate dehydrogenase (Beef heart)	glutaraldehyde	(11)
Urea	Urease (Jack bean)	glutaraldehyde	(12)
Glucose	Glucose oxidase (*Aspergillus niger*)	carbohydrate	(1,8)
	Peroxidase (horse radish)	carbohydrate	
Inorganic phosphate	Pyruvate oxidase (*Pediococcus* sp.	glutaraldehyde	(13)

MULTI-CHANNEL INSTRUMENT

Fig. 1 illustrates the flow diagram of a three-channel instrument we have designed and built under collaboration with Y. Saheki and H. Kagamiyama of Shiga University of Medical Sciences in Otsu, Japan (4). It has three immobilized enzyme reactors (R_1, R_2 and R_3), to which μL aliquots of the sample are supplied through computer-controlled electromagnetic valves. The enzymes used are oxidases for glucose, uric acid, and cholesterol. All produce hydrogen peroxide. The effluent from the enzyme column reactor is introduced into a photon-counting cell (C), where it is very rapidly mixed with a solution of luminol (L) and potassium ferricyanide (F). The intensity of chemiluminescence produced is determined by a photon-counting tube (T) attached to the cell. The special

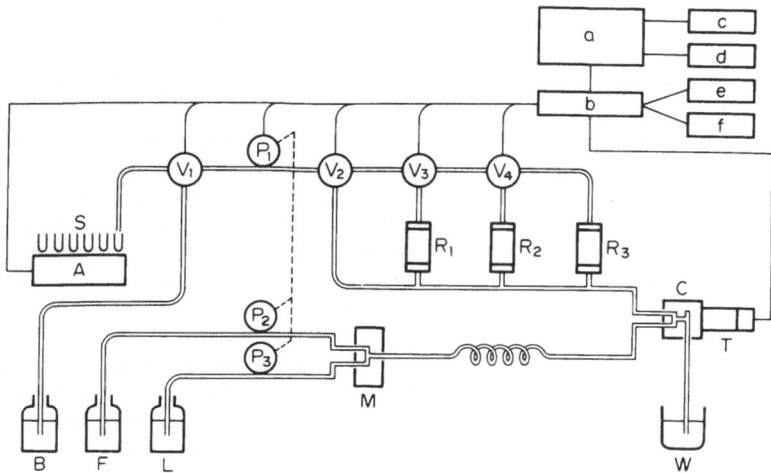

Fig. 1. Flow diagram for three-channel analyzer with immobilized enzyme columns. A autosampler; B buffer; C cell for chemiluminescence; F potassium ferricyanide; L luminol; M mixer; p pump; R immobilized enzyme column (reactor); S samples; T photon-counting tube; V valve; W waste; a computer; b interface; c printer; d key board; e display; f recorder

design of the cell and the direct attachment of the tube to the cell enable very efficient counting with high sensitivity to hydrogen peroxide (5).

A barrack model EBA-02, constructed under a Grant-in-Aid for Life Science Research from the Agency of Science and Technology of the Japanese Government, has performed one-min interval operation with standard and serum samples and showed at least 100-times higher sensitivity in analyses on glucose and uric acid compared with ordinary instruments on the market, with less satisfactory results on cholesterol (4).

The same principle can be applied to any assay system in which the product of the enzyme-catalyzed reaction is hydrogen peroxide. Creatinine in serum may be determined by combined use of creatininase, creatinase, and sarcosine oxidase (6). Similarly, the use of acyl-CoA synthetase and acyl-CoA oxidase for free fatty acids seems to be very promising (7).

REACTOR WITH COIMMOBILIZED ENZYMES

In many cases, two or more different kinds of enzymes are needed to complete the reaction sequence for enzymatic assays. For examples, one form of glucose assay needs glucose oxidase and peroxidase; while the three enzymes cholesterol ester hydrolase, cholesterol oxidase, and peroxidase are required for total cholesterol in serum. We reported that simultaneous use of glucose oxidase and peroxidase in the form of a coimmobilized enzyme column gave higher response to the concentration of glucose in serum than that in the form of a mixed-bed column (8).

We have compared three different types of enzyme reactors with immobilized cholesterol ester hydrolase and cholesterol oxidase: sequential, mixed-bed, and coimmobilized. The sequential enzyme column was prepared by packing the mini-column in two layers with two kinds of glass beads, the one bearing only the hydrolase and the other with only the oxidase. The mixed-bed enzyme column was prepared with a mixture of the hydrylase-glass beads and the oxidase-glass beads. The coimmobilized enzyme column contained glass beads to which the hydrolase and oxidase had been simultaneously immobilized. The size of the column was identical for each type; and the units of each enzyme in the different columns were

TABLE 2

ASSAY FOR TOTAL CHOLESTEROL USING IMMOBILIZED CHOLESTEROL
ESTER HYDROLASE AND CHOLESTEROL OXIDASE IN COLUMN FORM

Total Cholesterol (g/L)	Absorbance at 590 nm*		
	Sequential	Mixed-bed	Coimmobilized
1.0	0.070	0.130	0.168
2.0	0.134	0.250	0.335
3.0	0.198	0.366	0.515
4.0	0.256	0.492	0.674

*Hydrogen peroxide produced was determined using 3-methyl-2-benzothiazolinone hydrazone and N,N'-dimethylaniline.

made equal. A human serum sample containing 67% cholesterol ester
was used.

Table 2 summarizes the results of analyses obtained with the
three different types of reactor under otherwise identical conditions. The coimmobilized enzyme column gave the highest efficiency,
while the sequential type could oxidize only 40% of the total cholesterol present in the sample. This suggested that the ester hydrolase was almost ineffective in the sequential case (9).

RECYCLING OF NAD

We previously described the synthesis and coenzyme activity
of water-soluble, high MW NAD derivatives covalently bound to dextran and suggested that these derivatives could be used for enzymatic analyses as reusable coenzymes (10). Dextran T40-N^6-(NHCH$_2$
CO)$_3$-NHCH$_2$CH$_2$-NAD was synthesized from N^6-2-aminoethyl-NAD and
dextran T40-glycylglycylglycine by condensation with a water-
soluble carbodiimide reagent. The product had approximately 50%
coenzyme activity when assayed with lactate dehydrogenase (LDH)
from beef heart.

Using the dextran-NAD derivative, we have established a batchwise recycling system for the analysis of lactate with an immobilized LDH as shown in Fig. 2. Lactate was oxidized by immobilized
lactate dehydrogenase plus dextran-NAD in the column reactor; and
the amount of dextran-NADH produced was continuously monitored by
fluorescence emitted at 460 nm under excitation at 350 nm. The
reduced coenzyme was then re-oxidised with phenazine methosulfate;
and from the mixture the polymer NAD was recovered by gel filtration for recycling (11).

With lactate dehydrogenase at pH 9.0, dextran-NAD was found
to retain more than 89% of the original, specific coenzyme activity
after four cycles of regeneration. While in the assay for glutamate with immobilized glutamate dehydrogenase at pH 7.5, the
stability of dextran-NAD was much greater, so that it could be recycled seven times (11).

ACKNOWLEDGMENTS

This work was supported in parts by grants from the Ministry
of Education, Science and Culture and from the Agency of Science
and Technology, Japan, and by a research grant from the Japan-IBM,
Co., Tokyo.

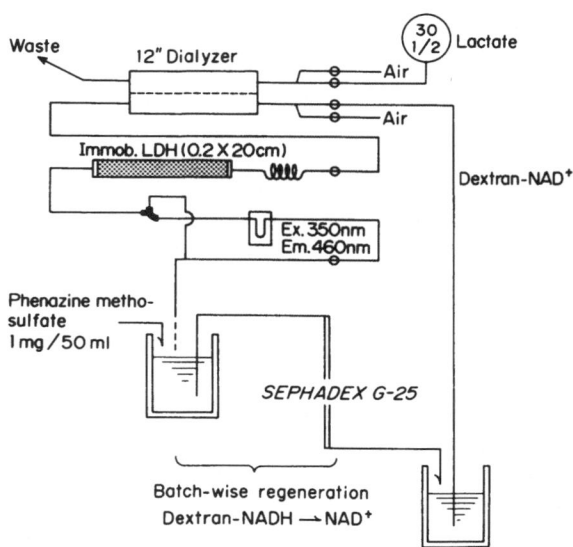

Fig. 2. Flow diagram for batch-wise recycling of dextran-NAD
with immobilized lactate dehydrogenase.

REFERENCES

1. ENDO, J., TABATA, M. & MURACHI, T. *Abstr. 10th Intern. Cong.
    Biochem.*, Hamburg, July (1976) p. 588.
2. ENDO, J., TABATA, M., OKADA, S. & MURACHI, T. *Clin. Chim.
    Acta 95:* 411 (1979).
3. MURACHI, T., SAKAGUCHI, Y., TABATA, M. & ENDO, J. in "Clinical
    Biochemistry, Principles and Practice" (A. W. Eng, and P.
    Garcia-Webb, eds.) Organizing Committee, 1st South East
    Asian and Pacific Congress of Clinical Biochemistry,
    Singapore (1980) p. 100.
4. SAHEKI, Y. & KAGAMIYAMA, H. in "Present and Future of Life
    Science," Institute for Chemical and Physical Research,
    Sozo, Tokyo, p. 366 (Japanes.).
5. SAHEKI, Y., NOZAKI, M. & OKANISHI, T. *Seikagaku 51:* 828
    (1979) (Japanes.).
6. RIKITAKE, K., OKA, I., ANDO, M., YOSHIMOTO, T. & TSURU, D.
    J. Biochem. (Tokyo) 86: 1109 (1979).
7. SHIMIZU, S., TANI, Y., YAMADA, H., TABATA, M. & MURACHI, T.
    Anal. Biochem. 107: 193 (1980).
8. MURACHI, T., SAKAGUCHI, Y., TABATA, M., SUGAHARA, M. & ENDO,
    J. *Biochimie 62:* 581 (1980).
9. TABATA, M., ENDO, J. & MURACHI, T. *J. Appl. Biochem. 3:*
    84 (1981).

10. SAKAGUCHI, Y. & MURACHI, T. *J. Appl. Biochem. 2:* 117 (1980).
11. SAKAGUCHI, Y., SUGAHARA, M., ENDO, J. & MURACHI, T. *J. Appl. Biochem. 3:* 32 (1981).
12. TABATA, M., ENDO, J., HARA, A. & MURACHI, T. *Abstr. 1st S. East Asian Pacific Cong. Clin. Biochem.* Singapore, October (1979) p. 150.
13. TABATA, M., IKEMOTO, M. & MURACHI, T. *J. Clin. Lab. Automation 6 Suppl.:* 10 (1981) (Japanes.).

IMMOBILIZED ENZYMES IN ANALYSIS: APPLICATIONS AND ECONOMIC ASPECTS

M. Gloger, M. Nelboeck, D. Doring*, and S. Klose

Boehringer Mannheim GmbH, Biochemical Research
Center, Tutzing, and Clinicon International
GmbH*, Mannheim, Federal Republic of Germany

Different requirements must be met for analytical and techno-
logical applications of enzymes. In the first case small amounts
of pure enzymes are used; while in the second case large scale
preparations of less purified enzymes can be applied to transform
pure substrates. At the Moment most of the practical analytical
work is carried out with soluble, native enzymes and not with immobilized enzymes; although there is a lot of information available
in the literature and some technical information from several companies concerning immobilized enzymes and their applications in
analysis.

The methods for preparing immobilized enzymes are well developed. For analytical applications covalently bound enzymes primarily are under investigation. Mechanical stability and rigidity
of the support also are of special interest; therefore, inorganic
supports or strong organic polymers are used as carriers.

Immobilized enzymes have advantages compared to soluble enzymes in analytical systems a) multiple use of a single batch of
enzyme is possible; b) immobilized enzymes exhibit increased stability under storage and also under reaction conditions; c) there
is no contamination of the sample by using immobilized enzymes;
d) it is easy to remove immobilized enzymes from the reaction mixture; e) immobilized enzymes are ready for use as reagents; f)
sequential reactions are possible with high efficiency by using
coimmobilized enzymes; and g) immobilized enzymes can be adapted
to different kinds of measuring systems. Besides the advantages
in handling and convenience, one can change the apparent properties
of an enzyme by immobilization in a more or less estimated way.
Changes in pH-optimum, K_m, or inhibition constants may result from

377

the new microenvironment which is built up during immobilization.
Some of these changes can be utilized to adapt an enzyme to specific test systems for kinetic sybstrate or inhibitor measurements
(1).

The most important analytical applications are in clinical
chemistry, food and environmental analysis, microbiology, and process control. In biochemical analysis discontinuous measurements
with single samples usually are carried out; but in process control
the measurements are done continuously. Here immobilized enzymes
can compete better with soluble enzymes, especially, if new equipment is developed for their application.

Immobilized enzymes are used or can be assayed in stirred tank
reactors (2) or in continuous flow systems (3). Continuous flow
analyzers (e.g., Auto Analyzer) can be adapted easily to immobilized enzyme preparations. The enzyme reactor is inserted into
the secondary loop; and there is no direct contact between the sample and the immobilized enzyme. In general enzyme tubes are used;
but there also have been good results with columns containing particle bound enzymes (4). The continuous flow system is very flex-
ible.

The so-called transducer-bound enzymes (5) are interesting
tools for analytical chemistry. They contain a thin layer of an
immobilized enzyme bound to a special transducer which can detect
an electrical or thermal signal triggered by a substrate in the
enzyme layer or in its surrounding. Electrodes as well as thermistors have been used as transducers until now. However new systems called CHEMFETs (chemical sensitive field effect transistors)
use integrated circuit technology in combination with chemical
sensitive membranes as sensors (6-9). They bring the advantages
of short response time, good base line stability, very small size,
and the possibility for signal amplification, as compared to classical transducer bound enzymes. Different types of CHEMFETS are
projected or are under development. One can expect their application in the future as cheap disposables for all kinds of selective
monitoring in biochemistry, medicine, and process control.

Clinical chemistry is an important application of immobilized
enzymes (5, 10-22) (Table 1). For each substrate parameter measured in a large hospital, several immobilized enzyme systems are
available for use in combination with different detectors. But
only one enzyme of clinical interest, creatin kinase (CK), uses
an immobilized material, luciferase, as indicated from the literature. Some of the enzyme tubes mentioned in Table 1 are available
commercially.

Food analysis (Table 2) and process control are other fields
for application of soluble and immobilized enzymes. Compared to

TABLE 1

APPLICATIONS OF IMMOBILIZED ENZYMES IN CLINICAL CHEMISTRY

Parameter	Immobilized Enzyme	Detector
glucose	nylon-tube: coimmobilization of hexokinase and glucose-6-phosphate dehydrogenase	photometer
	glucose oxidase column	H_2O_2-electrode
	glucose oxidase column	thermistor
	glucose oxidase membrane	H_2O_2 electrode
cholesterol	chol.-OD-membrane	H_2O_2-electrode
	chol.-OD/catalase-column	thermistor
triglycerides	lipoproteinlipase-membrane	pH-electrode
	nylon-tube: coimmobilization of phosphokinase, lactate dehydrogenase	photometer
	nylon-tube: glycerol dehydrogenase	photometer
urea	nylon-tube: urease	photometer
	urease-membrane	pH-electrode
	urease-column	thermistor
uric acid	nylon-tube: uricase	photometer
	uricase/aldehyde dehydrogenase	photometer
	uricase-column	photometer
	uricase-membrane	CO_2-electrode
creatine kinase	luciferase-BSA-gel	photomultiplier

clinical chemistry these fields are just in the early stages of their development. In food analysis there is no need for automated systems, because the number of samples to be handled in a given time is not as high as in the clinical laboratory. But there exists a wide variety of substances which can be measured only after a complex sample preparation. Here the application of en-

TABLE 2

APPLICATIONS OF IMMOBILIZED ENZYMES IN FOOD ANALYSIS

Parameter	Immobilized enzymes	Detector
L-amino acids	L-amino acid oxidase	H_2O_2-electrode
L-lysine	lysine decarboxylase	CO_2-electrode
L-aspartic acid	aspartate aminotransferase MDH	photometer
alcohols	alcohol-oxidase	H_2O_2-electrode
amygdalin	β-glucosidase	CN^--sensitive electrode
lactose	β-galactosidase + glucose oxidase and catalase	thermistor
sucrose, glucose, fructose	invertase, hexokinase, glucose-6-phosphate dehydrogenase (NADPH-regeneration by means of glutathion-reductase)	closed system, photometer

zymes is very promising. Procedures using immobilized enzymes are under development for estimation of carbohydrates, amino acids, and alcohols (Table 2). Selective electrodes, thermistors, and photometer units are used as detector systems. Continuous process control with enzymes is used in technology, health care, and environmental analysis. Immobilized enzymes can display advantages over soluble enzymes if complete systems can be developed. Such monitors can be used to control systems in the fermentation industry. Different types of penicillin sensors are described (8, 23, 24) that use immobilized β-lactamase. In some companies enzyme tubes and enzyme columns containing glucose oxidase and glycerol dehydrogenase are used for continuous fermentation control. A flow chart for this application is given in Fig. 1. Sampling is carried out with a dialyzer to prevent contamination of the fermentation broth.

Continuous monitoring in medicine and health care is another approach for application of immobilized enzymes. For diabetic

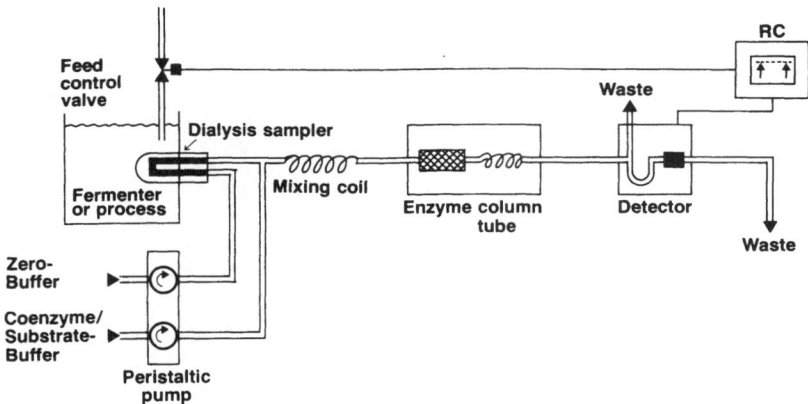

Fig. 1. Flow chart for continuous on-line measurement and
         control in fermentation processes.

patients a glucose sensor in combination with a computerized insu-
lin pump can be applied to insulin-dependent people (25). The on-
line glucose sensor for use in whole blood works with a glucose
oxidase membrane and an electrochemical cell to measure hydrogen
peroxide. The response is fast; and the glucose oxidase membrane
is very stable. But long time experiments are limited because ag-
gregated platelets block the membrane.

ECONOMIC ASPECTS

One of the advantages of immobilized enzymes compared to soluble enzymes is reusability. It is possible to reuse immobilized
enzymes in analytical systems 100-1000 times or more. In that respect the mechanical stability of the carrier as well as the general stability of the fixation product must be examined carefully.
Reuse is an economic factor if large amounts of enzymes are consumed. This is the case for technical application of enzymes but
not for analytical application. Even in clinical chemistry only
13 g of pure enzymes are consumed for about 120,000 measurements
in a large hospital. One can calculate an approximate annual requirement of 50-100 kg of highly purified enzymes worldwide for
clinical chemistry (26).

Cost reductions may arise not only through reuse of enzymes
but also from changes in technical equipment. Soluble enzymes
normally are used in batch reactors. With immobilized enzymes one
can work continuously in different types of reactors. For example,
the overall cost for the production of L-amino acids is reduced
to 60%; although only 25% of the enzyme cost is saved by reuse
(27).

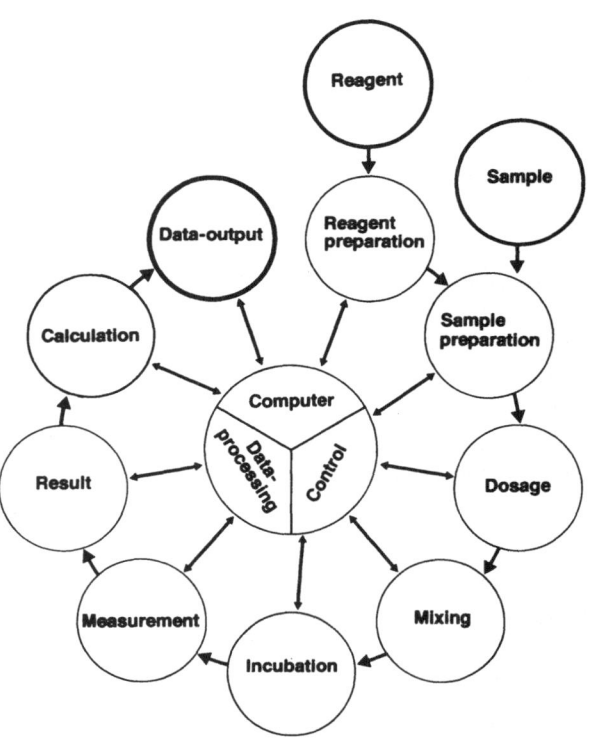

Fig. 2.   General processing-scheme for analysis in clinical
          chemistry.

        Since this field is dominated by applications in clinical
laboratories, the economic aspects will be limited to these laboratories. The analytical methods used in a clinical laboratory
utilize reagents, equipment, and personnel. Today the equipment
consists not only of mechanized or automated machines, but includes
computer systems which are essential for economic functioning of
the laboratories. As can be seen from Fig. 2 reagents represent
only a very small part of the complete system. Beside the technical equipment, a lot of personnel are necessary to do the work.
The costs for carrying out clinical tests were calculated carefully
(26). One can distinguish between fixed costs (staff, depreciation, services) and operational costs (reagents, disposables, other
costs). For our discussion the costs of the reagents and their
contribution to the total costs are important. These connections
are given in Fig. 3 for manual, semi-automatic, and fully auto-
matic testing. The numbers given are calculated as mean values
for various determinations on an average number of tests per year.
With increasing degrees of automation the costs for reagents and
staff decrease and the costs for disposables, depreciation, and

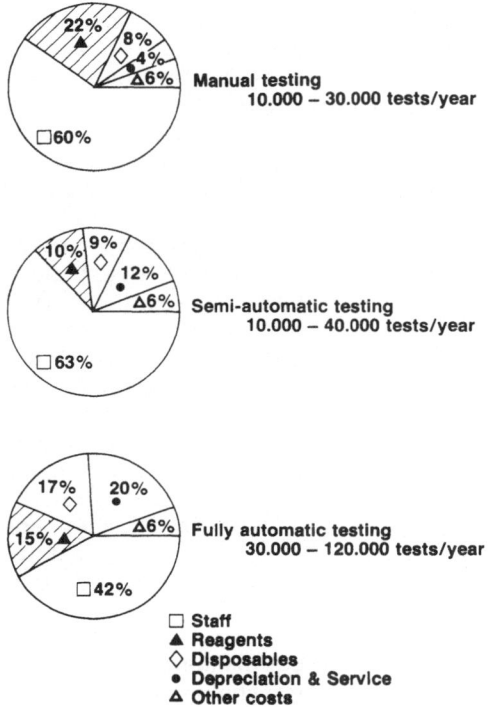

Fig. 3. Distribution of costs in different clinical laboratories.

service increase. Reagents contribute 10-22% of the total costs of a clinical laboratory. The enzyme cost can be calculated to be less than 2% of the total cost (only 50% of the tests use enzymes, and 20% of the reagent costs are enzyme costs).

These data demonstrate that the economy of a clinical laboratory is almost independent of the cost of enzyme and the degree of automation. Therefore, from a cost standpoint it matters little whether one uses soluble disposable or immobilized reusable enzymes in a clinical lab; however, in technology the economy of the process often depends on the cost of the reagents (enzymes). Similar calculations were done for other laboratories (28). Even for tests using radio-pharmaceuticals the cost of reagents is only about 25% of the total cost; again the largest single item is the cost for the technical staff.

In connection with the reusability and economics of immobilized enzymes in analysis it is necessary to mention some other important points. a) Application of immobilized enzymes in analysis requires the development of a more or less expensive module,

containing the immobilized enzymes. Such modules have limited
life spans; and their basic share of the total cost remains even
if the enzyme costs are reduced. b) During fixation there may
be a remarkable loss of enzyme activity due to inactivation during
the chemical reaction or due to the transition from a homogeneous
to a heterogeneous system. These losses must be compensated for
by additional reuse cycles. c) Enzymatic tests are generally done
with multicomponent systems containing substrates, enzymes, coenzymes, and buffers. Even if one component can be reused, the
others are consumed and must be added to every new sample. d)
Applications of immobilized enzymes in continuous flow systems must
compete with reagent mixes. These mixtures contain all the reagents necessary for one reaction in a stable form. The handling
and operation is simple; and reagents no longer are added consecu-
tively but as far as possible simultaneously. If the enzymes from
such a mixture are immobilized, the remaining components still require a mixture. The costs cannot be reduced significantly since
costs other than for the enzyme (i.e., stabilizers and packaging)
determine the price per test. There is another advantage of ready
to use stable reagent mixtures; they save operator time. This is
a more effective way to economize the costs of a clinical laboratory.

    We have seen that for discontinuous flow analyzers, which are
used in many laboratories today, the reusability of immobilized
enzymes or immobilized reagents is not an important factor. These
systems work better with reagent mixtures than with immobilized
enzymes. Quite different is the situation for closed system continuous flow analyzers (29). A permanent circulating buffer stream
makes the instrument continuously available. An absolute prerequi-
site for this mode of operation is localization of the enzyme,
which can be obtained only by immobilization. In this way economically long term runs without wasting enzymes are possible.
Closed circuit continuous flow apparatus are of special interest
in emergency diagnostics and for continuous measurements.

                        SUMMARY AND FUTURE ASPECTS

    For most of the important tests in clinical chemistry and
food analysis well developed methods with immobilized enzymes are
described in the literature. But their practical application is
rather limited at the moment. The reason for this is that only
small amounts of pure enzymes are necessary to perform all of the
analytical tests; and a reusable form has almost no advantage compared to soluble enzymes if open discontinuous flow analyzers are
considered. Instead, already to use stabilized reagent mixtures
better meet the analytical needs than do immobilized enzymes.
Furthermore, the economics of clinical laboratories is not influenced nearly so much by reagents and enzyme costs as by the

costs of personnel. Reusable enzymes, therefore, cannot reduce the costs to a marked degree.

The situation is quite different in the technical application of immobilized enzymes. For production processes large amounts of enzymes are consumed. Hence, the application of a reusable form of the catalyst together with an optimized equipment (reactor) leads to an important reduction in cost.

Therefore, it is expected that immobilized enzymes will find more practical uses in analytical applications that require the development of new equipment than in analyzers built for complete programs in clinical chemistry. Reagentless enzyme electrodes and closed flow analyzerswith immobilized enzymes will find their place for measurement of single parameters under special conditions, such as in the emergency laboratory or for bedside application. A dramatic change in analytical methods should be possible by the application of immobilized enzymes together with the new transducers to form enzyme transistors. Miniaturization and the possibility to build up multi-probe analyzers will influence very much the future developments in this field.

REFERENCES

1. GUILBAULT, G. G. J. Solid Phase Biochem. 2: 329 (1977).
2. MATTIASSON, B. & MOSBACH, K. Acta Chem. Scand. 24: 2093 (1970).
3. FILIPPUSSON, H., HORNBY, W. E., & MCDONALD, A. FEBS Lett. 20: 291 (1972).
4. ENDO, J., MASAYOSHI, T., SACHIKO, O., & MURACHI, T. Clin. Chim. Acta 95: 411 (1979).
5. BOWERS, L. D. & CARR., P. W. in "Advances in Biochemical Engineering, vol. 15,. (A. Fiechter, ed.), Springer Verlag, Berlin (1980) p. 89.
6. MOSS, ST. D. in "Roy. Soc. 1st Conf.: New Technologies in the Health Care Industry," London, June 11-12 (1979).
7. DANIELSSON, B., LUNDSTROM, I., MOSBACH, K., & STIBLERT, L. Anal. Lett. 12: 1189 (1979).
8. CARAS, S. & JANATA, J. Anal. Chem. 52: 1935 (1980).
9. United States Patent 4,020,830 (1977).
10. EVERSE, J., GINSBURGH, C. L., & KAPLAN, N. O. "Methods of Biochemical Analysis," vol. 25, Wiley, New York (1979) p. 135.
11. DANIELSON, B., GADD, K., MATTIASSON, B., & MOSBACH, K. Clin. Chim. Acta 81: 163 (1977).
12. BERTRAND, C., COULET, P. R., & GAUTHERON, D. C. Anal. Lett. 12: 1477 (1979).
13. SATOH, I., KARUBE, I. & SUZUKI, S. Anal. Chim. Acta 106: 369 (1979).

14. HINSCH, W. & SUNDARAM, P. V. *Clin. Chim. Acta 104:* 87 (1980).
15. LEON, L. P., SANSUR, M., SYNDER, R. L., & HORVATH, C. *Clin. Chem. 23:* 1556 (1977).
16. COLLISS, J. S. & KNOX, J. M. *Appl. Biochem. Biotechnol 6:* 15 (1981).
17. TABATA, M., ENDO, J., & MURACHI, T. *J. Appl. Biochem. 3:* 84 (1981).
18. CHUA, K. S. & TAN, I. K. *Clin. Chem. 24:* 150 (1978).
19. SALLEH, A. B. & LEDINGHAM, W. M. *Anal. Biochem. 116:* 40 (1981).
20. SUNDARAM, P. V., IGLOI, M. P., WASSERMANN, R., & HINSCH, W. *Clin. Chem. 24:* 1813 (1978).
21. RODRIGUEZ, O. & GUILBAULT, G. G. *Enzyme Microb. Technol. 3:* 69 (1981).
22. FRESENIUS, R. E., WONNE, K.-G., & FLEMMING, W. *Z. Anal. Chem. 271:* 194 (1974).
23. NILSSON, H. & MOSBACH, K. *Biotechnol. Bioeng. 20:* 527 (1978).
24. RUSLING, J. F., LUTTRELL, G. H., CULLEN, L. F., & PAPARIELLO, G. *J. Anal. Chem. 48:* 1211 (1976).
25. FOGT, E. J., LAWRENCE, M. D., JENNING, E. M., & CLEMENS, A. H. *Clin. Chem. 24:* 1366 (1978).
26. NELBOECK, M., DOERING, D., & KLOSE, S. in "Applied Biochemistry and Bioengineering," vol. 3 (L. B. Wingard Jr., E. Katchalski-Katzir, and L. Goldstein, eds.) Academic Press, New York (1981) p. 253.
27. CHIBATA, I. in "Immobilized Enzymes, Research and Development" (I. Chibata, ed.) Wiley, New York (1978) p. 175.
28. BARNARD, D. J., BINGLE, J. P., & GARRATT, C. J. *Brit. Med. J. 1:* 1463 (1978).
29. BERGMEYER, H.-U. & HAGEN, A. *Z. Anal. Chem. 261:* 333 (1972).

MICROBIAL SENSORS FOR GAS ANALYSIS

S. Suzuki and I. Karube

Research Laboratory of Resources Utilization
Tokyo Institute of Technology
Yokohama, Japan

Various kinds of biosensors have been developed and applied to clinical, fermentation, and environmental analyses (1-3). These biosensors are used for the determination of organic compounds dissolved in an aqueous solution. However, the determination of gaseous compounds in the atmosphere is also required for indistrial process and environmental control.

Most of the commercially available gas sensors use semiconductors and are based on the gases causing a change in the electrical conductivity of semiconductors. But the selectivity of these sensors is not sufficient. As previously reported, many microbial sensors consisting of immobilized whole cells and an oxygen electrode have been developed for determination of BOD (4), alcohols (5), acetic acid (6), and mutagens (7). These organic compounds were selectively determined from the respiration activity of immobilized whole cells with the oxygen electrode. Therefore, the principle of the microbial sensor can be applied to the determination of the gaseous compounds. In this paper, microbial sensors for ammonia, methane, and nitrogen dioxide are described.

MICROBIAL SENSOR FOR AMMONIA GAS

The determination of ammonia is important in clinical, environmental, and industrial process analyses. An ammonia gas electrode consisting of a combined glass electrode and a gas permeable membrane is usually used for this purpose (8). In this case, the determination must be performed under strong alkaline conditions (above pH 11). The ammonia electrode is based on potentiometric detection of ammonia. However, volatile compounds such as amines

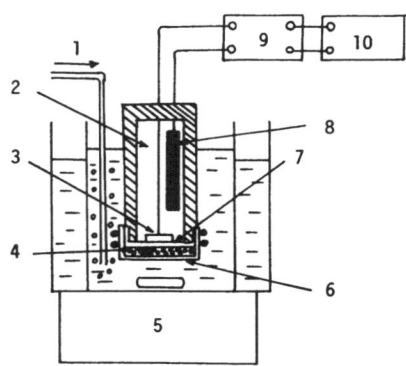

Fig. 1.   Schematic diagram of ammonia gas sensor: 1 = air at 280
          mL/min; 2 30% NaOH; 3 = Pt cathode; 4 = immobilized
          whole cells; 5 = magnetic stirrer; 6 = gas permeable
          membrane; 7 = teflon membrane; 8 = Pb anode; 9 =
          amplifier; 10 = recorder.

often interfere with the determination of ammonia.   Therefore, an
ammonia sensor based on amperometry is desirable.

     Nitrifying bacteria contain two genera of bacteria.   The
genus *Nitrosomonas* sp. utilizes ammonia as the sole source of
energy, with 1.5 mole O_2 used/mole NH_3 to give $NO_2^- + H_2O + H^+$.
The other genus *Nitrobacter* sp. oxidizes nitrite to nitrate, requiring 0.5 mole O_2/mole NO_2^-.   The oxidation of both NH_3 and NO_2^-
proceeds at a high rate; and oxygen uptake by the bacteria can
be determined directly by the oxygen electrode attached to the
immobilized bacteria.   The nitrifying bacteria isolated from
activated sludges were immobilized onto a porous acetylcellulose
membrane of 0.45 µm pore size.   The oxygen electrode consisted of
a teflon membrane, platinum cathode, lead anode, and 30% sodium
hydroxide electrolyte.   The porous membrane, retaining the immobilized bacteria, was fixed on the surface of the teflon membrane
of the oxygen electrode.   The bacterial membrane was covered with
a gas permeable teflon membrane of 0.5 µm pore size and fastened
with rubber rings (Fig. 1).

     The microbial sensor was inserted into a glycine buffered pH
10 sample solution saturated with dissolved oxygen; and the current was obtained at time zero.   This current corresponded to the
endogenous respiration level of the immobilized bacteria.   When
the ammonia solution was injected into the buffer solution, the
ammonium ion changed to ammonia gas.   Ammonia gas permeated through
the gas permeable membrane and was assimilated by the immobilized
bacteria.   Oxygen was then consumed by the bacteria so that the
concentration of dissolved oxygen around the membrane decreased.

The current decreased until it reached a steady state which indicated that the consumption of oxygen by the bacteria and the diffusion of oxygen from the sample solution to the bacteria membrane
were in equilibrium. The steady-state current depended on the
concentration of ammonia. The response time for the determination
of ammonia was within 4 min. The pH of the sample solution had
to be kept sufficiently above the pK for ammonia (9.1 at 30°C)
because ammonium ions could not pass through the gas permeable
membrane. When the sensor was inserted in tap water, the current
of the sensor returned to its initial level within 5 min.

A linear relationship was observed between the current decrease (the current difference between the initial and the steady
state) and the ammonia concentration below 42 mg/L. The minimum
concentration for the determination of ammonia was 0.1 mg/L. The
current decrease was reproducible within $\pm 4\%$ of the relative
error when a sample solution containing 21 mg/L of ammonium hydroxide was employed. The standard deviation was 0.7 mg/L in 20 experiments. Thus the amperometric determination of ammonia became
possible by the microbial sensor. The sensitivity of the microbial
sensor was almost at the same level as that of a glass electrode.

The selectivity of the microbial electrode for ammonia was
examined. The sensor did not respond to volatile compounds such
as acetic acid, ethyl alcohol, and amines (diethylamine, propyl-
amine, and butylamine), or to nonvolatile nutrients such as glucose,
amino acids, and metal ions (potassium ion, calcium ion, and zinc
ion). The long-term stability of the microbial sensor was examined
with a sample solution containing 33 mg/L of ammonia. The current
output of the electrode was almost constant for more than 10 days
and 200 assays.

The microbial sensor was applied to the determination of ammonia in human urine. The urine was diluted; and the concentration
of ammonia was determined by the electrochemical sensor and by a
conventional method. Good comparative results were obtained (correlation coefficient 0.9).

In conclusion, the microbial sensor appears to be quite
promising and very attractive for the amperometric determination
of ammonia.

MICROBIAL SENSOR FOR METHANE GAS

Methane is a clean fuel and major component of city gas (88%
methane) but forms an explosive mixture with air (5-14% methane).
A methane oxidizing bacterium, which grows well, has been isolated
from a natural source, grown in pure culture, and identified as
a new species, *Methylomonas flagellata* (9).

M. flagellata utilizes methane as its sole source of energy; and oxygen is consumed by respiration as follows; CH_4 + $NADH_2$ + O_2 → CH_3 OH + NAD + H_2O (10). *M. Flagellata* was immobilized in acetylcellulose filters with agar. The microbial sensor system, schematically illustrated in Fig. 2, is composed of an immobilized microorganism reactor (300 mg immobilized cells/ reactor), a control reactor, and two oxygen electrodes. Methane gas was introduced into both reactors by a pump at a controlled flow rate. The difference between the output currents of the two electrodes was related to the amount of methane in the flow lines. When sampl gas containing methane was transferred to the immobilized bacteria cells, methane was assimilated by the mocroorganisms. Oxygen was then consumed by the microorganisms so that the concentration of dissolved oxygen in the reactor decreased. The current decreased and reached a steady-state, which indicated that the consumption of oxygen by the microorganisms and the diffusion of oxygen from the sample gas to the immogilized bacteria were in equilibrium. The steady-state current depended on the concentration of methane. When air passed through the flow reactor, the current of the sensor returned to its initial level within 1 min. The response time required for the determination of methane gas was 1 min. The total time required for an assay of methane gas by this steady-state method was 2 min.

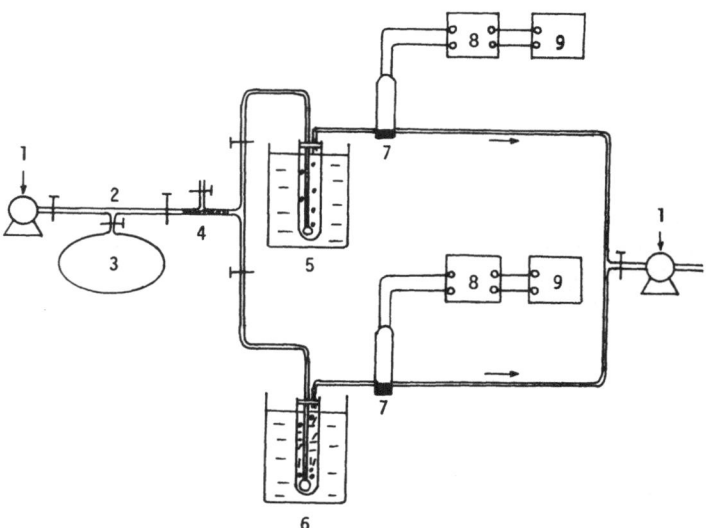

Fig. 2.   Schematic diagram of methane gas sensor: 1 = pump; 2 = gas sampler; 3 = sample gas; 4 = cotton filter; 5 = reference reactor; 6 = methane oxidizing bacteria reactor; 7 = oxygen electrode; 8 = amplifier; 9 = recorder.

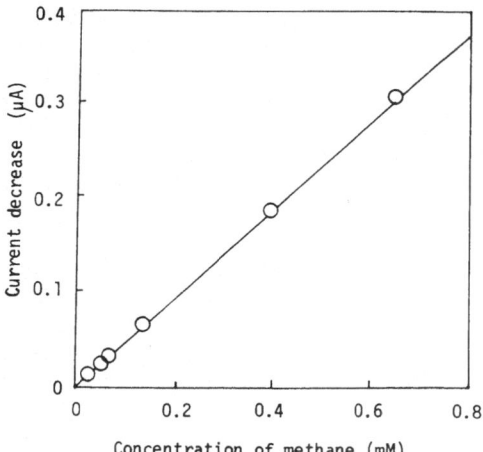

Fig. 3. Calibration curve for methane gas sensor. Conditions: 30°C, pH 7.2, sample gas flow 80 mL/min, and cell content 300 mg wet cells in 30 mL wet filter.

        Fig. 3 shows the calibration curve for this microbial sensor system. A linear relationship was observed between the current difference, between the electrodes, and the concentration of methane (below 6.6 mM). The minimum concentration for the determination was 13.1 µM. The current decrease was reproducible within ±5% in 25 experiments with sample gas containing 0.66 mM methane. The maximum current difference was observed at 30°C, which was used in the subsequent experiments. The current difference was almost constant below an air flow rate of 8.0 ml/min; but above this flow rate the current difference decreased. The long-term stability of the microbial sensor was examined with a sample gas containing 0.66 mM methane. The current output was almost constant for more than 20 days and ·500 assays. In the same experiment a good correlation was obtained between the methane concentrations determined by the electrochemical sensor and conventional gas chromatography methods (correlation coefficient 0.97). This sensor system can be used to determine the content of methane gas in the atmosphere. In conclusion, this microbial sensor system which used immobilized *M. flagellata* appears promising and a very attractive system for the rapid and continuous determination of methane.

                    MICROBIAL SENSOR FOR NITROGEN DIOXIDE GAS

        Nitrogen dioxide, formed during the combustion of all types of fossil fuel, is the most reactive of the gaseous oxides of nitrogen. It is a primary absorber of sunlight in the photochemical atmospheric reactions that produce photochemical smog (11).

Therefore, the determination of nitrogen dioxide is important in
environmental and industrial process analyses. The determination
of nitrogen dioxide is presently performed by spectrophotometric
methods (12). However, these require complicated procedures, a
long reaction time, and additional reagents. Gas-sensing probes
for nitrogen dioxide are already available; however, they are affected by carbon dioxide. Therefore, a new nitrogen dioxide sensor
is still required.

    Nitrogen dioxide reacts with water to give nitrous ions (13).
Nitrobacter sp. oxidizes nitrite to nitrate as described earlier.
As previously reported (4) oxygen uptake by the bacteria can be
directly determined by the oxygen electrode attached to the immobilized bacteria. The nitrifying bacteria, isolated from
activated sludges, were cultured in 0.3% $NaNO_2$ solution containing
various salts for 2 months. A schematic of the microbial sensor
is shown in Fig. 4. When the sample gas containing nitrogen dioxide was injected into the system for 2 min, nitrogen dioxide
passed through the gas permeable membrane, and changed to nitrite
in the bacterial layer. Nitrite was assimilated by the immobilized
bacteria. Consumption of oxygen by the microorganisms began and
caused a decrease in the dissolved oxygen around the membrane.
As a result, the current decreased markedly with time until a
steady state was reached. The steady-state current was obtained
within 3 min. When a sufficient quantity of bacteria was immobilized in the bio-sensor, the current depended mainly on the rate
of diffusion of nitrite from the sample solution to the immobilized bacteria. Therefore, the steady-state current depended on
the concentration of nitrogen dioxide. When buffer solution was

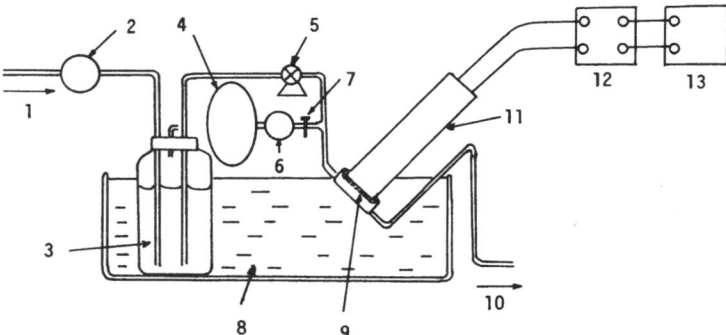

Fig. 4.   Schematic diagram of nitrogen dioxide gas sensor. 1 =
          air at 280 mL/min; 2 & 6 = pump; 3 = buffer at pH 2.0;
          4 = sample gas 5L; 5 = peristaltic pump; 7 = valve; 8 =
          incubator 30°C; 9 = immobilized whole cells; 10 = waste;
          11 = oxygen electrode; 12 = amplifier; 13 = recorder.

used, the current of the microbial sensor returned to its initial
level within 2 min.

A linear relationship was observed between the current decrease (the current difference between the initial and the steady-
state) and the nitrogen dioxide concentration below 255 ppm. The
minimum concentration for the determination of nitrogen dioxide
was 0.51 ppm. The current decrease was reproducible within $\pm$4%
of the relative error when a sample solution containing 51 ppm of
nitrogen dioxide was employed. The standard deviation was 2 ppm
in 25 experiments. The current output of the sensor was almost
constant for more than 24 days and 400 assays. The concentration
of nitrogen dioxide was determined by both the electrochemical
sensor and the conventional Saltzman's method with good correlation
(correlation coefficient 0.99).

REFERENCES

1. GUILBAULT, G. G., "Handbook of Enzymatic Methods of Analysis,"
     Marcel Dekker, New York (1976).
2. CHANG, T. M. S. (ed), "Biomedical Applications of Immobilized
     Enzymes and Proteins," vol. 2, Plenum, New York (1977).
3. SUZUKI, S. & KARUBE, I., "Immobilized Microbial Cells,"
     American Chemical Society, Washington D.C. (1979) p. 221.
4. KARUBE, I., MITSUDA, S., MATSUNAGA, T., & SUZUKI, S. *J.*
     Ferment. Technol. 55: 243 (1977).
5. HIKUMA, M., KUBO, T., YASUDA, T., KARUBE, I., & SUZUKI, S.
     Biotechnol. Bioeng. 21: 1845 (1979).
6. HIKUMA, M., KUBO, T., YASUDA, T., KARUBE, I., & SUZUKI, S.
     Anal. Chim. Acta 109: 33 (1979).
7. KARUBE, I., MATSUNAGA, T., NAKAHARA, T., SUZUKI, S., & KADA,
     T. *Anal. Chem. 53:* 1024 (1981).
8. BAILEY, P. L., "Analysis with Ion-Selective Electrodes,"
     Heyden Spectrum House, London (1978) p. 147.
9. MORINAGA, Y., YAMANAKA, S., OTSUKA, S., & HIROSE, Y. *Agric.*
     Biol. Chem. 40: 1539 (1976).
10. RIBBONS, D. W. *J. Bacteriol. 122:* 1351 (1975).
11. STERN, A.C., "Air Pollution," vol. 3, Academic Press,
     New York (1976).
12. Japanese Standards Assoc., J.I.S. K0516 (1976).
13. ENGLAND, C. & CORCORAN, W. H. *Ind. Eng. Chem. Foundament.*
     13: 373 (1974).

ANALYTICAL USES OF IMMOBILIZED ENZYMES

G. G. Guilbault

Department of Chemistry
University of New Orleans
New Orleans, Louisiana, USA

Although soluble enzymes can be used as excellent reagents
for the analysis of inorganic and organic compounds, they face a
serious challenge when attempts are made to utilize them in complex matrices, like blood or crude water. Problems center about
the effect of activators, inhibitors, other substrates, pH, and
temperature on the soluble enzyme. However, upon immobilization
most of these effects can be eliminated or minimized. For example,
an enzyme with a narrow pH range of 4-6 can be transformed upon
insolubilization to a more viable reagent with a broad pH range
of 4-10. Also, following immobilization the enzymes are much more
stable; they can be heated to 37, 40 or 50°C, with little loss of
activity; and the activity persists after several thousand analysis
are performed. However, the biggest advantage, analytically speaking, of immobilization, is that the insolubilized reagent becomes
a much more selective reagent. No longer do many activators and
inhibitors have an effect; only the most powerful can actually attack the enzyme.

In clinical analysis today electrolytes (Na^+, K^+, Ca^{++}) are
measured with ion-selective electrodes, in almost all instruments.
What then could be more natural than to place in these instruments
electrode probes for metabolites, such as glucose, urea, uric acid,
cholesterol, or creatinine. Such enzyme electrodes could be
fashioned by taking an ion-selective electrode for O_2, CO_2, or NH_3
and coating it with a layer of stabilized, immobilized enzyme.
The substrate to be measured diffuses into the enzyme layer where
it is converted to a product with the uptake or release of O_2,
CO_2, or NH_3. The latter is then measured with the appropriate ion
selective electrode, with the current produced linearly related
to the product concentration.

ENZYME ELECTRODE PROBES

The result of this perfect marriage of the enzyme, with its high selectivity and ultrasensitivity with the electrode sensor is a device that can measure directly either inorganic or organic substances in solution. The only reagent required is a buffer. The range determinable is generally 10^{-1} to 10^{-5} M, with a response time of about 1 min. A list of some of the enzyme electrode probes described in the literature for the anlaysis of substrates is given in Table 1. Over 50 substrate electrodes have been described, most totally specific for the analysis of their substrates. For several substrates many different types of electrode probes have been proposed, e.g. for glucose O_2, Pt, pH, and I^- electrodes have been described. Thus, almost 65 different enzyme electrodes have been proposed.

The stability of the electrode depends on the type of entrapment. Realistically, the immobilization characteristics and the stability should be defined in terms of dry storage and also use storage. Physically entrapped enzymes last about 3-4 weeks or 50-200 assays. For the chemically bound enzyme, 200-1,000 assays is a good range. Nylon tybes have been demonstrated for 10,000 assays. Interferences can be in the sensor itself or from other substrates for the enzyme. Using highly purified enzymes, the electrodes are quite specific for the primary substrate. Inhibitors of the enzyme also produce interferences; and here immobilization makes the enzyme much less susceptible to environmental factors.

D- and L-amino acid oxidases are less selective in their responses, as is alcohol oxidase. Hence, in using electrodes of these enzymes, a separation must be effected if two or more substrates are present. However, we have developed totally specific electrodes for the assay of L-amino acids such as L-methionine (31), L-lysine (29,30), L-tyrosine   (32) or L-phenyl-alanine (82), by inducing, isolating, separating, and purifying decarboxylative or deaminative enzymes, which react totally specifically with one L-amino acid. This same principle can be applied to many enzyme systems, which have previously been reported inaccurately to be non-specific, yet which after purification react with only one substrate. An example of this is creatininase, reported to react with several substrates; (43) yet, after purification the only substrate is creatinine. (44)

A key point to emphasize is that the immobilized enzyme is much less susceptible to inhibitors, especially weak or reversible inhibitors, due to the protection of the immobilization matrix. The user should always be aware of the potential dangers of activators or inhibitors, especially in assaying solutions containing metal ions and pesticides.

ANTIGEN-ANTIBODY PROBES

Another area of application of biological probes is the construction of sensor probes utilizing bound antibodies or antigens. We have, for example, successfully immobilized creatine kinase M antibody as a pretreatment for the detection of cardio-specific CK-MB isoenzyme. Goat anti-human CK-M IgG was immobilized on glass beads through glutaraldehyde coupling; and the immobilized carrier was packed into a magnetic stirring device, which was a rotating porous cell with a removable lid. (111) The bound antibody could be used for several hundred assays. It was regenerable; and excellent results in the use of this immunostirrer probe were obtained in the specific assay of CK-MB. An alternative approach was presented by Suzuki (112), who bound an antigen, and developed an assay for syphilis in blood. The contact potential was measured, with very low Δ mV changes of about 1-3 mV.

FUTURE APPLICATION AND COMMERCIAL PRODUCTS

Enzyme electrode probes will probably soon become an integral part of current commercial electrolyte analyzers. Indeed, they are already present in the Photovolt 4 + 2 (4 electrolytes and 2 metabolites, glucose and urea) and in the Medical Technology Associates Analyzer (glucose and urea). Owens-Illinois (Kimble) has designed a urea instrument using immobilized urase and an ammonia electrode probe, and a glucose instrument using insolubilized glucose oxidase and a Pt electrode. (109,110) Patent rights of this system have been purchased by Technicon, Inc. Yellow Springs Instrument Co. (Ohio) markets a glucose instrument with an immobilized glucose oxidase pad placed on a Pt electrode, and has other instruments available for other metabolites. Fuji Electric (Tokyo) markets instruments for glucose and uric acid which are similar in concept to the Yellow Springs system; and the Institute for Molecular Biology (DDR Berlin) markets a glucose instrument that uses bound glucose oxidase on a gelatin matrix.

Finally, Universal Sensors Inc. (New Orleans) markets self-contained enzyme electrodes for glucose, urea, amino acids (L-lysine, L-methionine, L-tyrosine, L-phenylalanine), lactate, and creatinine using a highly stable enzyme membrane of pig intestine. These electrodes, together with a glucose electrode from Tacussel (France), are the only ones available which are not part of large elaborate instruments.

TABLE 1

ENZYME ELECTRODE PROBES FOR SUBSTRATES

Type	Enzyme	Sensor	Ref.
Acetic & formic acids	Alcohol oxidase		
Acetylcholine	Acetylcholinesterase	Choline	(2,3)
Acetyl-β-methylcholine	Acetylcholinesterase	Acetyl-choline	(4-6)
Adenosine monophophate	5-Adenylate deaminase	NH_4^+	(7,8)
Alcohols	Alcohol dehydrogenase	Pt	(9)
	Alcohol dehydrogenase/diaphorase	Pt	(10)
	Alcohol oxidase	Pt	(11)
D-Amino acids*	D-Amino acid oxidase	NH_4^+	(12,13)
L-Amino acids** (general)	L-Amino acid oxidase	Gas (NH_3)	(14)
		NH_4^+	(15-19)
		Pt	(20)
	Decarboxylases	CO_2	(21)
L-Arginine	Arginase	NH_4^+	(22)
L-Asparagine	Asparaginase	NH_4^+	(17,23)
L-Cysteine	Proteus morganii	H_2S	(24)
L-Glutamine	Glutaminase	NH_4^+	(25)
L-Glutamic acid	Glutamate dehydrogenase	NH_4^+	(26)
	Glutamate decarboxylase	CO_2	(25,27)
L-Histidine	Histidinase	NH_4^+	(28)
L-Lysine	Lysine decarboxylase	CO_2	(29,30)
L-Methionine	Methionine ammonia lyase	NH_3	(31)
L-Phenylalanine	Phenylalanine ammonia lyase	NH_3	(82)
L-Tyrosine	Tyrosine decarboxylase	CO_2	(32)
	Tyrosinase	Gas (O_2)	(33)
Amygdalin	β-Glucosidase	CN^-	(34,35)
Butyryl thiocholine	Cholinesterase	Pt (SCh)	(36)
Cholesterol	Cholesterol esterase/cholesterol oxidase	Pt (H_2O)	(37)
		Pt (O_2)	(38-41)
	Cholesterol oxidase	Pt (O_2)	(42)
Creatinine	Creatininase	NH_4^+	(43)
	Creatininase (purified)	NH_3	(44)

TABLE 1 (CONT'D)

ENZYME ELECTRODE PROBES FOR SUBSTRATES

Type	Enzyme	Sensor	Ref.
Glucose	Glucose oxidase	pH	(45)
		$pt(H_2O_2)$	(46-50, (103-110)
		Pt(Quinone)	(51)
		Pt(DCIP)	(52,53)
		$Pt(O_2)$	(1, 54, 100, 102)
		I^-	(55-57, 99)
		$Gas(O_2)$	(58-63, 101)
	Glucose oxidase peroxidase	Pt	(64)
Lactic acid	Lactate dehydrogenase	$Pt-Fe(CN)_6^{4-}$	(51)
		C(NADH)	(65)
	$Cyt-b_2$	Pt	(52,53, 66-70)
Lactose	β-Galactosidase/ glucose oxidase	$Gas(O_2)$	(71,72)
Lectin	Glucose oxidase	$Gas(O_2)$	(118)
Maltose	Maltase/glucose/ oxidase	$Gas(O_2)$	(71-72)
NADH	Alcohol dehydrogenase	Pt	(74)
	Mitochandria	$Gas(O_2)$	(73)
Nitrate	Nitrate reductase/ nitrite reductase	NH_4^+	(75,96)
Nitrite	Nitrite reductase	$Gas(NH_3)$	(76)
Oxalic Acid	Oxalate decarboxylase	$Gas(CO_2)$	(77)
Penicillin	Penicillinase	pH	(45, 78-81)
Peroxide	Catalase	$Pt)O_2)$	(83)
Phenol	*Trichosporon cutaneum*	$Gas(O_2)$	(119)
Phosphate	Phosphatase/glucose oxidase	$Pt(O_2)$	(84)
Saccharose	Invertase/mutarotase/ glucose oxidase	$Pt(O_2)$	(85)
Succinic acid	Succinate dehydro- genase	$Pt(O_2)$	(86)

TABLE 1 (CONT'D)

ENZYME ELECTRODE PROBES FOR SUBSTRATES

Type	Enzyme	Sensor	Ref.
Sucrose	Sucrase/glucose oxidase	$Pt(H_2O_2)$	(87)
Sulfate	Aryl sulfatase	Pt	(88)
Thiosulfate	Rhodanase	CN^-	(89)
Urea	Urease	NH_4^+	(90-93, 129)
		pH	(45)
		$Gas(NH_3)$	(22,94-95)
		$Gas(CO_2)$	(97, 144)
Uric acid	Uricase	$Pt(O_2)$	(98)

*Responds to D-phenylalanine, D-alanine, D-valine, D-methionine, D-leucine, D-norleucine, and D-isoleucine.
**Responds to L-leucine, L-tyrosine, L-phenylalanine, L-tryptophan, and L-methionine.

REFERENCES

1. NANJO, M. & GUILBAULT, G., *Anal. Chim. Acta 75:* 169 (1975).
2. BAUM, G. *Anal. Biochem. 39:* 65 (1971).
3. BAUM, G. & LYNN, M. *Anal. Chim. Acta 65:* 385 (1973).
4. BAUM, G. *Anal. Lett. 3:* 105 (1970).
5. KOBOS, R. & RECHNITZ, G. *Arch. Biochem. Biophys 175:* 11 (1976).
6. KOBOS, R. & RECHNITZ, G. *Biochem. Biophys. Res. Comm. 71:* 762 (1976).
7. PAPASTHATOPOULOS, D. & RECHNITZ, G. *Anal. Chem. 48:* 862 (1976).
8. HJEMDAL-MONSEN, G., PAPASTHATOPOULOS, D., & RECHNITZ, G. *Anal. Chim. Acta 88:* 253 (1977).
9. SUZUKI, S., TAKAHASHI, F., SATOH, J., & SONOBE, N. *Bull. Chem. S oc. Japan 48:* 3246 (1975).
10. SMITH, M. & OLSON, C. *Anal. Chem. 47:* 1074 (1975).
11. NANJO, M. & GUILBAULT, G. *Anal. Chim. Acta 75:* 169 (1975).

12. GUILBAULT, G. & HRABANKOVA, E. *Anal. Chim. Acta 56:* 285 (1971).
13. LLENADO, R. & RECHNITZ, G. *Anal. Chem. 46:* 1109 (1974).
14. JOHANNSON, G., EDSTROM, K. & OGREN, L. *Anal. Chim. Acta 85:* 55 (1976).
15. GUILBAULT, G. & HRABANKOVA, E. *Anal. Lett. 3:* 53 (1970).
16. GUILBAULT, G. & HRABANKOVA, E. *Anal. Chem. 42:* 1779 (1970).
17. GUILBAULT, G. & HRABANKOVA, E. *Anal. Chim. Acta 56:* 285 (1971).
18. HSIUNG, C., KUAN, S., & GUILBAULT, G. *Anal. Chem. 40:* 45 (1977).
19. LLENADO, R. & RECHNITZ, G. *Anal. Chem. 46:* 1109 (1974).
20. GUILBAULT, G. & HRABANKOVA, E. *Anal. Lett. 3:* 53 (1970); *Anal. Chim. Acta 69:* 183 (1974).
21. CALVOT, C., BERJONNEAU, A. M., GELGF, G., & THOMAS, D. *FEBS Lett. 59:* 258 (1975).
22. ANFALT, T., GRANELLI, A., & JAGNER, D. *Anal. Lett. 6:* 969 (1973).
23. WAWRO, R. & RECHNITZ, G. A. *J. Memb. Sci. 1:* 143 (1976).
24. JENSEN, M. A. & RECHNITZ, G. A. *Anal. Chim. Acta 101:* 125 (1978).
25. GUILBAULT, G. G. & SHU, F. *Anal. Chim. Acta 56:* 333 (1971).
26. DAVIS, P. & MOSBACH, K. *Biochim. Biophys. Acta 370:* 329 (1974).
27. AHN, B. K., WOLFSON, S. K., & YAO, S. J., *Bioelectrochem. Bioenergetics 2:* 142 (1975).
28. NGO, Z. *Int. J. Biochem. 6:* 371 (1975).
29. GUILBAULT, G. G. & WHITE, C. *Anal. Chem. 50:* 1481 (1978).
30. SKOGBERG, D. & RICHARDSON, T. *Am. Assoc. Cereal Chem. 56:* 147 (1979).
31. GUILBAULT, G. G., KUAN, S. S., FUNG, K. W., & SUNG, H. Y. *Anal. Chem. 51:* 2319 (1979).
32. CALVOT, C., BERJONNEAU, A. M., GELLF, C., & THOMAS, D. *FEBS Lett. 59:* 258 (1975).
33. KUMAR, A. & CHRISTIAN, G. *Clin. Chem. 21:* 325 (1975).
34. LLENADO, R. & RECHNITZ, G. *Anal. Chem. 43:* 283, 1457 (1971).
35. MASCINI, M. & LIBERTI, A. *Anal. Chim. Acta 68:* 177 (1974).
36. BAUMANN, E., GUILBAULT, G. G., GOODSON, L., & KRAMER, D. *Anal. Chem. 37:* 1378 (1965).
37. HUANG, J., KUAN, S., & GUILBAULT, G. *Clin. Chem. 23:* 671 (1977).
38. DIETSCHY, J., WECHS, L., & DELENK, J. *Clin. Chim. Acta 73:* 407 (1976).
39. KAMERRO, J., NAKAMO, N., & BABA, S. *Clin. Chim. Acta 77:* 245 (1977).
40. KUMAR, A. & CHRISTIAN, G. *Clin. Chim. Acta 77:* 101 (1977).
41. NONNA, A. & NAKAJAMA, K. *Clin. Chem. 22:* 336 (1976).
42. SATOH, I., KARUBE, I., & SUZUKI, S. *Biotechnol. Bioeng. 19:* 1095 (1977).
43. THOMPSON, H. & RECHNITZ, G. *Anal. Chem. 46:* 246 (1974).

44. CHEN, B. & GUILBAULT, G. G. *Anal. Lett.* *13:* 1607 (1980).
45. NILSSON, H., ACKARLUND, A., & MOSBACH, K. *Biochim. Biophys.* *Acta 320:* 529 (1973).
46. CLARK, L. & LYONS, C. *Ann. N.Y. Acad. Sci.* *102:* 29 (1962).
47. GUILBAULT, G. *Digest X Intn. Conf. Med. Biol. Eng.* *2:* 74 (1973).
48. GUILBAULT, G. & LUBRANO, G. *Anal. Chim. Acta 64:* 439 (1973).
49. CLARK, L., FRG German Patent, Kl 421 3/04 Nr. 1598, 285 (1971).
50. MELL, L. & MALOY, J. *Anal. Chem.* *48:* 1597 (1976).
51. WILLIAMS, D., DOIG, A., & KOROSI, A. *Anal. Chem.* *42:* 118 (1970).
52. MINDT, W., RACINE, P., & SCHLAPFER, P., *Ber Bunsenges. Physk.* *Chem.* *77:* 805 (1973).
53. MINDT, W., RACINE, P., & SCHLAPFER, P., DDR Patentschrift, 100,556 (1973).
54. BESSMANN, S., SCHULTZ, R. *2nd Int. Conf. Med. Biol. Eng.* Dresden (1973).
55. NAGY, G., VON STORP, H., & GUILBAULT, G. *Anal. Chim. Acta* *66:* 443 (1973).
56. LLENADO, H. & RECHNITZ, G. *Anal. Chem.* *45:* 2165 (1973).
57. SAMBUCETTI, C., NETT, G., & ROTH, M. *Meth. Clin. Chem.* *1:* 118 (1970).
58. UPDIKE, S. & HICKS, G. *Nature 214:* 986 (1967).
59. TRAN-MINH, C. & BROUN, G. *Anal. Chem.* *47:* 1359 (1975).
60. OKUDA, J., OKUDA, G., & MUVA, I. *Chem. Pharmac. Bull (Tokyo)* *18:* 1945 (1970).
61. REITNAUER, P. *Med. Labortechnik 14:* 363 (1973).
62. *Ibid 16:* 284 (1975).
63. REITNAUER, P., DDR Patenschrift 101,299 (1973).
64. BLAEDEL, W. & OLSON, C., U.S. Patent 3367849 (1968).
65. THOMAS, L. & CHRISTIAN, G. *Anal. Chim. Acta 78:* 271 (1975).
66. RACINE, P., KLENKE, H.-J., & KOCHSIECK, K. *Z. Klin. Chem.* *Klin. Biochem.* *13:* 533 (1975).
67. DURLIAT, H., COMTAT, M., & BAUDRAS, A. *Clin. Chem.* *22:* 1802 (1976).
68. DURLIAT, H., COMTAT, M., MAHENC, J., & BAUDRAS, A. *Anal. Chim.* *Acta 85:* 31 (1976).
69. DURLIAT, H., COMTAT, M., MAHENI, J., & BAUDRA, A. *J. Electro-anal Chem.* *66:* 73 (1975).
70. GUILLOT, C., VANUXEM, D., & GRINAUD, C. *Path. Biol. Paris* *24:* 431 (1976).
71. CORDONIER, M., LAWNY, F., CHAPOT, D., & THOMAS, D. *FEBS Lett.* *59:* 263 (1975).
72. CHENG, F. & CHRISTIAN, G. *Analyst 102:* 124 (1977).
73. AIZAWA, M., WADA, M., & SUZUKI, S. *Anal. Chem.*, in press.
74. JAEGFELDT, H., TORSTENSSON, A., & JOHANSSON, G. *Anal. Chim.* *Acta 97:* 221 (1978).
75. KIANG, C. H., KUAN, S. S., & GUILBAULT, G. G. *Anal. Chem.* *50:* 1319 (1978).

76. KIANG, C., KUAN, S., & GUILBAULT, G. *Anal. Chim. Acta 80:* 209 (1975).
77. YAO, S., WOLFSON, S., & TOKARSKY, J. *Bioelectrochem. Bioenergetics 2:* 348 (1975).
78. PAPARIELLO, G., MUKHERJEE, A., & SHEARER, C. *Anal. Chem. 45:* 790 (1973).
79. RUSLING, J., LUTTRELL, G., CULLEN, L., & PAPARIELLO, G. *Anal. Chem. 48:* 1211 (1976).
80. OLLIFF, C. J., WILLIAMS, R. T., & WRIGHT, J. M. *J. Pharm. Pharmacol. 30:* 45p (1978).
81. NILSSON, H., MOSBACH, K., ENFORS, S., & MOLIN, N. *Biotechnol. Bioeng. 20:* 527 (1978).
82. HSIUNG, C. P., KUAN, S. S., & GUILBAULT, G. G. *Anal. Chim. Acta 90:* (1977).
83. AIZAWA, A., KARUBE, I., & SUZUKI, S. *Anal. Chim. Acta 69:* 431 (1974).
84. GUILBAULT, G. & NANJO, M. *Anal. Chim. Acta 78:* 69 (1975).
85. SATOH, I., KARUBE, I., & SUZUKI, S. *Biotechnol. Bioeng. 18:* 269 (1976).
86. GUILBAULT, G. G. & NANJO, M., unpublished results.
87. SAMBUCETTI, C., NETT, G., & ROTH, M. *Meth. Clin. Anal.* (1970).
88. GUILBAULT, G. G. & CSERFALVI, T., *Anal. Chim. Acta 84:* 259 (1976).
89. LLENADO, R. & RECHNITZ, G. *Anal. Chem 44:* 1366 (1972).
90. MONTALVO, J. *Anal. Chem. 41:* 2093 (1969).
91. GUILBAULT, G. & MONTALVO, J. *J. Am. Chem. Soc. 92:* 2533 (1970).
92. GUILBAULT, G. & MONTALVO, J. *Anal. Chem. 41:* 1897 (1969).
93. GUILBAULT, G. & MONTALVO, J. *J. Am. Chem. Soc. 91:* 2164 (1969).
94. JOHANNSON, G. & OLGREN, L. *Anal. Chim. Acta 84:* 23 (1976).
95. MASCINI, M. & GUILBAULT, G. *Anal. Chem. 49:* 795 (1977).
96. GUILBAULT, G. G., KIANG, C. H., & KUAN, S. S. *Anal. Chem. 50:* 1323 (1978).
97. GUILBAULT, G., SMITH, R. M., & MONTALVO, J. *Anal. Chem. 41:* 600 (1969).
98. NANJO, M. & GUILBAULT, G. *Anal. Chem. 46:* 1769 (1974).
99. GUILBAULT, G. G. & NAGY, G. *Anal. Chem. 45:* 417 (1973).
100. GUILBAULT, G. G. & LUBRANO, G. *Anal. Chim. Acta 60:* 254 (1972).
101. CLARK, L. C., U.S. Patent 3,539,455 (1970).
102. GUILBAULT, G. G. & LUBRANO, G. *Anal. Chim. Acta 97:* 229 (1978).
103. COULET, P. R., JULLIARD, J., & GAUTHERON, D. *Biotechnol. Bioeng. 16:* 1055 (1974).
104. THEVENOT, D. R., COULET, P. R., STERNBERG, R., GAUTHERON, D. C. in "Enzyme Engineering," vol. 4 (G. Broun, G. Manecke, and L. B. Wingard, eds.) Plenum Press, New York (1978) p. 221.

105.   COULET, P. R., GODINOT, C., & GAUTHERON, D. *Biochim. Biophys. Acta 391:* 272 (1977).

106.   ENGASSER, J. M., COULET, P. R., & GAUTHERON, D. C. *J. Biol. Chem. 252:* 7919 (1977).

107.   BRILLOUET, J. M., COULET, P. R., & GAUTHERON, D. C. *Biotechnol. Bioeng. 18:* 1821 (1976); *19:* 125 (1977).

108.   THEVENOT, D. R., COULET, P. R., STERNBERG, R., LAURENT, J., & GAUTHERON, D. C. *Anal. Chem. 51:* 96 (1979).

109.   GRAY, D. N., KEYES, M. H., & WATSON, B. *Anal. Chem. 49:* 1067A (1977).

110.   GRAY, D. N. & KEYES, M. H., *Chem. Tech. 7:* 642 (1977).

111.   YUAN, C., GUÍLBAULT, G., & KUAN, S. S. *Anal. Chim. Acta*, in press.

112.   SUKUKI, S. *J. Solid Phase Biochem 4:* 25 (1979).

COST ANALYSIS AND VIABILITY OF IMMOBILIZED ENZYMES

IN ROUTINE ANALYSIS

P. V. Sundaram

Medizinische Klinik Und Poliklinik
Abteilung Klinische Chemie
Goettingen,Federal Republic of Germany

Cost is not the only consideration on which to base the commercial viability of using immobilized enzymes in routine analysis. The factors to be considered are a) the operational and storage stability of the product, b) whether or not the products maintain specificity and precision, c) the cost of coenzymes whenever required, d) the ease of applicability, e) any requirement of special equipment, f) any requirement of more or special personnel to run the tests, g) the speed of analysis, h) the sample size, and i) the cost, especially of the enzyme.

Among the various developments in the field of enzyme technology in the last decade, analytical methods employing enzyme electrodes and immobilized enzyme nylon tube reactors are noteworthy. Not only has the theory behind the development and application of these devices been clearly spelled out, but especially in the case of the nylon tube reactors, detailed clinical trials have been carried out in various countries and the results have been reported (Table I). However, there still appears to be a lot of resistance on the part of industry in marketing these procedures. Are the reasons for this reluctance mainly technical or commercial? Based on over a decade of development experience with the nylon tube reactors, my view is that both technically and economically, routine testing using the nylon tube reactors should prove superior to testing with soluble enzymes.

All of the criteria listed above are satisfied by the nylon tube reactors. Special attention should be paid to points c, d, e, g, and i. In many cases partially purified enzymes can be used. Commercial kits available now for soluble enzyme tests use very high concentrations of cofactors; but our tests use only 10 to 20%

405

TABLE 1

COSTS OF CLINICAL TESTS*

Metabolite	Enzyme**	Cost per Test*** (US ¢)	Ref.
Urea	Urease	0.002	(3)
Uric acid	Uricase/dye	7.0	(4)
Glucose	Glucose dehydrogenase	90% cheaper than Merck's test kit	(5)
Lactate	LDH-ALT	1.6	(6)
Pyruvate	LDH	0.69	
Glycerol	GDH	0.2	
Triglycerides	GDH + lipase & esterase	1.4	(7)
Glucose, uric acid, & cholesterol	Aldehyde dehydrogenase & GOD, Uricase and cholesterol oxidase	Markedly cheaper than commercial test kits	(8-10)

*Tests for creatinine and creatine also were developed but not automated, so 50 to 60 tests/hr could not be made (2).

**LDH = lactate dehydrogenase, ALT = alanine aminotransferase, GOD = glucose oxidase, GDH = glycerol dehydrogenase

***1 US \$ = 100 US ¢

of that, which of course reduces the cost. A modular concept can be used in utilizing nylon tube reactors; and normally no special equipment is required, unless a UV spectrophotometer is to be added wherever needed. In all of our methods 50 - 60 tests/hr can be carried out.

Gloger (1) used statistics gathered by his company and stated that on an average 120,000 tests can be performed with 35 g of enzyme using soluble enzyme type tests. Our experience is that at least a 1,000 times greater number of tests can be made with the same quantity of enzyme when the enzymes are immobilized in nylon tubes. The nylon tube reactors are easily manufactured, stored, and shipped; and they should not cost more than US $50 or DM 100 each. Note that this estimate is based on the retail price for the enzymes from companies such as Boehringer, Merck, or Sigma. For companies having in-house facilities for enzyme manufacture, the cost of production and marketing of the nylon tube enzyme reactors should be much less. Market surveys that were conducted showed that consumer resistance to the nylon tube method was no more than one meets for any new product. Thus, what is required is a clear cut policy for active marketing and distribution. Only negative commercial reasons can keep these methods from gaining acceptance.

ACKNOWLEDGMENTS

The work was financed partly by the DFVLR and partly by the DFG. The author is on leave at the Indian Institute of Technology in Madras, India.

REFERENCES

1. GLOGER, M., this volume.
2. SUNDARAM, P. V., et al. Clin. Chim. Acta 94: 295 (1979).
3. SUNDARAM, P. V., et al. Clin. Chem. 24: 234 (1978).
4. SUNDARAM, P. V., et al. Clin Chem. 24: 1813 (1978).
5. SUNDARAM, P. V., et al. Clin. Chem. 25: 1436 (1979).
6. SUNDARAM, P. V., et al. Clin. Chem. 25: 285 (1979).
7. SUNDARAM, P. V., et al. Clin. Chim. Acta 104: 87, 94 (1980).
8. SUNDARAM, P. V., et al. Clin. Chem. 26: 1652 (1980).
9. SUNDARAM, P. V., et al. J. Clin. Chem. Clin. Biochem. 19: 307 (1981).
10. SUNDARAM, P. V., et al. Fres. Z. Anal. Chem., in press.

MULTIPURPOSE ENZYME-COLLAGEN MEMBRANE ELECTRODES

D. C. Gautheron, P. R. Coulet and C. Bertrand

LBTM-CNRS
Universite Clause Bernard (Lyon 1)
Villeurbanne, Cedex, France

Selective, multipurpose electrodes have been developed from a previously described glucose electrode. Several single or multienzyme systems can be covalently bound to collagen membranes by our patented coupling method (1,2). The enzyme membranes are closely associated to a platinum anode for monitoring the hydrogen peroxide generated during the specific oxidation of the enzyme substrate.

The same basic, two electrode, differential device can be used (Fig. 1); el_1 is mounted with a specific enzyme collagen membrane and el_2 with a collagen membrane treated in the same manner but without any enzyme. Electrochemical interferences in the mixtures are detected by this compensating electrode to provide an accurate measurement of the hydrogen peroxide generated enzymatically (3). The system allows for substitution of one enzymic membrane by another in a few minutes, thus changing the specificity

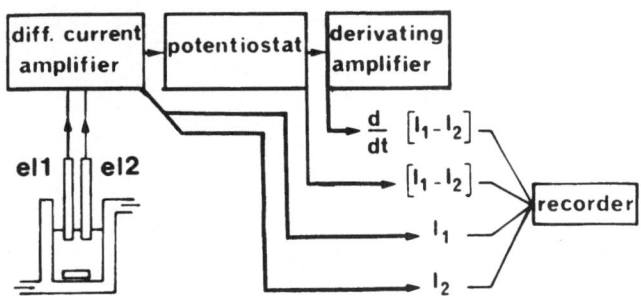

Fig. 1. Basic differential device.

of the device. Both mono and multienzyme electrodes have been developed (4). The glucose electrode (el_1 is glucose oxidase) has been extensively studied and is commercially available (Solea-Tacussel, Villeurbanne). Its main characteristics are: dynamic response time 45 sec, detection limit 10^{-8} M, linearity 10^{-7} – 5×10^{-3} M, reproducibility – 2%, stability more than 25 months at 22° C. An alternative galactose electrode (el_1 = galactose oxidase) allows the titration of galactose within the range 10^{-7} M – 2×10^{-3} M. Free cholesterol can be estimated in 1 min in the range 10^{-7} M to 10^{-4} M.

For disaccharide measurements a comparative study of membranes produced by random co-immobilization of enzymes, stacking of different enzyme membranes, and asymmetric coupling has recently been developed in our laboratory (5). For maltose determination, el_1 uses glucoamylase plus glucose oxidase with a titration range of 1.25×10^{-7} M – 10^{-3} M. The sucrose electrode uses invertase plus glucose oxidase for el_1; and the lactose electrose uses β-galactosidase plus glucose oxidase.

Many other systems can be used. Asymmetric coupling improved the electrode performances in every case; and the enzymatic sensors can be used for hundreds or thousands of assays, depending on the enzymes.

REFERENCES

1. COULET, P. R., JULLIARD, J. H. & GAUTHERON, D. C., French patent 2,235,133 (1973).
2. COULET, P. R., JULLIARD, J. H. & GAUTHERON, D. C. *Biotechnol. Bioeng. 16:* 1055 (1974).
3. THEVENOT, D. R., STERNBERG, R., COULET, P. R., LAURENT, J. & GAUTHERON, D. C. *Anal. Chem. 51:* 96 (1979).
4. BERTRAND, C., COULET, P. R. & GAUTHERON, D. C. *Anal. Chim. Acta 126:* 23 (1981).
5. COULET, P. R. & BERTRAND, C. *Anal. Lett. 12:* 581 (1979).

LONG-TERM STABILITY OF AIR-DRIED ENZYME ELECTRODES WITH

SELECTIVE ENZYMIC COLLAGEN MEMBRANES

P. R. Coulet, D. C. Gautheron, and G. Bardeletti

LBTM-CNRS
Universite Claude Bernard (Lyon I)
Villeurbanne, Cedex, France.

Enzyme collagen membranes have been prepared with the previously described coupling procedure developed in this laboratory (1, 2). Surface binding is obtained on both faces, allowing for asymmetric coupling with bienzyme systems, each enzyme being bound on only one face of the membrane (3). In our electrodes these membranes are associated with a platinum anode for the amperometric detection of H_2O_2, leading to very reliable enzyme electrodes for the detection of various species with selected mono or multienzyme systems that lead to hydrogen peroxide (4, 5). At present, the storage conditions of the enzyme electrodes need to be simplified for more practical use. Previously, the sensors were kept in buffer at $4°C$, and later at room temperature. Under these conditions the sensors were still usable after more than two years, provided a calibration was made for each set of measurements.

In the present work, our enzyme electrode for maltose determination with glucoamylase (GA) and glucose oxidase (GOD) bound to collagen in an asymmetric manner was used. This electrode can be calibrated with both maltose and glucose. This sensor was stored on the bench in a dry state at room temperature; and the main characteristics, i.e. detection limit and sensitivity, were checked periodically after rehydration. After two months of dry storage, a noticeable decrease in the slope of the calibration curves I/C appears; the residual values were 25% and 10% of the initial ones for glucose and maltose, respectively. The sensitivity, however, was sufficient to allow accurate measurements of the substrates in the range 10^{-7} M - 10^{-3} M with detection limits as low as 2.5 x 10^{-7} M and 1.5 x 10^{-7} M for glucose and maltose, respectively. For glucose, measurements of concentrations as low as 2.5 x 10^{-4} M were still possible with the same membrane after 160 day of dry

TABLE 1

EVOLUTION OF THE ELECTRODE CHARACTERISTICS WITH DRY STORAGE

Storage In Dry State* (day)	Sensitivity I/C (mA/M)		Detection Limit (M)	
	Glucose	Maltose	Glucose	Maltose
0	4.1	6.4	5×10^{-8}	2.5×10^{-8}
60	1.0	0.7	2.5×10^{-7}	1.5×10^{-7}

*GA/GOD assymmetric membrane

storage. The extension of such promising performances to newly
developed collagen-enzyme electrodes is now under investigation.

REFERENCES

1.   COULET,P. R., JULLIARD, J. H., & GAUTHERON, D. C., French
        Patent 2,235,133 (1973).
2.   COULET, P. R., JULLIARD, J. H., & GAUTHERON,D. C. *Biotechnol.
        Bioeng. 16:* 1055 (1974).
3.   COULET, P. R. & BERTRAND, C. *Anal. Lett. 12:* 581 (1979).
4.   THEVENOT, D. R., STERNBERG, R., COULET, P. R., LAURENT, J., &
        GAUTHERON, D. C. *Anal. Chem. 55:* 96 (1979).
5.   BERTRAND, C., COULET, P. R., & GAUTHERON, D. C. *Anal. Chim.
        Acta 126:* 23 (1981).

ENZYME ELECTRODES BASED ON INSOLUBILIZED ENZYME MEMBRANES

COUPLED WITH AN ELECTROCHEMICAL TRANSDUCER

J. L. Romette, N. D. Tran, P. Durand, J. L. Boitieux, and J. L. Navarro

Laboratoire de Technologie Enzymatique, UTC
Compiegne, France

A glucose sensor, based on the amperometric measurement of oxygen consumed by immobilized glucose oxidase, has been successfully and widely used in measuring glucose in very small samples (50 μl) of whole blood or industrial broths. The glucose electrode owes its success to lack of interference from other electroactive species in whole blood and to the high specificity of the immobilized enzyme. Studies (1) on the enzyme support have permitted a solution to the difficult problem of the variability of the oxygen content of the sample. The sensor also can be used with immobilized invertase and mutarotase to measure sucrose or with immobilized lactase to measure lactose.

An L-lysine sensor (2), based upon the potentiometric measurement of CO_2 produced by immobilized L-lysine decarboxylase, has been used successfully for actual time control of industrial production of amino acids. The device provides a measurement every two min. Studies on pyridoxal phosphate have shown the cofactor concentration dependence of the electrode response. L-alanine concentrations have been determined in the same way by coupling a pNH_3 sensor with an L-alanine dehydrogenase enzyme membrane. The influence of the NAD cofactor on the response of this electrode also has been studied. Measurements of hormone concentrations (4) have been realized in the same way (Fig. 1).

The determination of antigen, hormone, or hapten concentrations have been carried out by a potentio-enzymo-immunological sensor (3). The technique has been tested on Australia Antigen, which is responsible for hepatitis B, detection in whole blood. It consists of an Elisa test coupled with a specific iodide sensor. The procedure gave the same sensitivity as did the radio-immuno

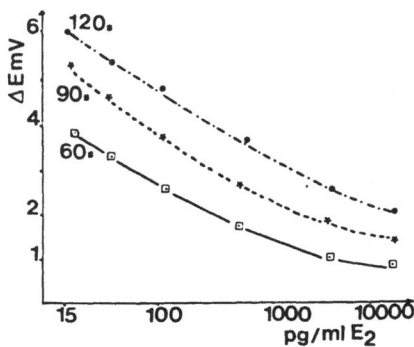

Fig. 1. Calibration curve for β-oestradiol determination with
         different cofactor concentrations.

assay technique. Inhibitors of enzyme reactions also can be determined with the enzyme electrode. An example is the pesticide
sensor (6), based on the inhibition of acetylcholine (5) by the
pesticide. This electrode has been used successfully for organo-
phosphate or carbamate pesticide measurements in water, milk, and
foodstuffs.

REFERENCES

1. ROMETTE, J. L., FROMET, B., & THOMAS, D. *Clin. Chim. Acta 95:*
    249 (1979).
2. TRAN, N. D., ROMETTE, J. L., & THOMAS, D. *Anal. Chem.*, in
    press.
3. BOITIEUX, J. L., ROMETTE, J. L., DESMET, G., & THOMAS, D.
    ITBM 2: (1981).
4. BOITIEUX, J. L., DESMET, G., & THOMAS, D. *Clin. Chim. Acta*,
    in press.
5. DURAND, P., DAVID, A., & THOMAS, D. *Biochim. Biophys. Acta*
    527: 277 (1978).
6. DURAND, P. & THOMAS, D. *J. Environm. Toxicol. Pathol.*, in
    press.

POTENTIOMETRIC GLUCOSE SENSOR: ENZYMATIC ACTIVITY AND

POTENTIOMETRIC MEASUREMENTS

L. B. Wingard Jr. and J. F. Castner

Department of Pharmacology, School of Medicine
University of Pittsburgh, Pittsburgh
Pennsylvania, USA

The development of a sensor to continuously monitor *in vivo* glucose levels, without the need to remove samples, is at present a major area of diabetes research. *In vitro* studies in our laboratory, have shown that changes in glucose concentrations can be measured potentiometrically by using either glucose oxidase (GOD) (1) or GOD/catalase (2) enzyme electrodes. However, reproducibility of these electrodes has been a problem. Recently, we examined several parameters associated with the fabrication of the enzyme electrodes. This paper presents some of the data obtained in this investigation.

The work described here is limited to the glutaraldehyde enzyme immobilization technique. In previous work (2) GOD, catalase, bovine serum albumin, and glutaraldehyde in 0.1 M sodium phosphate buffer pH 7.4 were mixed and poured over a platinum screen or wire, which was resting on a glass plate. After the glutaraldehyde had crosslinked to form a membrane around the platinum, the unit was peeled off the glass, washed with buffer, and tested for potentiometric response in buffered glucose solutions. The thickness of each membrane was a variable by this procedure. The use of a polymethacrylate mold to cast enzyme membranes over platinum foils (*ca.* 0.005 cm thick) proved to be a more successful approach in obtaining reproducible membrane thickness.

Measurements of immobilized GOD activity in the membranes were done by the o-dianisidine/peroxidase spectrophotometric assay. Membrane GOD activity in the mold was assayed in the absence of catalase and platinum since these materials are both known to catalyze the decomposition of H_2O_2, and hence would interfere with

the coupled enzyme spectrophotometric procedure. The assay was
done at 25°C and with 1800 mg/dl glucose at pH 7.4. Suitable agita-
tion was provided to minimize external diffusion. The immobilized
GOD activity was defined here as the rate of change of absorbance
at 460 nm, with the same abs/min assumed to equal a unit of
either immobilized or soluble glucose oxidase. The stability of
GOD activity in several electrodes fabricated using the mold-foil
technique was as follows (mean $\pm$ SD shown as Δ abs/min): at 0 hr
(n = 4) $2.38 \times 10^{-3} \pm 2.2 \times 10^{-4}$, at 3.5 - 4.0 hr (n = 4) $2.32 \times$
$10^{-3} \pm 3.2 \times 10^{-4}$, at 24 hr (n = 2) $2.13 \times 10^{-3} \pm 4.2 \times 10^{-5}$, at
53 hr (n = 2) 1.99×10^{-3}, at 55 hr (n = 2) 1.74×10^{-3}, and at
73 hr (n = 2) 2.04×10^{-3}. Thus, the reproducibility was good
after a few hr use.

Potentiometric testing of the GOD/catalase/platinum electrodes
in the mold showed better reproducibility in potential response
(mean of 39.0 mv $\pm$ 8.0 SD) at 100 mg glucose/dl than did electrodes prepared without a mold and with platinum screens (mean of
67.3 mV $\pm$ 19.9 SD).

A concurrent study of the effect of dissolved oxygen concentration on potentiometric response of plain platinum in
phosphate buffer suggested that the variations in potentiometric
response with the enzyme electrodes may be due to platinum-oxygen
surface reactions. This surface effect is currently under study
and will be reported elsewhere, along with the details of the above
work.

This work was supported by grant 1R01AM26370 from NIH. The
technical assistance of L. Cantin is greatly appreciated.

REFERENCES

1. WINGARD JR., L. B., ELLIS, D., YAO, S. J., SCHILLER, J. G.,
    LIU, C. C., WOLFSON JR., S. K. & DRASH, A. L. *J. Solid
    Phase Biochem.* *4:* 253 (1979).
2. WINGARD JR., L. B., SCHILLER, J. G., WOLFSON JR., S. K., LIU,
    C. C., DRASH, A. L. & YAO, S. J. *J. Biomed. Matl. Res. 13:*
    921 (1979).

MICROBIAL SENSOR FOR PRELIMINARY SCREENING OF MUTAGENS

I. Karube and S. Suzuki

Research Laboratory of Resources Utilization,
Tokyo Institute of Technology, Yokohama, Japan

The recombination of deficient strain B. subtilis M45 Rec$^-$ and the wild strain B. subtilis H17 Rec$^+$ were employed for developing a sensor for testing for mutagens (1-4). Immobilization of the bacteria was performed by dropping a bacteria suspension containing 2.7×10^8 cells onto a porous acetylcellulose

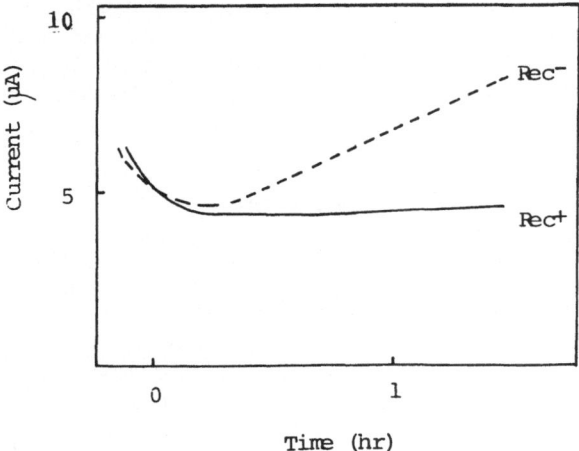

Fig. 1. Response curves of microbialsensor to AF-2, when 1.6 µg/ml of AF-2 was added to the system.

membrane (Millipore, Type HA, 0.45 µm pore size) with slight
suction. The microbial sensor system consisted of two microbial
electrodes: a) an electrode of *B. subtilis* Rec⁻ (Rec⁻ electrode) and b) an electrode of *B. subtilis* Rec⁺ (Rec⁺ electrode)
plus two oxygen electrodes. When the Rec⁻ and Rec⁺ electrodes
were inserted into 0.3 g/L buffered glucose solution, steady-
state currents were obtained. Then, the mutagen 2-(2-furyl)-3-
(5-nitro-2-furyl) acrylamide (AF2) was added. The time course of
electrode current is shown in Fig. 1. The rate of current increase was a measure of the mutagen concentration and was most
easily measured as the linear slope at the midpoint of the sig-
moidal curve. When chemical mutagens, such as AF-2, mitomycin,
captan, 4-nitroquinoline-N-oxide, N-methyl-N'-nitro-N-nitroso-
guanidine, aflatoxin B, 2-aminoanthracene, and 2-acetylamino-
fluorene, were added to the glucose-buffer solution, the currents of the Rec⁻ electrode increased markedly; and the mutagenicity of these chemicals could be estimated with the electrochemical system.

The relationships between the rate of current increase of
the Rec⁻ electrode and the concentration of AF-2 and mitomycin
were linear for the range 1.6-2.8 g/mL for AF-2 and 2.4-7.3 g/mL
for mitomycin. The responses of the Rec⁻ and the Rec⁺ electrodes
to 10 and 50 g/ml streptomycin, which is a known inhibitor of
protein synthesis in bacteria, gave a current increase with both
Rec⁺ and Rec⁻ electrodes of 5∿6 µA/hr. The current of neither
electrode increased with 10 g/mL chloramphenicol. Therefore,
chloramphenicol and streptomycin were not mutagens.

Other tests showed that the sensor had higher sensitivity
than the rec-assay (4) or the Ames test. The minimum measurable
mutagen concentration was 1.6 µg/mL by the microbial sensor and
5.0 µg/mL by the rec-assay and 10 µg/mL by the Ames test for
AF-2.

REFERENCES

1. BRIDGES, B. A. *Nature 261:* 195 (1976).
2. AMES, B. N., LEE, F. D. & DURSITON, W. E. *Proc. Natl. Acad.
     Sci. U.S.A. 70:* 782 (1972).
3. BRIDGES, B. A., MOTTERSHEAD, R. P., ROTHWELL, M. A. & GREEN,
     M. H. L. *Chem. Biol. Interact. 5:* 77 (1972).
4. KADA, T., TSUCHIKAWA, K. & SADAIE, Y. *Mutat. Res. 16:* 167
     (1972).

APPLICATION OF MICROBIAL ELECTRODE TO ANALYSIS OF WASTE WATER

M. Hikuma, H. Suzuki, T. Yasuda, I. Karube* and
S. Suzuki*

Central Research Laboratories, Ajinomoto Co. Inc.
Kawasaki and Tokyo Institute of Technology,*
Yokohama, Japan

The biochemical oxygen demand(BOD) test and determination of
nitrogenous compounds are required to control the pollution of
water. Microbial electrodes consisting of immobilized microorganisms and electrochemical devices have been developed for
rapid estimation of BOD (1) and determination of ammonia (2),
nitrite (3), and nitrate with response times of 4-20 min. They
were prepared according to procedures previously reported (1-3).

The microbial electrode was inserted into a jacketed flow
cell; and buffer solution was pumped through the flow cell continuously together with the air or nitrogen gas (for the nitrate
electrode). When a sample solution (BOD, NH_3 or NO_2^-) was transferred to the flow cell and contacted the electrode, the substrate diffused into the layer of immobilized microorganisms and
was assimilated by the microorganisms. The respiration of the
microorganisms was activated; and the dissolved oxygen around the
oxygen probe decreased. The current of the electrode decreased
until it reached a steady state, which corresponded to the concentration of the sample solution. *P. aeruginosa* produced carbon
dioxide from nitrate under anaerobic conditions in the presence
of organic compounds such as ethyl alcohol. Therefore nitrate was
determined by measuring carbon dioxide with the CO_2-probe.

The microbial electrodes were applied to analysis of untreated and pretreated waste waters of a fermentation factory.
The assay could be done in a shorter time by using a 3-10 min
sample introduction period. The waste waters were also assayed
by conventional methods (4). As shown in Table 1, good agreement
was obtained between the concentrations determined by the
microbial electrodes and those determined by the conventional

methods. The microbial electrodes could be used for more than 3 weeks. In conclusion the microbial electrodes appear quite attractive for analysis of waste waters.

TABLE 1

COMPARISON OF ELECTROCHEMICAL AND CONVENTIONAL METHODS

Substate	Microbial Electrode	Conventional Method	Difference
	(mg/L)	(mg/L)	(%)
BOD	27.3	28.8*	-5.5
	33.1	29.8	10
NH_3	0.558	0.585**	-4.8
	1.43	1.42	0.7
NO_2^-	4.02	3.66***	9.0
	5.88	5.46	7.1
NO_3^-	92.5	100****	-8.1
	183	175	4.4

*5-day method, **distillation-acidimetry, ***diazotization method, ****xylenol method.

REFERENCES

1. HIKUMA, M., SUZUKI, H., YASUDA, T., KARUBE, I. & SUZUKI, S. *Eur. J. Appl. Microbiol. Biotech. 8:* 289 (1979).
2. HIKUMA, H., KUBO, T., YASUDA, T., KARUBE, I. & SUZUKI, S. *Anal. Chem. 52:* 1020 (1980).
3. SUZUKI, H., HIKUMA, M., YASUDA, T., KARUBE, I. & SUZUKI, S. *Eur. J. Appl. Microb. Biotechnol.* (in press).
4. RAND, M. C., GREENBERG, A. E. & TARAS, M. J. "Standard Methods for the Examination of Water and Wastewater," 14th Edit., American Public Health Association (1975).

BIOSENSORS BASED ON ENZYME AMPLIFICATION AND IMMUNOCHEMICAL

SELECTIVITY

M. Aizawa and S. Suzuki*

Institute of Materials Science, University of
Tsubuka, Ibaraki and Tokyo Institute of Technology,*
Yokohama, Japan

To improve the sensitivity of non-labeling immunosensors,
an enzyme has successfully been incorporated as a label and used
to determine IgG. The sensor depends upon an antibody-bound membrane for its selectivity, a label enzyme for its sensitivity,
and a Clark-type oxygen electrode for its electrochemical performance (1-9).

Cellulose triacetate (250 mg) was dissolved in 5 ml dichloro-
methane; and 0.05 ml 50% glutaraldehyde was added followed by
0.15 ml 1,8-diamino-4-aminomethyl octane. The solution was cast
on a glass plate and allowed to stand at room temperature to complete intermolecular crosslinking of the 1,8-diamino-4-amino-
methyl octane and glutaraldehyde. The membrane was peeled off
and placed in 0.1% glutaraldehyde at $30^{\circ}C$ and pH 8.0 for 2 hr;
then it was placed in contact with phosphate buffer containing
anti-IgG antibody around $4^{\circ}C$ for 16 hr. The membrane was reduced
with 0.1 M $NaBH_4$ for 3 min. After thorough washings, the membrane
was used for the immunosensor. The sensor consisted of an antibody-bound membrane, oxygen permeable teflon membrane, platinum
cathode, lead anode, and alkaline electrolyte solution. The sensor output current, resulting from electrochemical reduction of
oxygen, was displayed on a recorder. The test solution was
stirred during measurements. IgG (10 mg) was dissolved in 1 ml
0.1 M phosphate buffer, pH 7.5, containing 0.08% glutaraldehyde.
Glucose oxidase (GOD) (10 mg) was added to the solution. After
2 hr incubation the preparation was fractionated by Sepharose CL-
6B chromatography.

Competitive enzyme immunoassays were performed with the
immunosensor. The immunosensor was soaked in an IgG containing

test solution, to which a known amount of the GOD labeled IgG was added. Both nonlabeled and labeled IgG competitively adsorbed on the membrane-bound antibody due to immunochemical specific affinity. After sufficient immunochemical reaction, the sensor was rinsed to remove unreacted IgG and was then placed in pH 7.0 phosphate buffer. In order to quantitate the labeling enzyme, specifically adsorbed on the sensor membrane, the sensor membrane was assayed for GOD activity. Addition of glucose caused the sensor output to decrease due to oxygen consumption by the GOD reaction. The sensor output reached a steady state within 15 sec. The extent of the sensor output change reflected the amount of labeled IgG specifically adsorbed on the sensor membrane. In case a test solution contained no IgG, the sensor output change was the maximum. The change decreased with increase in IgG concentration of the test solution. Under the optimum conditions, a calibration curve was obtained for determination of IgG with this labeling immunosensor. It showed that the lower and upper detection limits of the sensor were 10^{-9} and 10^{-5} g/ml, respectively.

The lower detection limit of a non-labeled immunosensor was reported as 10^{-5} g/ml for the determination of human serum albumin (9). The sensitivity of the labeled immunosensor is thus a marked improvement. The catalytic function of the labeling enzyme may account for the improved sensitivity. The amount of the labeling enzyme is comparable to that of the antigen to be determined. However, the enzyme converts glucose very rapidly with consumption of oxygen. The oxygen molecules decrease in a manner to give chemical amplification.

REFERENCES

1. AIZAWA, M., KATO, S. & SUZUKI, S. *J. Membrane Sci. 2:* 125 (1977).
2. AIZAWA, M., SUZUKI, S., NAGAMURA, Y., SHINOHARA, R. & ISHIGURO, I. *J. Solid-Phase Biochem. 4:* 25 (1979).
3. AIZAWA, M., SUZUKI, S., NAGAMURA, Y., SHINOHARA, R. & ISHIGURO, I. *Chem. Lett. 1977:* 779 (1977).
4. AIZAWA, M., KATO, S. & SUZUKI, S. *J. Membrane Sci. 7:* 1 (1980).
5. AIZAWA, M., MORIOKA, A., MATSUOKA, H. SUZUKI, S. *J. Solid Phase Biochem. 1:* 319 (1976).
6. AIZAWA, M., MORIOKA, A. & SUZUKI, S. *J. Membrane Sci. 4:* 221 (1978).
7. AIZAWA, M., MORIOKA, A., SUZUKI, S. & NAGAMURA, Y. *Anal. Biochem. 94:* 22 (1979).
8. AIZAWA, M., MORIOKA, A. & SUZUKI, S. *Anal. Chim. Acta 115:* 61 (1980).
9. AIZAWA, M. & SUZUKI, S. *Jpn. J. Appl. Phys. 21:* 219 (1982).

APPLICATION OF CHEMILUMINESCENCE OF *CYPRIDINA* LUCIFERIN ANALOG

TO IMMOBILIZED ENZYME SENSORS

T. Kobayashi*, K. Saga, S. Shimizu, and T. Goto

Department of Food Science and Technology, and
Department of Chemical Engineering*
Nagoya University, Nagoya, Japan

In recent years, various sensors with immobilized enzyme have been developed (1). In most of the sensors based on electrochemical methods, a substrate reacts in the immobilized enzyme membrane, and the decrease of substrate or increase of product is measured by a detector which is attached to the inner surface of the membrane. As it takes several minutes for the substrate or the product to diffuse through the membrane, the stable output signal is dependent on the thickness of the immobilized enzyme membrane and is usually obtained after more than 5 min.

To develop an immobilized enzyme sensor with a higher sensitivity and rapidity, the chemiluminescence of a *Cypridina* luciferin analog, 2-methyl-6-phenyl-3,7-dihydroimidazol [1,2-α] pyrazin-3-one, was applied to immobilized enzyme sensors. In this case, photons were emitted as soon as the luciferin analog and a substrate met at the outer surface of the immobilized enzyme membrane. As a result, the response time was independent of the thickness of the membrane; and it was not necessary to prepare a thin membrane. Xanthine oxidase, peroxidase, glucose oxidase, uricase, and cholesterol oxidase were immobilized by using photo-crosslinkable resin prepolymer or porous glass beads. The immobilized enzyme sensor system was composed of a photoncounter and a test tube in which the immobilized enzyme membrane or particles were placed. A linear relation between the concentration of substrates and luminescence rate was obtained on log-log scales (Fig. 1). This immobilized enzyme sensor system could be used repeatedly. The detection limits for xanthine, hypoxanthine, hydrogen peroxide, glucose, cholesterol, and uric acid were 0.02, 0.02, 0.2, 0.4, 2, and 2 μM, respectively. Concentrations of hypoxanthine in tuna muscle and glucose and cholesterol in serum, measured using this

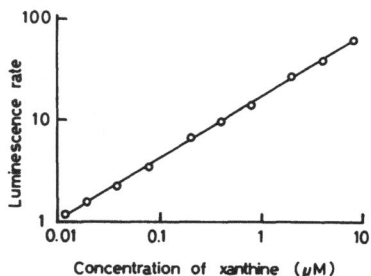

Fig. 1.   Relation between the luminescence rate and the concen-
          tration of xanthine.

sensor system, were comparable with those measured by the standard
methods.

     The optimum concentration of the luciferin analog was 4 μM
and 5 μM and the pH 6.0 and 9.0 for xanthine oxidase and peroxidase, respectively. In the batch system, one sample could be
measured within 30 sec; forty samples could be analyzed within
1 hr in the flow system. Compared with luminol, the luciferin
analog had the advantages of a) needing a smaller amount (4-5 M)
of analog than that of luminol (200 μM) (2), b) use at or near
neutral pH (luminol must be used at pH 10.5), and c) a ten times
lower detection limit for hydrogen peroxide.

REFERENCES

1.   SUZUKI, S. & KARUBE, I. *Ann. N. Y. Aca. Sci. 326:* 255 (1979).
2.   BOSTICKS, D. T. & HERCULES, D. M. *Anal. Chem. 47:* 447 (1975).

USE OF HYDROGEN SENSITIVE Pd-MOS COMPONENTS IN BIOCHEMICAL ANALYSIS

B. Danielsson, F. Winquist, K. Mosbach and
I. Lundstrom

Pure & Applied Biochemistry, University of Lund
and Applied Physics, Linkoping Institute of
Technology,* Lund and Linkoping, Sweden

Various hydrogen and ammonia sensitive palladium coated
semiconductor components have been applied to the determination
of hydrogen (1) and ammonia (2) evolved or consumed in biochemical reactions. As a common feature all components studied have
been of the metal oxide semiconductor type (MOS). This is a silicon chip with a 100 nm silicon oxide layer coated with a very thin
(100 nm) film of a catalytically active metal, palladium (3).
Pd-MOS capacitors as well as Pd-MOS field effect transistors
(Pd-MOSFETs) have been used. The Pd-MOSFETs used for monitoring
reactions catalyzed by urease and creatininase (2) had an integrated heater and a temperature control circuit to keep the semiconductor chip at a suitable temperature.

When hydrogen or some gases containing hydrogen, such as
NH_3, H_2S, or CH_4, are present in the atmosphere surrounding the
component, hydrogen molecules are readily dissolved in the Pd-
layer, where they dissociate. Hydrogen atoms migrating towards
the Pd-SiO_2 interface become polarized in the electric field
applied across the structure. The voltage drop thus created
affects the properties of the device in such a way that can be
related to the hydrogen pressure. In order to speed up the time
response and to prevent water molecules from sticking to the
metal surface, the device is heated to 100-150°C, either in the
way just mentioned or by mounting it in a heated chamber. The
time response will then be about one min and the sensitivity at
least 1 ppm of H_2 or 10 ppm of NH_3 in air.

The high temperature presently required, precludes direct
attachment of the enzyme to the surface of the sensor. Therefore

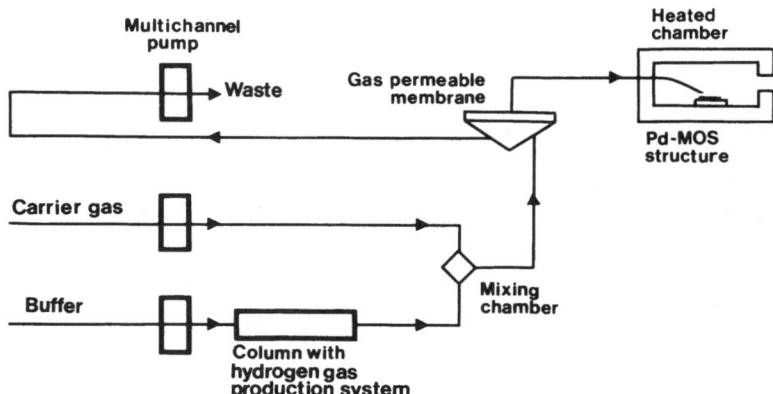

Fig. 1. Experimental arrangement. The hydrogen producing system
         could be a small column with immobilized *Clostridium*
         Acetobutylicum or hydrogen dehydrogenase.

the gas to be measured must be transferred to a carrier gas, e.g.
air or N_2, and transported past the sensor. This can be done as
shown in Fig. 1, where a multichannel peristaltic pump is used to
maintain the liquid and gas flows required. This system proved
to work quite well when tested with hydrogen producing mocroorganisms and with hydrogen dehydrogenase immobilized in small
columns (1). The hydrogen dehydrogenase could be used to measure
the concentration of $NADH/NAD^+$ in a solution, thus opening a
general route to following dehydrogenase reactions with use of
hydrogen sensors.

Finally, we have found that, under certain circumstances, the
electrical resistance across a Pd-MOS-component is affected by
the NH_3-concentration in the gas tested. The utilization of this
phenomenon is examplified here by the determination of urea using
immobilized urease. The measuring range was linear up to 100 mM
or more with a 0.8 ml urease column containing about 250 units
bound to controlled pore glass and operated at a flow rate of 0.3
ml/min. The sensitivity was such that the method could be used
for clinical urea determination. Since the NH_3 formed is transferred to a carrier gas in about the same way as shown in Fig. 1,
crude samples can be applied without any pretreatment. It should
also be noted that the components described here operate in the
gas phase in contrast to similar FET devices which operate on
ionic species in solution (4).

REFERENCES

1. WINQUIST, F., DANIELSSON, B., LUNDSTROM, I. & MOSBACH, K.
    Appl. Biochem. Biotechnol. 7: 135 (1982).
2. DANIELSSON, B., LUNDSTROM, I., MOSBACH, K. & STIBLERT, L.
    Anal. Lett. 12: 1189 (1979).
3. LUNDSTROM, I. *Physica Scripta* 18: 424 (1978).
4. CARAS, S. & JANATA, J. *Anal. Chem.* 52: 1935 (1980).

COMPUTER CONTROLLED MASS SPECTROMETER MONITORING OF FERMENTATIONS

E. Pungor Jr., C. L. Cooney, and J. C. Weaver

Massachusetts Institute of Technology
Cambridge, Massachusetts, USA

Previous work has shown that computer controlled mass spectrometer (MS) monitoring in real time of dissolved aqueous phase volatile compounds and the same compounds in the gas phase appears feasible (1-3). If multicompound monitoring capability is realized, a significant problem in fermentation monitoring will have been solved. The mass spectrometer and its interface consist of a system whose output is proportional to the gas or liquid phase concentrations of volatile compounds of interest, but with linear interferences. For example, the area, S_{28}, of the peak at mass 28 is due to contributions from both N_2 and CO_2, while that at mass 44 is due solely to CO_2, as shown in the two equations below.

$$S_{28} = A_{28,N_2} [N_2] + A_{28,CO_2} [CO_2]$$

$$S_{44} = A_{44,CO_2} [CO_2]$$

These equations can be easily solved in real time for continuous determination of $[N_2]$ and $[CO_2]$. The coefficients, A, in the equations are first determined by calibration.

In addition common volatile species such as nitrogen or water can be monitored since they are present at high concentrations in almost all cases, and can serve as useful internal standards. Typically, gas phase concentrations can be measured with respect to the gas phase nitrogen concentration, while aqueous concentrations can be measured with respect to water in the liquid phase. For example, ethanol dissolved in the liquid phase can be monitored by using the ratio of S_{31}/S_{18}, which is equivalent to $(A_{31,ETHANOL} [ETHANOL])/ (A_{18,H_2O} [H_2O])$. Internal standards provide significant

429

reduction in drift, which is necessary for long term monitoring. This approach has been used to monitor gas and liquid phase concentrations of carbon dioxide, oxygen, and alcohol; possible extension are shown in Table 1. For volatile acids and bases it is necessary to adjust the pH at the MS membrane interface to be either low or high (2). Immobilized enzymes may also be used to convert nonvolatile species to volatile species (4). The challenge for the future is to determine the extent of measurement error propagation within the real time computer calculations which are used to determine the concentrations.

TABLE 1

DISSOLVED VOLATILE COMPOUNDS TESTED

VOLATILE COMPOUNDS	MW	B.Pt. ($^{\circ}$C)
Ammonia	17	-33
H_2O	18	100
DHO	19	101
Formaldehyde	30	-20
Oxygen	32	-193
Methanol	32	65
Carbon Dioxide	44	
Acetaldehyde	44	21
Formic Acid	46	100
Ethanol	46	79
Acetone	58	57
Acetic Acid	60	118
Ethylene Glycol	62	198
Acrylic Acid	72	141
Propionic Acid	74	141
Butyric Acid	88	164

REFERENCES

1. REUSS, M., PIEHL, H., & WAGNER, F. *Euro. J. Appl. Microbiol.* *1:* 323 (1975).
2. WEAVER, J. C. & ABRAMS, J. H. *Rev. Sci. Inst.* 50: 478 (1979).
3. PUNGOR JR., E., PERLEY, C. R., COONEY, C. L., & WEAVER, J. C. *Biochim. Biophys. Acta.* 438: 296 (1976).
4. WEAVER, J. C., MASON, M. K., JARRELL, J. A., & PETERSON, J. W. *Biochim. Biophys. Acta.* 438: 296 (1976).

Session VII

MEDICAL APPLICATIONS OF ENZYME TECHNOLOGY

Chairmen: K. Mosbach and T. Murachi

IMMOBILIZED HEPARINASE: PRODUCTION, PURIFICATION, AND

APPLICATION IN EXTRACORPOREAL THERAPY

R. Langer, R. J. Linhardt, C. C. Cooney,
D. Tapper and M. D. Klein

Department of Nutrition and Food Science, MIT
Cambridge, and Department of Surgery
Children's Hospital Medical Center
Boston, Massachusetts, and University of Michigan
Ann Arbor, Michigan, USA

There are nearly 20,000,000 perfusions involving extracorporeal medical machines performed each year (1,2). The artificial
kidney is employed several times weekly by over 100,000 persons
each year (1). Open-heart operations employing cardio-pulmonary
bypass are an everyday occurrence at many medical centers (2).
Less often employed is the membrane oxygenator for pulmonary support in the critically ill infant (3). Still other applications
of extracorporeal devices, such as the artificial liver, are in
a conceptual stage (4).

In every case, systemic levels of heparin, which fully anti-
coagulate the patient, are required for the operation of extracorporeal medical devices. However, the resulting circulating
concentration of heparin results in a high incidence of hemmorhagic
complications, particularly in patients at risk of bleeding and
for certain longer term perfusions. The severity of hemmorhage
may vary from mild mucosal oozing to massive intracranial, gastro-
intestinal, genitourinary, and intrathoracic bleeding (5,6). The
incidence of hemmorhage during heparinization is reported to be
from 8 to 33% (7). Six to 10% of patients develop coagulation
abnormalities with excessive bleeding following open-heart surgery;
and this percentage increases drastically with the use of long-
term pulmonary support with membrane oxygenators (3). Bleeding
complications occured 10-19% of the time using either low dose
heparin or regional heparin regimens in hemodialysis of patients
with increased risk of hemorrhage (8); and nearly 25% of all
patients suffering from acute renal failure are subject to increased bleeding risk. The incidence of partial clotting in the

433

dialyzer was 3-5% with either low dose heparin or regional heparinization (8). In addition to hemorrhage, there are a number of other complications associated with heparinization, particularly when the drug is administered over a long period. Some of these complications include alopecia (9) and interference with bone repair (10), leading in some cases to severe decalcifying bone disease (11).

Because of the problems associated with systemic heparinization, a number of approaches have been explored to solve this problem. Attempts at regional heparinization, by infusing heparin into the machine and trying to reverse it with a heparin antagonist such as protamine, have met with little clinical success (8). Protamine only reversibly neutralizes heparin probably by occupying its highly charged regions. It causes hypotension as well as anticoagulation; but most importantly its anti-heparin affect is quite variable (12). Some investigators have used very low dose heparin successfully to minimize the amount of anticoagulation to which the patient is subject (8). There has also been a great deal of research devoted to the creation of blood compatible materials (13), including materials with heparin (14) or other substances (15) bonded to their surface. Machines could be fabricated from these materials; and theoretically no anticoagulant would be needed. The development of heparin substitutes has also been explored (16). At present however, there is no ready solution to the problem of using extracorporeal devices requiring blood perfusion without anticoagulating the patient.

We are proposing a new approach which may enable heparin to be removed specifically from the blood after it has served its purpose in the extracorporeal device and before it is returned to the patient. This approach consists of a blood filter, containing immobilized heparinase, which could be placed at the effluent of any extracorporeal device. Theoretically, this would allow the extracorporeal machine to be anticoagulated while the patient was not. The significance of the proposed heparinase filter is that it may open new possibilities in extracorporeal treatment. Among the potential advantages include a) the possibility of obviating the need of potentially toxic neutralizing substances such as protamine (17) (i.e., the filter could be used at the termination of the perfusion in place of current protamine reversal techniques) and b) the opportunity of heparinizing the extracorporeal system without simultaneous heparinization of the patient (i.e., the filter could be used continuously in a manner analogous to regional heparinization by replacing the protamine influx with the filter). The filter would also permit the use of higher heparin concentrations in extracorporeal devices.

The primary objective of our research is to determine if immobilized heparinase is capable of removing heparin from blood.

The amount of data on heparinase has until now been limited and
the methods for producing it are inadequate for large scale use.
Therefore, the principal focus of our research has been on conduct-
ing feasibility studies on heparinase production, purification,
immobilization, and initial trials of filters of immobilized heparinase with human blood and in dogs. These studies are reported
here.

HEPARINASE PRODUCTION

The limitations of earlier procedures (18) to produce heparinase were a) non reproducability, b) extreme expense, c) not
conducive to high yields, and d) not suitable for scale up. In
order to solve these problems, we have conducted two types of
studies: a) one aimed at the kinetics of microbial growth and
heparinase production and at optimizing the conditions (e.g.,
harvest time) of fermentation and b) one aimed at developing a
simple defined low cost medium for heparinase fermentation.

The wild-type strain of *Flavobacterium heparinum* produces
a cell associated heparinase during growth only when heparin is
supplied to the growth medium as an inducer. The inducer was provided at the time of innoculation of the sterile medium. Our findings were that growth was initially exponential and heparin
was rapidly taken up by the cell at a rate of 1.1 g/g cell/hr.
Enzyme specific activity began to increase just as heparin uptake
was finishing and increased at a volumetric rate of 375 units/L/hr.
One unit is defined as one mg of heparin degraded/hr. At the onset
of the stationary growth phase, enzyme production stopped; and a
deactivation was observed resulting in a 86% loss of total activity within 4 hr. Thus, understanding the kinetics of enzyme production, and implementing timely harvest, were critical to obtaining highly active heparinase. By implementing the conditions of
induction and harvest outlined above, nearly a 100 fold increase
in maximum volumetric productivity was obtained compared to previous studies. Many fermentations have been performed yielding an
enzyme level on the average of 9600 units of heparin/L fermentor
broth, demonstrating the reliability of this method (19).

In earlier procedures a complex media involving trypticase-
soybroth, vitamins, heparin, and mineral salts was used. Results
in this complex medium were difficult to reproduce and difficult
to interpret. We conducted a step-by-step investigation of the
factors critical to the fermentaiton. Our findings were as follows: a) nineteen compounds (other than heparin) were tested for
their ability to induce heparinase. Of these, only hyaluronic acid,
heparin monosulfate, maltose, and N-acetyl D-glucosamine induced
Flavobacteria heparinum to produce heparinase. None induced at
levels higher than heparin. b) The nutritional requirements in-

clude an absolute requirement for histidine and growth stimulation
by methionine. No other amino acids and no vitamins were required.
c) A defined growth medium was developed consisting of glucose,
ammonium sulfate, potassium and sodium phosphate, magnesium sulfate, trace minerals, heparin, histidine, and methionine. d) The
nitrogen source (ammonium sulfate) optimum concentration for growth
was 0.5 g/L. e) The optimum temperature for growth was 27°C. f)
The optimum concentration of initial glucose (carbon source) for
growth was 8 g/L. g) Above a sodium chloride concentration of
4.8 g/L (as sodium) the growth rate was diminished by 0.04 hr^{-1}/g
sodium added. h) The optimum phosphate concentration for growth
was 20 mM (19).

Using the defined media developed in these studies, the volumetric heparinase production, cell growth rate, and densities all
increased. A typical production run resulted in a 10 fold increase
(compared to our optimized complex medium) in total enzyme obtained
to 96,000 units/L fermentor broth. This fermentation has been repeated many times demonstrating the reliability of the method.
Furthermore, there is not a rapid loss of the enzyme at the end of
the fermentation in this medium (19).

PURIFICATION AND CHARACTERIZATION OF HEPARINASE

The objectives of our work on the purification of heparinase
were two-fold: a) to adapt previous purification schemes (20) to
large scale production, and b) to purify heparinase to homogeneity. The cell pellet produced from centrifugation of the fermentation broth at 10,000 x g was resuspended at 100 mg/ml protein in
0.01 M phosphate buffer pH 7.0 and disrupted soncially; the nucleic acids were precipitated with 12.5 mg/ml protamine sulfate; and
the protein solution was added to 4 g hydroxylapatite/g protein.
The hydroxylapatite bound protein was then washed stepwise in a
batch procedure with increasing concentrations of sodium chloride
and sodium phosphate (from 0 M and 0.01 M to 0.50 M and 0.25 M,
respectively). The resulting enzyme preparation (HA) was obtained
in 0.125 M NaCl and 0.07 M sodium phosphate wash (21).

For the second goal, we tried affinity chromatography. In
preliminary experiments, we found that a heparin-Sepharose column
failed to bind heparinase. We therefore searched for a competitive
and reversible heparinase inhibitor to act as a ligand. Over 30
sulfated substances were screened; and three synthetic heparin
substitutes polyvinylsulfate (PVS), polyanethole sulfonate (PAS),
and polystyrene sulfonate (PSS) were found to meet these requirements. These are the first heparinase inhibitors to be discovered.
The inhibitory effect of PVS ($K_i = 3.0 \times 10^{-8}$M; MW $\sim$ 10,000) was
lost as PVS was hydrolyzed. An affinity column was prepared by

immobilizing partially hydrolyzed PVS on epoxy-activated Sepharose
(22). Heparinase (HA purified) was bound to this column and released at either pH 11 or 4 to give 500% enrichment in specific
enzyme activity with 5-10% total activity recovery.

Isoelectric focusing (IEF) also was tried for obtaining
highly pure heparinase. The enzyme was loaded onto prefocused
acrylamide gel at pH 7.0. After isoelectric focusing the enzymatic
+ 0.5 to give a specific activity of about 5000 units/mg protein
and an enrichment of 50 fold.

The purification of heparinase was followed by SDS-gel electrophoresis. The crude sonicate gave >20 major bands, the HA
purified enzyme 3 major bands, and the IEF purified enzyme 2 major
bands. Heparinase has a MW of 51,000 $\pm$ 6,000 as judged by Sephadex G-200 gel exclusion chromatography. SDS gel electrophoresis
of isoelectric focused heparinase showed only two bands, one at
80,700 MW and one at 45,700 $\pm$ 1,600. The latter band fell within
the range of that obtained by Sephadex chromatography, further
suggesting the accuracy of the 51,000 $\pm$ 6,000 value obtained by
that method.

The enzyme is very specific, acting only on heparin (Km =
4.2×10^{-5} M) and slightly on heparin monosulfate. Over 30 very
similar polysaccharides, including other glycosaminoglycans, were
tested. We also found that heparinase acts endolytically as an
α-1,4-eliminase cleaving heparin (M.W. $\sim$ 10,000) at 9-10 sites. A
computer model has been developed and used successfully to predict
the degradation kinetics of heparin by heparinase (23).

IMMOBILIZED HEPARINASE

Immobilization of heparinase should prevent the enzyme from
circulating and thereby reduce or eliminate immunological problems.
Heparinase was immobilized to Sepharose using a variation (21) of
the procedure of March *et al.* (24) at a yield of 91%. The hepa-
rinase-Sepharose had enhanced thermal stability, which was especially noticeable at low temperature storage of this enzyme. At
$4°C$ the immobilized enzyme had a half life of denaturation of
>3600 hr compared with 125 hr for the native enzyme at the same
temperature. The greater stability of the immobilized enzyme is
also seen at higher temperatures (21).

In initial trials with blood we took a 100 cc Bentley blood
transfusion filter, packed it with 50 cc of Sepharose-heparinase,
and fluidized the Sepharose by pumping it through a U-tube inter-
connected to the filter at a flow rate 6 times higher than blood
was being pumped through the same filter (Fig. 1). This

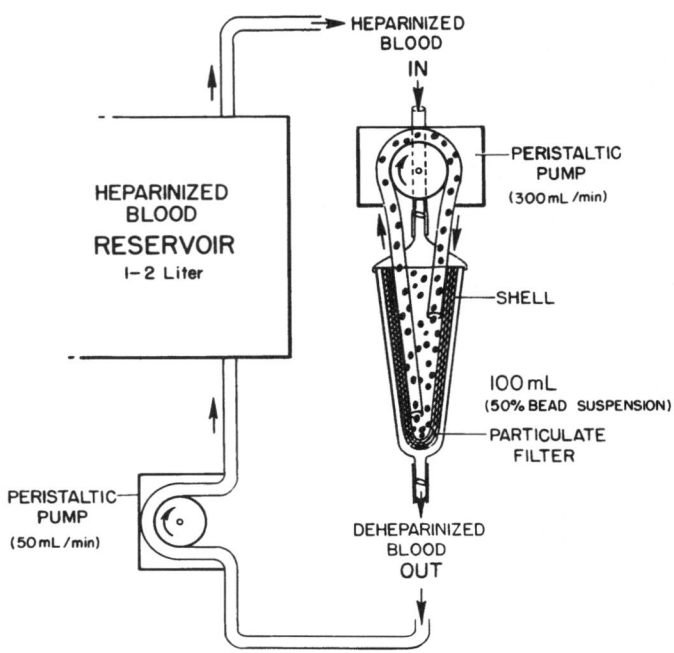

Fig. 1. Heparinase reactor.

fluidization prevented the Sepharose from packing and preventing blood flow. Using this reactor, two sets of experiments were conducted. In one set, whole human blood (2 units) was obtained from Boston's Children's Hospital Blood Donor Service and anticoagulated at 100 units heparin/ml. This was much higher than clinical levels; but gave a more difficult test. The blood was stored at 37°C during use. Blood was passed through the filter at 50 ml/min. Samples were taken after each of 6 passes and prior to heparinization, centrifuged at 4500 rpm for 20 min; and the plasma was assayed by the APTT (25), and Azure A (26) methods. Sixty percent of heparin anticoagulant activity was gone by 6 min or 3 passes by the APTT. The Azure A test was performed as a control to insure that heparin was actually being lost and that the results obtained were not merely due to damage to coagulation factors. Azure A measures the concentration of heparin for all heparin segments of hexasaccharide or larger (27). The reason the Azure A test does not show zero heparin activity is that a few of hexasaccharide or larger are present, as shown by both HPLC and gel permeation chromatography and by digestion of heparin by heparinase in buffer. These segments show Azure A binding activity but do not possess anticoagulant activity as measured by APTT. Control filters containing Sepharose without heparinase showed no effect as judged by either assay.

We performed a similar test on an 11 kg dog, which was anes-thetized and catheterized through the carotid artery and vein with a Scribner Shunt fitted with silastic tubing, which interfaced directly with the heparinase reactor. The dog was administered 4500 units of heparin. Blood samples were taken and anlayzed by 3 assays: APTT (25), whole blood recalcification time (26), and Azure A (26). As shown in Fig. 2, within 2 min or 1 pass through the filter, nearly all anticoagulant activity was lost. The Azure A test again was used as a control assay. The controls showed no detectable effect (Fig. 2). An *in vivo* half life of 2 hr for hep-arin was measured in the dog. The dogs appeared healthy and were still alive 3 month after the deheparinization experiments.

We have also taken the heparinase degradation products from these filters and have subjected them to toxicity tests using the *S. typhimurim* mutagenesis assay (28). No mutagenicity was ob-served even with concentrations 1000-fold in excess of those we would anticipate clinically. Heparin was also negative in this assay. Some platelet loss was noted using this filter; this may be circumvented, however, by a different choice of biomaterials for the heparinase filter, co-immobilization of other compounds (e.g., prostacylin) to the filter, or a different filter design.

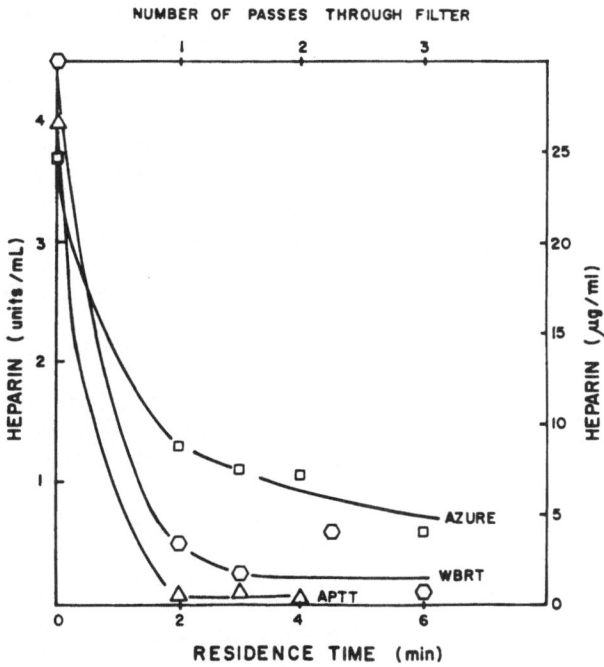

Fig. 2. Heparin levels in blood of dogs.

Although the filter is at an early stage of development, it is clear that such a filter could have broad generality. At present there exist blood filters, in some cases as large as 2000 cc, that are routinely used at the termination of extracorporeal procedures to remove aggregates formed during these procedures. It is possible that heparinase could be bound to these filters. In fact, one attractive feature of the application of immobilized heparinase is that the filter would only be used in situations where the blood must leave the patient and where existing biomaterials already interface with blood at the desired location. Thus, the eventual application of this process might not require any additional inconvenience to the patient or additional apparatus for the physician.

REFERENCES

1. SCHREINER, G. E. *Trans. Am. Soc. Art. Int. Org. 26:*   9 (1980).
2. KENNEDY, R. H., KENNEDY, M. A., MCGOON, D. C., PLUTH, J. R. & NOBREGA, F. T. *J. Thorac. Cardiovasc. Surg. 80:*   702 (1980).
3. FLETCHER, J. R., MCGEE, A. E., MILLS, M., SNYDER, K. C. & HERMAN, C. M. *Surgery 80:*   214 (1976).
4. GLABMAN, S., GERONEMUS, R., VONALBERTINI, B., KAHN, T., MOUTOSSIS, G. & BOSCH, J. P. *Trans. Am. Soc. Art. Int. Org. 25:*   394 (1979).
5. BASU, D., GALLUS, A., HIRSCH, J. & CADE, J. *New Eng. J. Med. 287:*   324 (1972).
6. GILSON, S. *Throm. Diath. Hemorh. 6:*   152 (1961).
7. GERVIN, A. S. *Surg. Gyn. Obstet. 140:*   789 (1975).
8. SWARTZ, R. D. & PORT, F. K. *Kidney Int. 16:*   513 (1979).
9. HIRSCHBOECK, J., MADISON, F. & PISCIOTTA, A. *Am. J. Med. Sci 227:*   279 (1954).
10. STINCHFIELD, F., SANKAVAN, B. & SAMILSON, R. *J. Bone Joint Surg. 38:*   270 (1956).
11. GRIFFITH, G., NICHOLS, G., ASHER, J. & FLANAGEN, B. *J. Am. Med. Assoc. 193:*   91 (1965).
12. JASTRZEBSKI, J., HILGARD, P. & SYKES, M. K. *Card. Res. 9:* 691 (1975).
13. KOLFF, W. J. & STELLWAG, F. *Ann. N.Y. Acad. Sci. 283:* 443 (1977).
14. LEININGER, R. I., FALB, R. D. & GRODE, G. A. *Ann. N.Y. Acad. Sci. 146:*   11 (1968).
15. KUSSEROW, B. K., LARROW, R. & NICHOLS, J. *Trans. Am. Soc. Artif. Int. Org. 17:*   1 (1971).
16. BERGLIN, E., HANSSON, M. A., TGER-NILSSON, A. C. & WILLIAM-OLSSON, G. *Thromb. Res. 9:*   81 (1976).
17. JACQUES, L. B. *Canad. Med. Ass. J. 108:*   1291 (1973).
18. HOVINGH, P. & LINKER, A. *J. Biol. Chem. 245:*   6170 (1970).

19. GALLIHER, P. M., COONEY, C. L., LANGER, R. & LINHARDT, R. *Appl. Envir. Micro. 41:* 360 (1981).

20. LINKER, A. & HOVINGH, P. *Biochemistry 11:* 53 (1972).

21. LANGER, R., LINHARDT, R. J., GALLIHER, P. M., FLANAGAN, M., KLEIN, M. & COONEY, C. L.in "Biomaterials: Interfacial Phenomena and Applications" (S. Cooper, A. Hoffman, B. Ratner, and N. A. Peppas, eds.), Advances in Chemistry Series, in press.

22. VRETBALD, P. *Biochim. Biophys. Acta. 434:* 169 (1976).

23. LINHARDT, R. J., FITZGERALD, G. L., COONEY, C. L. & LANGER, R. *Biochim. Biophys. Acta.*, in press.

24. MARCH, S. C., PARIKH, I. & CUATRECASAS, P. *Anal. Biochem. 60:* 149 (1974).

25. LANGDELL, R. D. in "Thrombosis and Bleeding Disorders," (N. U. Bank, F. K. Beller, E. Deutsch, E. F. Mammen, eds.) Academic Press, New York (1971).

26. HOFFBERG, S., thesis, M.I.T. (1981).

27. DIETRICH, C. P. *Biochem. J. 108:* 647 (1968).

28. SKOPEK, J. R., LIBER, H. L., KADEN, D. A. & THILLY, W. G. *Proc. Natl. Acad. Sci. USA 75:* 4465 (1978).

CLINICAL UTILITY OF UROKINASE-TREATED POLYMER FOR

ANTITHROMBOGENIC MATERIAL

T. Ohshiro

Second Department of Surgery
Osaka University Medical School
Osaka, Japan

There is an increasing need to develop better medical materials with high antithrombogenecity. Attempts have been made to improve such materials by the following methods: a) to develop a material biologically or physiochemically compatible with tissue, b) to coat a material with albumin or globulin, c) to immobilize an anticoagulant agent or anti-platelet agent on the surface of a material, and d) to immobilize a fibrinolytic enzyme on the surface. Among these methods, the immobilization of a fibrinolytic enzyme is especially intriguing; and immobilized urokinase has been shown to maintain fibrinolytic ability (1-5). Described herein are the results of clinical tests of urokinase-treated polymer as an antithrombogenic material.

Urokinase was immobilized by etching polymer with HCl or NaOH for one hr, followed by dipping in 10 wt% polyethyleneimine at 30°C for two hr in the presence of 5 wt% dicyclohexyl carbodiimide, dipping in 4 wt% Gantrez (maleic anhydride/methylvinyl ether copolymer) for five hr, and finally dipping in 600 U/ml urokinase solution at 30°C for one hr. The key to obtaining strong fibrinolytic activity of the urokinase-treated polymer is to form a grafted copolymer layer between the urokinase and polymer in order to increase considerably the number of binding sites. Polyethyleneimine and Gantrez gave good results. In measuring the effective activity of the immobilized urokinase, the weight of immobilized urokinase and the enzyme activity were determined separately. The amount of immobilized urokinase, obtained from radioactivity, was 1.5×10^{-3} mg/cm^2. This was equivalent to a fibrinolytic activity of 68 IU/cm^2. The fibrinolytic activity calculated from the zymolytic activity by the MCA method was 20 IU/cm^2. The ratio

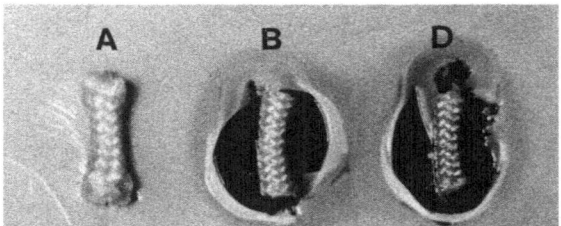

Fig. 1. Fibrinolytic ability of urokinase-treated nylon threads
         before and after disinfection with ethylene oxide gas
         (EOG) on standard human fibrin plate. A = untreated,
         B = urokinase treated (before EOF), D = urokinase treated
         (after EOG).

of the former to the latter was 7:2. The effective enzyme activity was approximately 30% (Fig. 1).

    The possibility that urokinase-treated polymer had antithrom-
bogenecity *in vivo* was foreseen (6-8). Hence, we prepared
urokinase-treated evatate catheters from ethylene vinyl acetate
copolymer. The catheter, 1.5 mm in outer diameter and 40-70 cm in

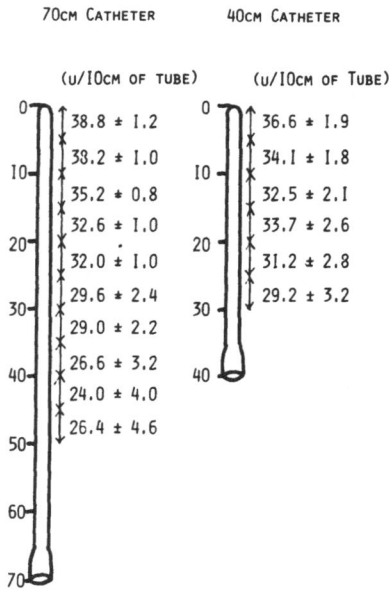

Fig. 2. Fibrinolytic activity of urokinase-treated evatate
         catheter measured by MCA method.

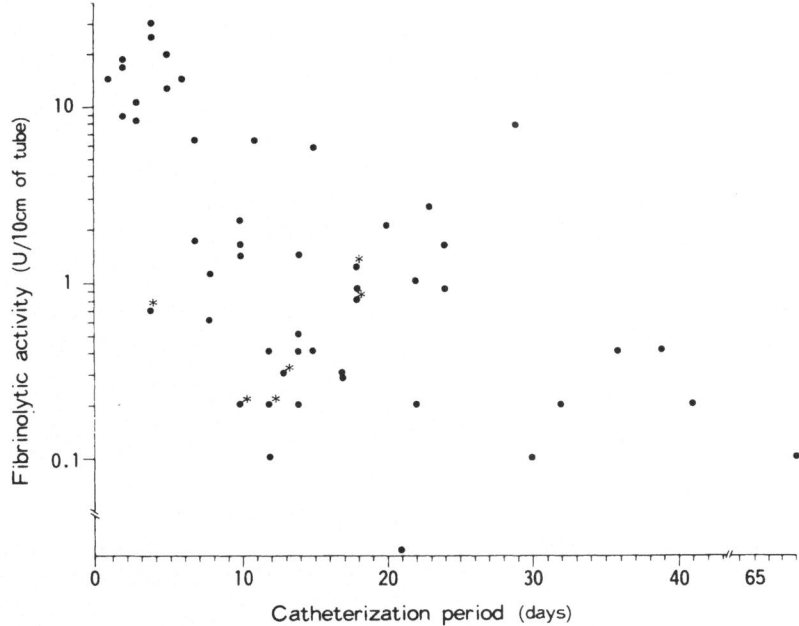

Fig. 3. Fibrinolytic activity of individual catheters after use
          in fasting patients. * = thrombus formation.

length, had the fibrinolytic activity of 38.8 - 26.4 U/10 cm of
tube (ca 4 U/cm^2) as shown in Fig. 2. The catheter was placed in
the superior caval vein through the right subclavian vein. We
used 55 catheters in 55 patients. Thrombi were formed in 6 cases
(10.9%), while 49 (89.1%) were free from thrombus formation; this
compares favorably with 40-95% thrombi formed with conventional
catheters. The fibrinolytic activity of the catheter decreased
with extension of the catheterization period, as shown in Fig. 3;
but it still showed approximately 1.7 U/10 cm of tube 7 day after
insertion and 0.3 U/10 cm of tube 21 day after insertion. Of the
six catheters that failed to prevent thrombus formation, five gave
a value less than 1 U/10 cm of tube, and accounted for 20.8% of
the 24 catheters whose fibrinolytic activity was under 1 U/10 cm
of tube. On the other hand, there were 27 catheters which gave a
value above 1 U/10 cm of tube; and only one (3.7%) of these was
associated with thrombus formation. The new catheter did not
undergo qualitative changes even during two months retention.

     In thoracic surgery, the failure to draw off hemorrhagic effu-
sions causes serious complications, such as atelectasis, pneumonia,
abscess, or sepsis. We devised urokinase-treated polyvinyl
chloride tubes for liquid drainage and used them 29 times in 19

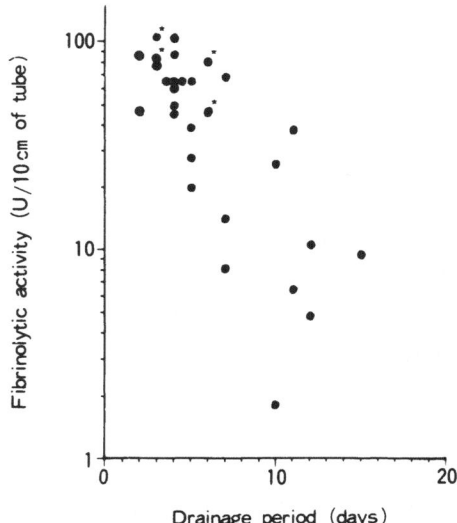

Fig. 4. Fibrinolytic activity of individual drain tubes after use in thoracic surgery in patients. * = thrombus formation

patients. The mean drainage period was 5.9 day (3-15 day).
Thrombi, resembling buffy-coat clots, were observed in 4 tubes
(13.8%), while none were seen in 25 tubes (86.2%). None of the
patients suffered any complications. The fibrinolytic activity of
individual drain tubes after insertion is shown in Fig. 4. The
initial activity of about 200 U/10 cm of tube (*ca* 8 U/cm^2) decreased to about 40 U/10 cm of tube 6 day after insertion.

One of the surgical procedures to treat lymphedema is
Handley's method; this involves enclosing lymph drainage materials
in the subcutis of affected limbs. In this method, stagnant lymph
is led to normal tissues through the interspace within threads,
tubes, or wicks of silk, nylon, or teflon. The passage of lymph
through the interspace is smooth soon after embedding; but the
flow becomes obstructed after several months due to clotting of
lymph and invasion of connective tissues. As a result, lymphedema
develops again. We prepared urokinase-treated nylon treads made
up of more than a thousand fibers and tested them clinically in
12 patients (Table 1). In moderate cases, only Handley's method
was performed. In severe cases, Thompson's method was employed
in combination with Handley's method. Out of 12 cases, 4 were
primary and 7 were secondary, followed by radical uterine or
breast cancer operations. Eight patients showed excellent or good
effects. The tension of firm subcutaneous tissue was weakened,
even in 3 patients where a reduction of circumference did not

TABLE 1

LYMPH DRAINAGE OF UROKINASE-TREATED NYLON THREAD*

Category	Number of Cases
Sex	
male	1
female	11
Classification	
primary	4
secondary	7
other	1
unknown	0
Localization	
upper extremity	3
lower extremity	9
Clinical effect	
excellent	2
good	6
moderate	3
poor	1

*Mean age 41.5 yr

occur. One patient became worse from an infection of the embedded threads. In one patient threads tentatively withdrawn after 19 month of embedding possessed about 5% of the original activity. The histological observations showed that connective tissues had invaded part of the interspace between the fine nylon fibers, composing each thread, to occlude the capillary lumens, but the extent of invasion was mild. It was assumed that the thread was still functioning as a lymphatic pathway.

In microvascular surgery, minute thrombi formed on suture materials become the cause of arterial re-occlusion. Then urokinase-treated nylon filaments of 3-0 to 5-0 size were used for suturing the small vascular vessels. This procedure was attempted 59 times in a total of 28 patients needing reconstruction of the iliac to popliteal artery due to angiitis, arteriosclerosis, aneurysm, or trauma (Table 2). During the 2 year 8 month mean follow up period, favorable blood flow was maintained in 20 cases (71.4%). The prosthetic grafts were occluded in 8 cases (28.6%), consisting of 3 thromboangiitis obliterans and 5 arteriosclerosis obliterans. Although the real factor responsible for the re-

TABLE 2

VASCULAR SUTURE MATERIAL OF UROKINASE-TREATED NYLON FILAMENT*

Category	Number of Cases	Notes
Sex		
male	24	
female	4	
Purpose		artery-prosthetic graft anastomosis
Prosthetic graft		
bifurcated graft	3	
straight graft	25	
Total number of anastomosis		
abdominal aorta	3/59	
common-external iliac artery	18/59	
superficial femoral artery	38/59	
Occlusion of graft	8	
Cause of occlusion		unknown

*Mean age 58.1 years.

occlusion was not known, the urokinase-treated suture material was thought to have been useful.

The antithrombogenic mechanism of the urokinase-treated poly-mer may be interpreted, as follows: when blood contacts the sur-face of the material, Factor XII is activated; this leads to the formation of a lot of fibrin monomer. Then, plasminogens, which have an affinity for fibrins, gather around each fibrin monomer; urokinase then converts the plasminogens into plasmins, which lyse the fibrin monomer directly. This mechanism has three points: a) fibrinolysis as seen on the surface of the material occurs directly at the site of contact between the material and the fibrin monomer, so that it is not affected by antiplasmins abun-dantly present in blood; b) the situation thus created on the surface of the material has a close resemblance to the antithrom-bogenic condition of the endothelium in a vascular vessel, and c) fibrinolysis uninterruptedly taking place on the surface of the material is nothing but a localized phenomenon and never becomes generalized.

CONCLUSION

Urokinase-treated polymer can become a good antithrombogenic medical material. The necessity for antithrombogenic medical devices will increase further in the future.

REFERENCES

1. AMBRUS, C. M., ROHOLT, O. A., & MEYER, B. K. *Fed. Proc. 31:* 267 (1972).
2. DEUTH, D. G. & MERTZ, E. T. *J. Med. 3:* 224 (1972).
3. KUSSEROW, B. K. & LARROW, R. W. *Circulation 2:* 54 (1972).
4. WIMAN, B. & WALLEN, P. *Eur. J. Biochem. 36:* 25 (1973).
5. CAPET-ANTONINI, F. C., GRIMARD, D., & TAMMENASSE, J. *Thromb. Res. 2:* 479 (1973).
6. KUSSEROW, B. K., LARROW, R. W., & NICHOLS, J. E. *Trans. Am. Soc. Art. Int. Organs 19:* 8 (1973).
7. OHSHIRO, T., MONDEN, M., KOSAKI, G., & TAKAGI, K. *Artif. Organs (Jpn) 6:* 237 (1977).
8. OHSHIRO, T. & KOSAKI, G. *Artif. Organs 4:* 58 (1980).

ARTIFICIAL CELL IMMOBILIZED MULTIENZYME SYSTEMS AND COFACTORS

T. M. S. Chang, Y. T. Yu, and J. Grunwald

Artificial Cells and Organs Research Centre
McGill University, Montreal, Canada

The biomedical applications of immobilized enzymes have already been demonstrated using single enzyme systems (1-3). However, most metabolic functions are carried out in the body by complex multienzyme systems with cofactor requirements. As a result, research has been carried out here for microencapsulation of multienzyme systems with cofactor regeneration incorporated. Artificial cells containing hexokinase and pyruvate kinase could recycle ATP for the continuous conversion of glucose into G-6-P (4,5). Artificial cells containing alcohol dehydrogenase and malic dehydrogenase can recycle NADH (4,5). Multienzyme systems (urease, glutamate dehydrogenase, glucose-6-phosphate dehydrogenase) inside semipermeable microcapsules can convert urea sequentially into ammonia and glutamate (6). Glucose-6-phosphate dehydrogenase is used to recycle the cofactor. In order to allow for the use of blood glucose, artificial cells containing urease, glutamine dehydrogenase, and glucose dehydrogenase have been developed to convert urea or ammonia into glutamate (7,8). Further development of this approach has been carried out to prepare artificial cells containing urease, glutamate dehydrogenase, transaminase, and glucose dehydrogenase (9). This sequential reaction includes the conversion of urea into ammonia then to glutamic acid and finally to alanine (9). The required cofactor can be recycled by glucose dehydrogenase. These studies demonstrate the feasibility of using artificial cells containing multienzyme systems for the sequential conversion of substrates with cofactor recycling. This is now ready for *in vitro* application. However, for *in vivo* use it would be much more desirable to retain the cofactor within the artificial cells for continuous recycling. The following work is being carried out to study this feasibility.

COFACTOR LINKED TO MACROMOLECULES FOR IMMOBILIZATION OF COFACTOR
IN ARTIFICIAL CELLS (10,11)

One approach, carried out in this laboratory, is to link cofactors to macromolecules to form soluble macromolecules. In this
way the macromolecules can be retained within the artificial cell
and do not cross the semipermeable membrane. Typical examples are
given (10,11).

NAD^+-N^6-[N-(6-aminohexyl)-acetamide] was coupled to dextran
T-70 by a modification of an earlier technique (12). The resulting
soluble dextran-NAD^+ contained up to 40 µmol of coupled cofactor
per g (dry weight) of dextran. This is equivalent to three cofactor molecules per dextran molecule. The native and the immobilized
NAD^+ showed very good stability in Tris buffer but were rapidly
de-activated in crude hemoglobin (11). It is likely that NAD^+ deactivation in the presence of crude hemoglobin was caused by enzymes, such as NAD^+-glycohydrolases and poly(ADP-ribose) polymerase
which are known to be present in red blood cells. The deactivation
of NAD^+ was eliminated after purification of the hemoglobin by affinity chromatography on NAD^+-Sepharose. Dextran-NAD^+ was recycled
within the microcapsules by the sequential reaction of yeast alcohol dehydrogenase and malic dehydrogenase (11). The microcapsules
were placed in a shunt to form a continuous flow system; so that
65% of the oxaloacetic acid was converted to malic acid in one
passage through the shunt. The dextran-NAD^+ present within the
microcapsules was regenerated two times each min during the reaction. The shunt showed good stability, the reaction rate remaining constant for the first hr and 83% of the original activity being retained even after 3 hr of continuous reaction. When microcapsules were crosslinked with glutaraldehyde by the technique described (13), there was complete stability during 3 hr of continuous reaction; although the activity was only about 10% that of
the untreated microcapsules. The same shunt could be stored and
reused several times.

In certain studies, one may want to avoid the use of hemoglobin with its other possible enzyme contaminants. The method of
preparing artificial cells (13) was modified earlier to substitute
for hemoglobin with polyethyleneimine (14). This was used here
for the preparation of artificial cells containing dextran-NAD^+,
alcohol dehydrogenase, and malic dehydrogenase (11) (Fig. 1). The
recycling of dextran-NAD^+ within the nylon-polyethyleneimine microcapsules was measured in a continuous flow shunt (11). The dex-
tran-NAD^+ within the microcapsules was regenerated twice each min
during the continuous reaction. No significant decrease in activity was observed during the 3 hr of reaction. After this reaction,
each shunt was stored at 4°C. In reusing this after 7 and 12 day
of storage, 63% of the original activity was retained after 7 day;
and 41% was still retained after 12 day. The microcapsules could

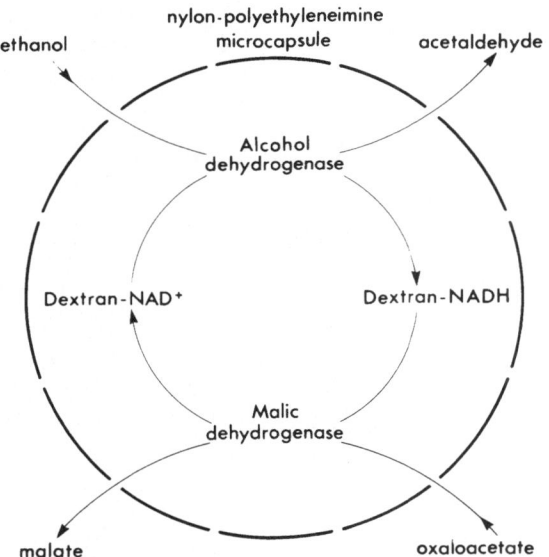

Fig. 1. Artificial cells containing alcohol dehydrogenase, malic
         dehydrogenase, and NAD^+ -dextran.

be further stabilized by crosslinking with glutaraldehyde; but this
crosslinking caused a sharp decrease in the initial recycling activity within the microcapsules.

Dextran-NAD^+ is an excellent model for studying the feasibility of cofactors linked to macromolecules within artificial cells.
However, the cofactor activity was not sufficient for many multienzyme systems. As a result, we have looked at another system,
polyethylene-glycol linked cofactor (15).

LIPID-POLYMER MEMBRANE ARTIFICIAL CELL FOR THE IMMOBILIZATION OF MULTIENZYME SYSTEMS, COFACTORS, AND SUBSTRATES (16)

Another approach for retaining cofactors inside artificial
cells for continuous recycling is being studied for a system which
can act on lipid-soluble metabolites. The detailed study and analysis of a lipid-polymer complex membrane artificial cell system
(13) has recently been completed (17). In this system, the artificial cells with the polymer membrane is complexed with a lipid material similar to lipids in biological cells. In this way the membrane has a very low permeability to nonlipid soluble materials;
but lipid-soluble substrates can cross the membrane very freely.

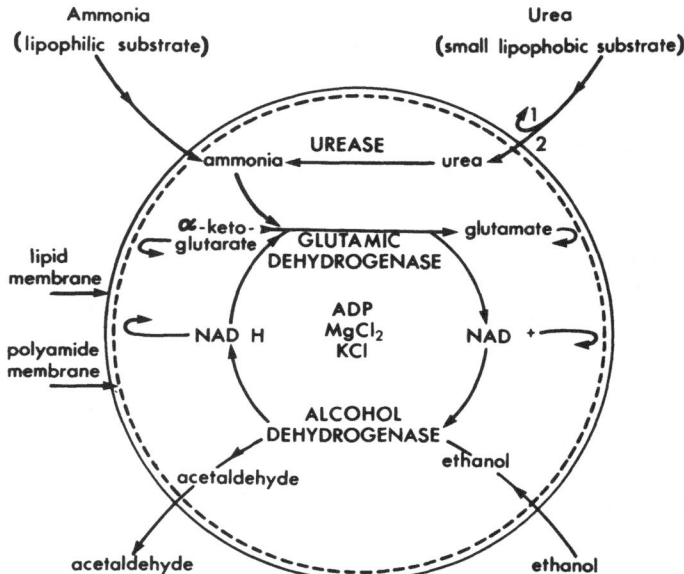

Fig. 2.   Artificial cells formed from lipid-polyamide ultrathin
          membrane. Enzymes, free cofactors, and α-ketoglutarate
          are retained inside to act on lipid soluble substrate
          diffusing into the artificial cells.

         A study was carried out to see whether ultrathin lipid-nylon
membrane microcapsules could retain enzymes, NAD$^+$, NADH, and α-
ketoglutarate; and whether external ammonia and alcohol could cross
the lipid membrane to take part in the multistep reactions shown
in Fig. 2. The results showed that the enzymes, NAD$^+$, and α-
ketoglutarate did not leak out from the microcapsules despite extensive washings. When microcapsules containing glutamic dehydrogenase, α-ketoglutarate, KCl, MgCl$_2$, ADP, and 10 μmol NADH were
added to ammonia-alcohol substrate solution at pH 9; the ammonia
levels fell rapidly. Ammonia (10 μmol) was converted into glutamate by the microcapsules. Here, NADH was retained in the microcapsules, but no recycling enzyme was included. When microcapsules containing 0.25 μmol NAD$^+$, alcohol dehydrogenase, glutamic
dehydrogenase, α-ketoglutarate, MgCl$_2$, KCl, and ADP were added to
the substrate solution, the change in ammonia levels indicated
that 10 μmol ammonia was converted into glutamate. Thus, the
0.25 μmol NAD$^+$ retained in the microcapsules could be recycled 40
times to convert 10 μmol ammonia. By increasing the NAD$^+$ from
0.25 μmol to 0.50 μmol, the rate of conversion was comparable to
10 μmol of NADH with no recycling.

These results show that the recycling cofactors and substrates required for multienzyme reactions can be retained within ultrathin lipid-polyamide microcapsules to act on permeant external substrates. This system can be useful for lipophillic substrates or toxins. The lipid-polymer membrane artificial cell is more analogous to that of biological cells; and further basic studies and developmental studies will be carried out. Further study is being continued to modify the permeability of the membrane by changing the composition of the cholesterol and lecithin in the membrane. Other multienzyme systems for the conversion of lipid soluble metabolites and substrates will be studied.

DISCUSSION

In addition to artificial cells containing simple enzymes, we now have two artificial cells systems containing multienzymes with cofactor recycling for possible biomedical applications. The one involving the use of cofactors covalently linked to macromolecules, is suitable for use with large water soluble substrates. The other, involving the use of lipid-polymer membranes, to retain the free cofactor, is suitable for use with lipid soluble substrates.

REFERENCES

1. CHANG, T. M. S. *Science 146:* 524 (1964).
2. CHANG, T. M. S., MACINTOSH, F. C., & MASON, S. G. *Can. J. Physiol. Pharmacol. 44:* 115 (1965).
3. CHANG, T. M. S., "Biomedical Applications of Immobilized Enzymes and Proteins", vols. 1 & 2, Plenum Press, New York (1977).
4. CAMPBELL, J. & CHANG, T. M. S. *Biochem. Biophys. Res. Commun. 69:* 562 (1976).
5. CAMPBELL, J. & CHANG, T. M. S. in "Enzyme Engineering", vol 3 (E. K. Pye and H. H. Weetall, eds.) Plenum Press, New York (1978). p. 371.
6. COUSINEAU, J. & CHANG, T. M. S. *Biochem. Biophys. Res. Commun. 79:* 24 (1977).
7. CHANG, T. M. S. & MALOUF, C. *Trans. Am. Soc. Artif. Intern. Organs 24:* 18 (1978).
8. CHANG, T. M. S. & MALOUF, C. *Artif. Organs 3:* 38 (1979).
9. CHANG, T. M. S., MALOUF, C. & RESURRECCION, E. *Artif. Organs (Supp.) 3:* 284 (1979).
10. GRUNWALD, J. & CHANG, T. M. S. *J. Applied Biochem. 1:* 104 (1979).
11. GRUNWALD, J. & CHANG, T. M. S. *J. Mol. Cat. 11:* 83 (1981).
12. LARSSON, P. O. & MOSBACH, K. *FEBS Lett. 46:* 119 (1974).

13. CHANG, T. M. S., "Artificial Cells", Charles C. Thomas
     Publisher, Springfield, IL. (1972).
14. AISINA, R. B., KAZANSKAFA, N. F., LUKASHEVA, E. V., &
     BEREZIN, V. *Biokhimiya 41:* 1656 (1976).
15. BUCKMANN, A. F., KULA, M. R., WICHMANN, R., & WANDREY, C.
     J. Applied Biochem, in press.
16. YU, Y. T. & CHANG, T. M. S. *FEBS Lett. 125:* 94 (1981).
17. ROSENTHAL, A. M. & CHANG, T. M. S. *J. Memb. Sci. 6:* 329
     (1980).

ANTITHROMBOGENIC ACTIVITY OF ARTIFICIAL MEDICAL MATERIALS

IMPROVED BY ENZYME IMMOBILIZATION TECHNIQUES

Y. Miura, S. Aoyagi and K. Miyamoto

Department of Biochemical Engineering,
Faculty of Pharmaceutical Sciences, Osaka
University, Suita, Osaka, Japan

The immobilization of proteins onto the surface of artificial polymers was investigated in order to prepare antithrombotic medical materials which can inhibit blood coagulation, activate the fibrinolytic system, and block platelet aggregation.

The thrombus formation time (TFT) was measured by Chandler's method using 0.8 ml samples of native rabbit blood. Immobilization of the proteins was done using Sepharose 4B as the carrier. When the TFT was determined in the presence of the immobilized proteins, 5 mg of immobilized antithrombin III-heparin (I-AT III·Hep) and 10 mg of immobilized urokinase (I-UK) gave slight prolongations of the TFT; but these materials could not block completely the thrombus formation.

Furthermore, both the antithrombin III-heparin complex and urokinase were coimmobilized on Sepharose 4B (I-AT III • Hep-UK) and showed the highest antithrombotic activity of these immobilized preparations. I-AT III • Hep-UK, 5 mg, blocked thrombus formation in all runs of 0.8 ml rabbit whole blood; while 5 mg of I-AT III·Hep blocked thrombus formation in only 4 out of 11 runs. I-AT III·Hep-UK showed much higher antithrombogenecity than I-AT III·Hep and I-UK, even when I-AT III·Hep-UK had only half the anticoagulant activity of I-AT III·Hep and only half the fibrinolytic activity of I-UK per unit dry weight of carrier.

We also investigated the mechanism of platelet aggregation, using citrated rabbit blood. The formation of platelet aggregates in a shear field was followed in an artificial blood circulation device. The platelet aggregation was induced by the adenosine diphosphate, which was liberated from the blood cells injured in the shear field.

457

Heparin failed to prevent platelet aggregation in sheared whole blood. The enzymatic elimination of adenosine diphosphate by pyruvate kinase and phosphoenolpyruvate or apyrase was effective in the prevention of the formation of platelet aggregates. Immobilized apyrase showed stable activity for adenosine diphosphate decomposition and thereby prevented platelet aggregation in the sheared blood.

APPLICATION OF IMMOBILIZED ENZYMES FOR BIOMATERIALS USED IN THE FIELD OF THORACIC SURGERY

S. Watanabe and T. Teramatsu

Department of Thoracic Surgery, Chest Disease
Research Institute, Kyoto University, Kyoto, Japan

With a view to developing biomedical materials for reconstructive surgery, we have studied a composite material constituted with collagen and a synthetic polymer, which possesses high tissue compatibility(1). We have been also investigating the replacement of various organs with this new material experimentally and clinically, for example for reconstruction of the trachea, chest wall, or diaphragm(2). This collagen-synthetic polymer composite material was applied as a support for immobilization of enzymes to provide the material surface with biological functions. Enzymes were successfully bound to the collagen membrane layer by activation of its carboxyl groups. We have already described the *in vitro* and *in vivo* behavior of urokinase immobilized on this composite material(3). When artificial organs and biomedical materials are implanted in the body, bacterial infection may cause serious complications. We now report a new method for producing antibacterial biomedical materials by immobilization of lysozyme and a polypeptide antibiotic Polymyxin B on this composite material. Lysozyme-bound collagen-polypropylene mesh composite material showed good bacteriolytic activity against *Micrococcus lysodeikticus* on agarose plates. The activity remained for about 19 months after the preparation. Pieces of this lysozyme-bearing composite material were implanted into rabbit dorsal subcutaneous tissue. Although various histological reactions were seen soon after the implantation, the lysozyme-bound composite material was adapted to the subcutaneous tissue of a different species after prolonged implantation (Fig. 1). Hematological examination showed no abnormal findings. Polymyxin B-bound composite material also showed antibacterial activity against *Bordetella bronchiseptica* on agar plates. The activity was maintained sufficiently during 6 months storage. Pieces of this Polymyxin B-bearing composite

material were implanted in the rabbit dorsal subcutaneous tissue.
Some inflammatory reactions were seen soon after the implantation;
but the composite material was later adapted to the subcutaneous
tissue. Any side effect such as renal and neural toxic reaction
of soluble Polymyxin B were not seen.

We wish to thank T. Murachi for his helpful advice and support during the course of this investigation.

REFERENCES

1. SHIMIZU, Y., ABE, R., TERAMATSU, T., OKAMURA, S., & HINO, T.
   Biomat. Med. Dev. Artif. Organs 5: 49 (1977).
2. SHIMIZU, Y., MIYAMOTO, Y., TERAMATSU, T., OKAMURA, S., &
   HINO, T. *Biomat. Med. Dev. Artif. Organs 6:* 375 (1978).
3. WATANABE, S., SHIMIZU, Y., TERAMATSU, T., MURACHI, T., &
   HINO, T. *J. Biomed. Mater. Res. 15:* 553 (1981).

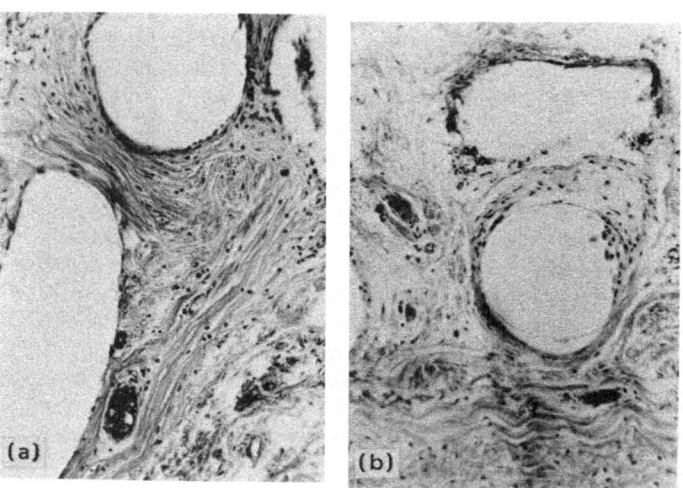

Fig. 1. Histological findings of collagen-polypropylene mesh
composite materials with immobilized lysozyme (a) and
without immobilized lysozyme (b), implanted in rabbit
dorsal subcutaneous tissue for five months (X 100).

USE OF IMMOBILIZATION PRINCIPLES FOR THE CONSTRUCTION OF

DRUG TARGETING SYSTEMS

V. P. Torchilin, A. L. Klibanov, V. R. Berdichevsky,
V. G. Omelyanenko, and V. N. Smirnov

National Cardiology Research Center, Moscow, USSR

A most interesting therapeutic idea is that of targeting of drugs, using the specific affinity of liposomes for affected organs and tissues. The most obvious method for production of these affinity liposomes is the modification of the liposomal surface with affinity macromolecules possessing specific affinity for characteristic components of the target organ or tissue (Fig. 1).

Effective covalent binding of proteins can be achieved when phosphatidylethanolamine-containing liposomes are first activated with dialdehyde or diimidate; and then protein is added. In such

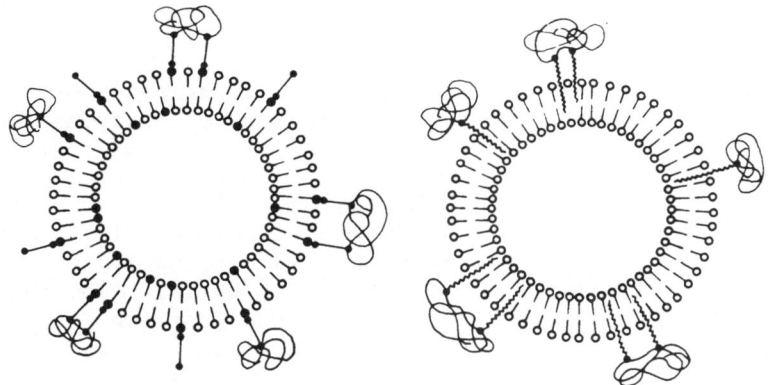

Fig. 1. Binding of proteins to liposomes. Left: covalent binding via spaces; right: hydrophobic binding. O— lecithin, ●—phosphatidylethanolamine, ●—● bifunctional reagent, ●—∿∿ hydrophobic modifyer.

461

a way we were able to bind up to 7.1×10^{-5} moles of active enzyme (α-chymotrypsin) per mole of lipid; 70% of the bound enzyme could be inhibited by pancreatic protease inhibitor. This method had a low binding efficiency and different liposomes could be cross-linked to cause aggregation of liposomes. In another approach the same enzyme was modified with palmitoyl chloride and then bound to liposomes by the cholate dialysis technique. Binding of 3×10^{-4} moles protein/mol of lipid was achieved when ·3 NH_2- groups, from 15 titrated, were modified. About 80% of the immobilized enzyme was still available for pancreatic inhibitor. Results of fluorescence and EPR studies showed that incorporation of modified proteins into the liposomes occurred rather slowly; the immobilized protein was located above the liposomal surface. It was supposed that the protein modification with natural phospholipid that had two hydrophobic tails would require a smaller degree of modification, and thus the enzyme activity for interaction with macromolecular ligands would be preserved. The modifyer phosphatidylinositol was activated by periodate oxidation, reacted with protein amino groups, and reduced by sodium borohydride. The modified enzyme was bound to liposomes via the cholate dialysis technique to give 2.4×10^{-3} moles of active chymotrypsin per mole of lipid. Almost 90% of the protein bound preserved its affinity towards protease inhibitor.

Antibodies against canine cardiac myosin, possessing specific affinity towards the necrotic area in myocardial infarction, were immobilized on liposomes via glutaraldehyde activation (1-2 moles protein/10^4 mol lipid were bound). The experiments carried out *in vivo* in dogs showed that [111]$lnCl_3$- containing liposomes with immobilized antibodies were targeted into necrotic tissues after experimental myocardial infarction.

The initial step of different vessel injuries is mainly the deendothelization with subsequent exposition of the subendothelial layer, causing platelet activation and adhesion Antibody against collagen, the major constituent of the subendothelial layer, and fibronectin, a protein possessing specific affinity towards collagen, were immobilized on the surface of liposomes via the glutaraldehyde method. The conjugates effectively recognized the bund to exposed subendothelial and not endothelial cells. From these results we concluded that these liposomes may function *in vivo* as microcontainers for a diffusable drug or for biologically active compounds capable of promoting secondary endothelium growth.

REFERENCES

1. TORCHILIN, V. P., GOLDMACHER, V. S., & SMIRNOV, V. N. *Biochem. Biophys. Res. Commun. 85:* 983 (1978).

2. TORCHILIN, V. P., KHAW, B. A., SMIRNOV, V. N., & HABER, E. *Biochem. Biophys. Res. Commun. 85:* 1114 (1979).
3. TORCHILIN, V. P., BERDICHEVSKY, V. R., BARSUKOV, A. A., & SMIRNOV, V. N. *FEBS Lett. 111:* 184 (1980).
4. KOELSCH, R., LASCH, J., KLIBANOV, A. L., & TORCHILIN,V. P. *Acta Biol. Med. Germ. 40:* 331 (1981).
5. TORCHILIN, V. P., OMELYANENKO, V. G., KLIBANOV, A. L., MIKHAILOV, A. I., GOLDANSKII, V. I., & SMIRNOV, V. N. *Biochim. Biophys. Acta. 602:* 511 (1981).
6. CHAZOV, E. I., ALEXEEV, A. V., ANTONOV, A. S., KOTELIANSKY, V. E., LEYTIN, V. L., LYUBIMOVA, E. V., REPIN, V. S., SVIRIDOV, D. D., TORCHILIN, V. P., & SMIRNOV, V. N. *Proc. Natl. Acad. Sci. USA,* in press.
7. TORCHILIN, V. P. & KLIBANOV, A. L. *Enz. Microb. Technol.* in press.

APPLICATION OF BIOREACTORS WITH IMMOBILIZED L-ASPARAGINASE

G. Mazzola, C. Giordano*, R. Longhi, G. Vecchio, and
R. Esposito*

Instituto di Chimica degli Ormoni del C.N.R., Milano
and Cattedra di Nefrologia,* I Facolta di Medicina
Universita, Napoli, Italy

L-asparaginase is one of the most important therapeutic agents
for acute lymphoblastic leukemia (1). To reduce the side effects
of intravenous injection of free enzymes, the use of extracorporeal
circulation using reactors with immobilized L-asparaginase (2) has
been studied. A hollow fiber dialyser can be utilized to support
the enzyme with the enzyme on the outer surface to allow for com-
partmentalization from immune system, high MW inhibitors, and proteolytic enzymes.

Hollow fiber reactors with immobilized L-asparaginase showed
the same kinetic characteristics both in buffer solution and in
whole blood. The apparent K_m value in pH 8 phosphate buffer was
1.738×10^{-2} M and in whole blood enriched with a fixed quantity
of L-asparagine (L-ASN) was 2.19×10^{-2} M. These reactors show
bulk diffusion controlled kinetics; and the maximum apparent activity could not be reached with blood because the pressure drop became too high. With the flow rate usually employed in extracorporeal circulation, the 21 cm long reactors showed a loss of apparent
activity, probably due to the formation of a substrate gradient
in the longitudinal direction. Higher apparent activities could
be obtained using shorter reactors with a higher flow rate and the
same pressure drop.

Reactors made with 50 hollow fibers, 21 cm long with an internal volume of 300 µl, were able to lower the L-ASN to zero in
3-4 hr during *in vivo* extracorporeal circulation in rats. The
level of L-ASN remained at zero for about 6 hr; but then rose to
its original level 24 hr after the end of the treatment. *In vitro*

experiments with the same conditions and 50 ml of whole blood
gave the same results.

L-ASN depletion from blood plasma occured after 1 hr; and the
level remained at zero after the treatment. However, in the
blood cells the L-ASN concentration decreased slowly and only to
about 300 nmole/ml; after the end of treatment the concentration
rose at once; when it reached the initial level in the blood cells,
the L-ASN reappeared in the plasma.

Extracorporeal perfusion over L-asparaginase in the therapy
of human acute lymphoblastic leukemia did not produce improvements;
although the *in vitro* and *in vivo* data showed an effective decrease of the L-ASN level. In fact, even if the clearance levels
obtained were accurate, the *in vivo* experiments demonstrated that
the L-asparaginase activity overcame the *ex novo* production of
L-ASN only during the extracorporeal circulation. The L-ASN depletion was limited, however, to a few hr after the suspension of
the treatment. This was too little time to obtain a therapeutic
effect. This behavior has been confirmed in our data concerning
a leukemic patient treated with extracorporeal therapy. Thus at
present, extracorporeal use of L-asparaginase is not advantageous
over the use of free enzyme for the rapid reappearance of L-ASN.
This suggests that for therapeutical use, other metabolites impli-
cated in the metabolic pathway should be removed.

REFERENCES

1. BURCHENAL, J. H. *Cancer Res. 33:* 350 (1970).
2. JACKSON, J. A., HALVORSON, H. R., FURLONG, J. W., LUCAST, K.
    D., & SHORE, J. D. *J. Pharmacol. Exper. Therap. 209:* 271
    (1979).
3. MAZZOLA, G. & VECCHIO, G. *Int. J. Artif. Organs 3:* 120 (1980).

ACYL-CoA SYNTHETASE AND ACYL-CoA OXIDASE FOR DETERMINATION OF SERUM FREE FATTY ACIDS

S. Shimizu, Y. Tani, and H. Yamada

Department of Agricultural Chemistry, Kyoto University
Kyoto, Japan

Quantitative determination of serum free fatty acids (FFA) has been carried out by titrimetric determination of total acidity or by colorimetric analysis based on the transfer of metal soaps from a copper or nitrate triethanolamine reagent, after extraction of FFA into organic solvents. These methods have the disadvantages of limited sensitivity, many sources of error, and complexity of procedure. We have recently developed new enzymatic methods (1, 4) for the determination of serum FFA using microbial acyl-CoA synthetase (E.C. 6.2.1.3) and acyl-CoA oxidase (E.C. 1.3.3.x), isolated and characterized from *Pseudomonas aeruginosa* (1,2) and *Candida tropicalis* (3), respectively. We utilize new UV and colorimetric methods.

The principal reactions for the UV method are: a) FFA + CoA + ATP + acyl-CoA synthetase to form Acyl-CoA + AMP + PPi; b) AMP + ATP + myokinase to give 2 ADP; c) 2 ADP + 2 PEP + pyruvate kinase to give 2 ATP + 2 pyruvate; and d) 2 pyruvate + 2 NADH + lactate dehydrogenase to give 2 lactate + 2 NAD. To 0.63 ml of incubation mixture containing 50 μmol Tris-HCl pH 8.0, 6 μmol MgCl$_2$, 3.2 μmol ATP, 7.4 μmol phosphoenolpyruvate, 0.28 μmol NADH, 0.4 μmol EDTA, 0.8 mg Triton X-100, 21.6 U myokinase, 12 U pyruvate kinase, and 12 U lactate dehydrogenase was added 10-25 μl serum (1-20 nmol as FFA) and 3-6 mU acyl-CoA synthetase. After a brief incubation, the reaction was initiated by addition of 0.26 μmol CoA in a total volume of 0.68 ml; and the rate or total decrease in absorbance at 340 nm was followed at 25°C. As a standard, a mixture of myristic-palmitic-stearic-oleic acids (1:1:1:1 molar ratio) in 0.25% Triton X-100 was used. Usually one assay was completed within a few min. Strict proportionality was observed between the concentration of FFA and the rate or total decrease in absorbance, provided the final concentration of FFA in the mixture did not exceed 15 μM. Recovery of palmitic acid added to serum was essentially quantitative. Multiple determination on a single serum gave 0.53 $\pm$ 0.02 μmol/ml (n=7, value $\pm$ SD) by the rate procedure and 0.57 $\pm$ 0.02 μmol/ml (n=10) by the end point procedure.

The principal reactions for the colorimetric method are: a) FFA + CoA + ATP + acyl-CoA synthetase to form Acyl-CoA + AMP + PPi; b) Acyl-CoA + O_2 + acyl-CoA oxidase to give Enoyl-CoA + H_2O_2; and c) H_2O_2 + phenol + 4-aminoantipyrine + peroxidase to form quinoimine dye. To 0.42 ml of the reaction mixture containing 9 μmol Tris-HCl pH 8.0, 0.34 μmol CoA, 0.9 μmol ATP, 0.23 μmol $MgCl_2$, 0.023 μmol EDTA, 0.045 mg Triton X-100, 138 mU acyl-CoA synthetase, and 4 U miokinase were added 30 μl serum. After incubation at 37°C for 5 min, 0.05 ml of N-ethylmaleimide (12.5 mmol phenol in 100 ml 12 mM N-ethylmaleimde) was added to stop the reaction. To the mixture was added 0.1 ml of color reagent containing 6 μmol FAD, 500 μmol 4-aminoantipyrine, 200 U acyl-CoA oxidase, and 3000 U peroxidase in 100 ml of 300 mM potassium phosphate pH 7.4. After further incubation at 37°C for 10 min, the color produced was measured at 500 nm against a reagent blank. This method gave an almost identical response for fatty acids with carbon chain lengths of 6 - 18. The degree of saturation did not affect the response. Using native human sera, we found a within-day precision of 2.9% for a mean FFA concentration of 0.67 μmol/ml (SD 0.020 n=10) and a day to day precision of 3.1% (n=6). FFA values by this method correlated well with the UV method and with a commercially available kit method based on a chemical procedure (5). The correlation coefficient and regression curves for the UV method and for the chemical method were r=0.991 and y=0.870x + 0.081 (n=26), and r=0.989 and y=0.932x + 0.004 (n=23), respectively.

The present enzymatic method has the advantages of being rapid, simple, and good sensitivity. The new method should be easily automatable and a new key measurement technique for the clinical and biochemical areas.

REFERENCES

1. SHIMIZU, S., INOUE, K., TANI, Y., & YAMADA, H. Anal. Biochem. 98: 341 (1979).
2. SHIMIZU, S., MORIOKA, H., INOUE, K., YASUI, K., TANI, Y., & YAMADA, H. Agric. Biol. Chem. 44: 2659 (1980).
3. SHIMIZU, S., YASUI, K., TANI, Y., & YAMADA, H. Biochem. Biophys. Res. Commun. 91: 108 (1979).
4. SHIMIZU, S., TANI, Y., YAMADA, H., TABATA, M., & MURACHI, T. Anal. Biochem. 107: 193 (1980).
5. ITAYA, K. & UI, M. J. Lipid Res. 6: 16 (1965).

ENZYME IMMUNOASSAY FOR FREE THYROXINE

H. H. Weetall, W. Hertl, F. B. Ward, and L. S. Hersh

Research and Development Division
Corning Glass Works
Corning, New York, USA

The determination of free thyroxine (FT_4) concentrations in serum is of great importance because it represents the active physiological fraction of the total thyroxine (TT_4) in the serum (1). The serum concentration of FT_4 is determined by the concentrations of the TT_4 present and the various thyroxine (T_4) binding proteins present in association with the binding affinitives of these proteins for T_4. The fraction of FT_4 In the serum is usually less than 0.03% (1, 2). While investigating the characteristics of horseradish peroxidase-labeled T_4 (T_4/HRP) we observed that the conjugated T_4 did not appear to bind to the thyroxine binding globulin (TBG) present in serum. This phenomenon was similarly observed and reported by others (3, 4). By taking advantage of this unusual phenomenon we have devised a direct method for measurement of FT_4 using EIA and equilibrium conditions. In the presence of serum to which T_4/HRP has been added one finds two distinct immunologically active fractions, the FT_4 and that of the added T_4-HRP conjugate. The concentration of the FT_4 is determined by the serum constituents; the FT_4/HRP concentraiton is determined only by the albumin. When a small quantity of antibody specific for the T_4 is added, it samples the variable FT_4 and the constant FT_4/HRP, thus giving a measure of the FT_4 present in the sample.

In all cases reagents were combined in the following order: a) assay buffer with BSA, b) antigen standard or patient serum, c) tracer conjugate, and d) immobilized antibody (400 µg IMA/ 0.15 ml). After reagent addition, the mixture was incubated at 37°C for two hr, centrifuged for five min at about 1300 x g, and the supernatant liquid was decanted. Substrate solution (2 ml) was added, followed by incubation at room temperature, color

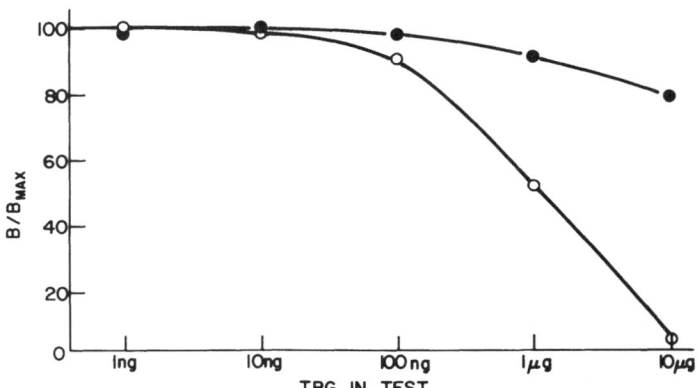

Fig. 1.   Non-binding of T_4/HRP to TBG.   (O) tubes contained anti-
          body and radiolabelled T_4; (●) tubes contained antibody
          and T_4/HRP conjugate.

development quenched by addition of 1 ml 1 M citric acid, centrifuged, and read spectrophotometrically at 455 nm.

   The results of experiments in which increasing amounts of TBG
were added to the assay tubes are shown in Fig. 1. With increasing
TBG decreasing amounts of the radio-labeled thyroxine were bound
to the antibody due to the interaction of the thyroxine with the
TBG; whereas with T_4/ HRP there was little difference in the amount
bound to the antibody with increasing TBG. These data demonstrate
that there was little interaction between the T_4/HRP conjugate and
the TBG. Dose-response curves for the free T_4 assay, using typical
quantities of constituents but various incubation times, indicated
that after 60 min incubation the reaction was nearly at equilibrium.

   The rationale for this assay is that there are two populations of free antigens. One (serum T_4) binds to TBG and TBPA; the
other (T_4/HRP tag) does not bind appreciably to TBS or TBPA. Both
populations do bind to albumin (both HSA and BSA) and to the antibody. In the assay the total albumin concentration (HSA plus BSA)
is almost constant, so that the concentration of free T_4/HRP is
also essentially constant. The T_4 and the TBG introduced by the
serum is variable and unknown; and the free T_4 is a function of
these TBG and albumin concentrations. If a small quantity of antibody is added, the ratios of the populations binding to the antibody will be the same as the ratios of the free antigen populations. Since the amount of T_4HRP antigen on the antibody can be
measured, we have a measure of the dilution of the unknown free
T_4 by the known free T_4HRP. Thus, the lesser the free T_4 the
greater the ratio of free T_2/HRP to free T_4.

The performance of the free T_4 EIA was compared to a commercial free T_4 RIA kit. The data are given in Table I. Various human serum samples with free T_4 RIA values ranging from 0.1 to 6.8 ng/dl were assayed by both EIA and RIA methods, total T_4 and TBG were determined by RIA kits. A critical test of the free T_4 EIA is the accuracy obtained with patient samples containing higher than normal TBG and normal free T_4. The EIA values (Table I) for five sera whose TBG ranged from 54 to 75 µg/ml displayed excellent agreement with RIA values for free T_4. Intra assay variations and inter assay variations never exceeded 4%.

We have demonstrated that we do indeed have a method capable of measuring the concentration of FT_4 in a liquid sample. This approach gives satisfactory values for the determination of FT_4 values in a number of patient sera which include hyperthyroid, and euthyroid samples. It is also obvious that this approach may possibly be extended to the measurement of other free hormones and drugs which are found in serum normally bound to one or more bind-

TABLE I

COMPARISON OF FREE T_4 EIA WITH RIA VALUES

Sample	FT_4-EIA (ng/dl)	FT_4RIA (ng/dl)	TT_4RIA (ng/ml)	TBG-RIA (µg/ml)
M1	0.39	0.28	34	35.5
M9'	0.54	0.73	64	38.5
M11'	0.24	0.13	14	37.5
SM-10	0.58	0.66	33.5	26
SM-22	0.54	0.43	40	26
M15	1.30	1.70	98	28.5
M18	1.53	1.69	117	29
AO-6	1.31	1.23	96	26
AO-10	091	1.08	54	18
KS-6	1.04	1.18	121	28.2
AO-5	0.88	1.19	147	53.5
C-10	0.70	0.98	178	75
C-11	0.71	0.88	131	54
C-13	0.84	0.88	125	56
C-17	1.02	0.99	150	56
M-39	6.1	6.1	>300	22.5
T-15	6.8	6.8	177	16
T-17	3.75	4.5	190	24
V-13	3.4	3.5	178	18.7

ing proteins. Whether one can design a conjugate with the specific properties of retaining antibody binding capacity and loss of binding protein affinity remains to be determined experimentally.

REFERENCES

1. ROBBINS, J. & RALL, J. E. *Physical Rev. 40:* 415 (1960).
2. HAMADA, S., NAKAGAWA, T., MORI, T., & TARIGAKA, K. *J. Clin. Endocrinol. Metab. 31:* 166 (1970).
3. STERLING, K. & BRENNER, M. A. *J. Clin. Invest. 45:* 153 (1966).
4. KLEINHAMMER, G., DEITSCH, G., LINKE, R., & STAEHLER, F. *Abst. Ann. Meet. Clin. Chem.* (1978).

ROUTINE DETERMINATION OF HYDROGEN PEROXIDE IN CLINICAL CHEMISTRY

WITH IMMOBILIZED ALDEHYDE DEHYDROGENASE

P. V. Sundaram

Medizinische Klinik Und Poliklinik
Abteilung Klinische Chemie
Goettingen, Federal Republic of Germany

Nylon tube reactors, containing multienzyme systems in which aldehyde dehydrogenase (E.C. 1.2.1.5) is always the final enzyme, have been developed. They facilitate the estimation of stoichiometrically produced hydrogen peroxide in *in vitro* reactions. They have been put into routine use in clinical chemistry by linking the reactors to flow-through autoanalyzer systems. In these systems H_2O_2 is produced from substrates such as glucose, uric acid, or cholesterol by specific enzymes such as glucose oxidase, uricase, or cholesterol oxidase. Ethanol, which is supplied in excess along with catalase and the unknown substrate, is converted to CH_3CHO and subsequently to acetate by aldehyde dehydrogenase in the presence of NAD. Thus, the appearance of NADH gives a measure of the original H_2O_2 producing substrate. The three steps in the sequence H_2O_2 to acetate have a 1:1 stoichiometry. This is the first time that such a mixed system, of immobilized and enzymes in solution, have been combined to form a multienzyme sys-

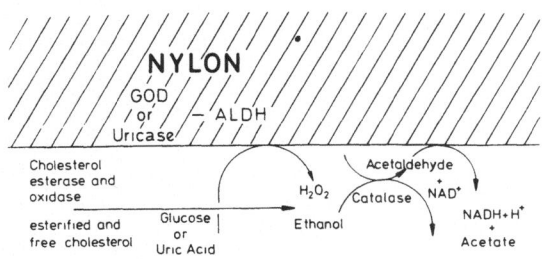

Fig. 1. Analysis scheme.

473

tem.   Cholesterol oxidase and esterase could not be immobilized in stable form, so that samples were pretreated with these enzymes in solution (Fig. 1).

Routine analysis conducted in a 1,000 bed hospital compared very well with conventional methods upon statistical analysis for the three metabolites.   Glucose was compared with the GOD/PAP or glucose dehydrogenase methods (r = 0.994), uric acid with the SMA 12/DMA-MBTA or SMAC-phosphotungstate method and ACA-UV methods (r = 0.993 to 0.997), and cholesterol with the cholesterol oxidase-PAP method (r = 0.993).   Reactors containing either aldehyde de-hydrogenase (ALDH) alone or those containing GOD-ALDH and uricase-ALDH were stable enough to carry out a minimum of 4,000 tests each. This combined with the fact that the NAD concentrations were kept low for these tests, unlike the commercial test kits, reduced the cost of testing quite dramatically(1-3).

This project was financed by the Deutsche Forschungsgemein-schaft.   Thanks to A. Antonijevic for his expert technical assist-ance.   The author is on leave at The Indian Institute of Technol-ogy, Madras, India.

REFERENCES

1. HINSCH, W., ANTONIJEVIC, A. & SUNDARAM, P. V. *Clin. Chem. 26:*
   1652 (1980).
2. HINSCH, W., ANTONJIEVIC, A., & SUNDARAM, P. V. *J. Clin Chem.*
   Clin. Biochem., 19: 307 (1981).
3. HINSCH, W. & SUNDARAM, P. V. *Fres. Z. Anal. Chem.* in press.

ENZYME ELECTRODES FOR SIMULTANEOUS DETERMINATION OF CREATININE

AND CREATINE IN SERUM OR WHOLE BLOOD

T. Tsuchida and K. Yoda

Katata Research Center, Toyobo Co., Ltd.
Honkatata, Otsu, Japan

Many of the currently used assay procedures for creatinine
and creatine are based on the Jaffé alkaline picrate method, which
is not specific and is subject to interferences. Several enzymatic
methods have been investigated recently to increase the specificity
(1-3); but these methods are laborious and less economic. In this
paper we describe a rapid, specific, and economic enzyme electrode
method for the analysis of creatinine and creatine.

The three enzymes used in the electrode method were creatinine
amidohydrolase (E.C. 3.5.2.10, *Pseudomonas* sp.) (CN), creatine
amidinohydrolase (E.C. 3.5.3.3, *Pseudomonas* sp.) (CI), and sarcosine oxidase (E.C. 1.5.3.1, *Corynebacterium* sp.) (SO). Two kinds
of immobilized enzyme membrane were prepared, CN/CI/SO and CI/SO,
by co-immobilization onto the porous side of a cellulose acetate
membrane with asymmetric structure. The membrane was treated with
γ-aminopropyl triethoxy silane prior to coupling of the enzymes
by glutaraldehyde. The skin layer of the asymmetric membrane had
selective permeability for hydrogen peroxide. Two types of enzyme
electrodes were constructed, using combinations of the enzyme membranes and polarographic electrodes for sensing hydrogen peroxide.
The multi-enzyme reactions and electrochemical oxidation of the
resulting hydrogen peroxide were as follows:

$$\text{Creatinine} + H_2O \xrightleftharpoons{\text{CN}} \text{Creatine}$$

$$\text{Creatine} + H_2O \xrightarrow{\text{CI}} \text{Sarcosine} + \text{Urea}$$

$$\text{Sarcosine} + O_2 + H_2O \xrightarrow{\text{SO}} \text{Formaldehyde} + \text{Glycine} + H_2O_2$$

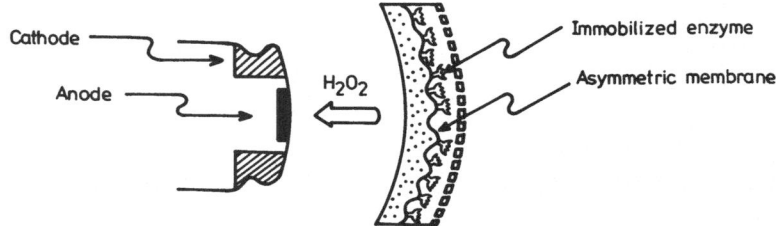

Fig. 1. Schematic diagram of enzyme electrode.

$$2H_2O_2 \longrightarrow 4H^+ + 2O_2 + 4e^- \qquad \text{(Anode)}$$

$$4H^+ + O_2 + 4e^- \longrightarrow 2H_2O \qquad \text{(Cathode)}$$

A schematic diagram of the enzyme electrode is shown in Fig. 1. The creatinine-creatine sensory responded linearly to creatinine and creatine from zero to 10 mg/dl, with response time of 20 sec and a detection limit of 0.1 mg/dl for whole blood. It was possible to test 60 samples/hr without deproteinization. Recovery of added substrates was 104% for creatinine and 102% for creatine. The within-day and between-day precision in serum was 1.3% and 8.4% at 2.2 mg/dl (n = 55) for creatinine and 5.7% at 1.1 mg/dl and 11.5% at 1.0 mg/dl (n = 21) for creatine; for whole blood the within day precision was 30% at 0.4 mg/dl for creatinine and 21% at 0.2 mg/dl for creatine. Only 25 μl of sample was required. Correlation with the kinetic Jaffé model was excellent (r = 0.985, Y = 1.078 x - 0.26) for creatinine and (r = 0.962, Y = 1.101 x - 0.25) for creatine. The immobilized enzymes were stable for more than 500 assays.

REFERENCES

1. MOSS, G. A., BONDER, R. J. L., & BUZZELLI, D. M. *Clin. Chem. 21:* 1422 (1975).
2. MEYERHOFF, M. & RECHINITZ, G. A. *Anal. Chim. Acta 85:* 277 (1976).
3. KINOSHITA, T. & HIRAGA, Y. *Chem. Pharm. Bull. 28:* 3501 (1980).

Session VIII
GENETIC ENGINEERING FOR ENZYME
(OR IMPORTANT BIOLOGICAL SUBSTANCES)
PRODUCTION
Chairmen: D. Fink and K. Sakaguchi

CONSTRUCTION OF VARIOUS HOST VECTOR SYSTEMS AND THE VARIATION

OF ENZYME LEVELS

K. Sakaguchi

Laboratory of Microbiological Chemistry, Mitsubishi-
Kasei Institute of Life Sciences
Tokyo, Japan

The *E. coli* gene engineering system is neither the best nor
universal one. The host vector systems of each bacteria, yeast,
mold, and even plants have advantages, especially in regard to developing their original specificities. These systems should be
and will be investigated further for applications and for construction of safer systems. This paper describes various microorganism
and plant systems that have been studied in our laboratory. For
a general review see (1).

BACILLUS SUBTILIS

Bacillus subtilis is a Gram-positive, rod-shaped spore-
forming bacterium, which in many respects belongs to another
world from *E. coli*. Foreign genes which are introduced but fail
to form proteins in *E. coli* cells are more likely to be expressed in *B. subtilis* cells. Gram-Positive and -negative
bacteria are different not only in their cell wall construction,
but also in their genetic and biochemical nature. This group of
bacteria and related species are not parasitic to animals and are
non-pathogenic to both animals and plants; they are therefore
candidates for experimental host-vector systems which might be
safer than the *E. coli* system. It is important to note that
Gram-positive bacteria do not produce endotoxin.

To establish recombinant systems in *B. subtilis*, the presence of plasmids was investigated by the cesium chloride-ethidium
bromide density centrifugation technique. Four strains containing
plasmids were obtained from nineteen *B. subtilis* strains from
type culture collections in Japan. Further, twenty plasmids were

479

isolated from seventeen strains of *Bacillus subtilis (natto)* (IFO3009-IAM1075) (Table 1), which is taxonomically and genetically identical with *B. subtilis*. Some strains contained one large plasmid and several smaller plasmids in a single strain.

Recombinant plasmids composed of *B. subtilis* 168 leucine genes and a *B. subtilis (natto)* plasmid pLS28 were constructed in a recombinant-deficient (*recE4*) mutant of *B. subtilis* 168. The process involved *Eco*RI fragmentation and ligation of pLS28 DNA and RSF2124-B. *leu*, which contained *B. subtilis* leucine genes and propagated in *E. coli*. The mixture was transformed into *B. subtilis* RM125 arg15 leuA8 thr5 r_m^- m_m^- recE4 cells. The *recE* mutation was absolutely necessary in order to avoid integration of the leucine gene into the host chromosomal DNA. This is particularly important when *B. subtilis* chromosomal genes are to be cloned on plasmids in *B. subtilis* 168.

A series of *Bacillus subtilis* plasmids was constructed which carried either the *leu* region or both the *leu* and the dihydrofolate reductase (DHFR) regions of the *B. subtilis* chromosome. The DHFR-coding gene was derived from a trimethoprim resistant (Tmp^r) *B. subtilis* strain; and cells harboring the DHFR plasmid showed resistance to trimethoprim (Tmp). One such leu^+ tmp^r plasmid, pTL12, was found to be useful for cloning DNA fragments at the *Bam*HI, *Eco*RI, *Bgl*II and *Xma*I sites. It was also shown that insertion of DNA fragments at the *Bam*HI and *Xma*I sites of pTL12 inactivated the *leu*A gene function (insertional inactivation) but not tmp^r, indicating that cells carrying recombinant plasmids can be detected easily by selecting Leu^- Tmp^+ colonies. Combination of *B. subtilis* 168 and plasmid pTL12 should serve as an efficient homologous cloning system in *B. subtilis*.

SPECIES BARRIERS TO THE MAINTENANCE AND EXPRESSION OF FOREIGN DNA

Some of the barriers to genetic materials from foreign DNA are due to nuclease, replication unit, transcription, translation, proteinase, and intervening genes. Nuclease barriers arise from surface degradation and release, intracellular inactivation (SP82 G DNA suffers about 40 lesions when transfected into *Bacillus subtilis*), and restriction endonucleases (Over 10^5 times more DNA is necessary in transforming *E. coli* r_k^+ m_k^+ cells compared to the restriction-deficient cells). The replication unit barrier is examplified by the inability to introduce *E. coli* plasmids PSC101, RSF1010 into *B. subtilis* r^- m^- cells and the degradation of phage T7 DNA after penetration into Syrian hamster embryonic cells.

TABLE 1

BACILLUS SUBTILIS PLASMIDS

Strain	Plasmid	Plasmid type	MW (x 10⁶)	Copies/ Chromosomes	No. of Cleavage Sites EcoRI	BamNI	HindIII
IFO3022	pLS11	1	5.4	7	1	1	5
IFO3215	pLS12	1	5.4	5	1	1	5
IAM1232	pLS13	2	4.6	10	4	0	3
IAM1261	pLS14	3	5.0	7	3	0	4
IFO3009	pLS15	4	3.6	5	1	2	4
IFO3013	pLS16	6	33	0.5	16	5	—
	pLS17	4	3.9	6	1	2	4
	pLS18	6	33	0.9	15	3	—
IFO3335	pLS19	4	3.6	4	1	2	4
	pLS20	6	34	0.4	17	4	—
IFO3936	pLS21	4	3.6	4	1	2	4
IFO13169	pLS22	4	3.4	4	1	2	4
IAM1143	pLS23	6	34	0.7	16	4	—
	pLS24	4	3.3	3	1	2	4
IAM1207	pLS25	6	31	0.7	16	4	—
	pLS26	4	3.7	4	1	2	4
	pLS27	6	33	0.6	16	4	—
IAM114	pLS28	5	3.6	5	2	1	5
IAM1114	pLS28	5	3.6	5	2	1	5
IAM1168	pLS29	6	35	0.6	17	5	—
	pLS30	5	4.1	3	2	1	5
IAM1163	pLS31	6	32	0.7	16	4	—
	pLS32	7	46	0.9	20	9	—
IAM1113	pTA1030	8	4.6	16	3	0	2
IAM1075	pTA1031	8	4.6	16	3	0	2

Transcription barrier, as denoted by promoter specificity, occurs with molecular differences of RNA polymerase in bacteria, phage, and eukaryotic cells due to exchange of σ factor with $\sigma\,\lambda$ in phage λ-infected *E. coli* or to differences of the DNA sequence on each promoter. Termination point specificity also occurs. Translation barrier arises from the binding specificity or ribosomes (*Bacillus brevis* or *Caulobacter crescentus* 30S subunit does not bind to MS2 RNA). The proteinase barrier can be due to intracellular proteinase destroying unfamiliar or poorly conformed proteins produced in the cell; for instance, the peptides produced from synthesized DNA or eukaryotic protein in bacteria. And finally, many eukaryotic organisms have a gene construction with intervening genes. Such genes do not form normal protein when they are introduced into prokaryotes.

SELECTIVE DISADVANTAGE OF HOSTS HARBORING RECOMBINANT MOLECULES

Nagahari found that RSF1010 trp hybrid plasmid carrying *E. coli trp* operon produced over two hundred times more *E. coli*-type tryptophan synthetase in *Pseudomonas* cells. However, after overnight culture, 90% of the surviving cells had lost the plasmid. A point to be stressed is that host microorganisms which carry a composite plasmid constructed *in vitro* should not be able to become predominant in a natural environment because they are forced to produce unnecessary enzymes (in this case, antibiotic-inactivating enzymes and tryptophan-synthesizing enzymes), suppressing the production of normal enzymes necessary for rapid growth under natural circumstances. They thus have a selective disadvantage. Microorganisms that have acquired composite plasmids by conjugation also have the same disadvantages under natural conditions.

PSEUDOMONAS

Gene engineering with *Pseudomonas* cells has several important aspects. *Pseudomonas* and related bacteria have the ability to decompose various strange compounds which are not usually metabolized by other microorganisms, for instance phenol, toluene, hydrocarbons, haloacetates, or even industrially formed new substrates such as nylon oligomers. These capabilities provide a strong efficiency for combating pollution and for future use as biocatalysts in the chemical industry. However, cloning genes of *Pseudomonas* or related bacteria in *E. coli* results in very poor expression of the enzymes. But, their expression is realized in the *P. putida* system. The *Pseudomonas* gene engineering system is more potent than the *E. coli* system. Striking fortification of *E. coli* enzymes is possible by introducing *E. coli* genes into *Pseudomonas* (3). RSF1010 and its derivatives exhibit convenient

TABLE 2

STRIKING FORTIFICATION OF *E. COLI* TRYPTOPHAN SYNTHETASE
IN *PSEUDOMONAS AERUGINOSA* USING RSF1010 PLASMID AS VEHICLE
AND THE DECREASE OF THE SPECIFIC ACTIVITY OF TSase β BY ONE WEEK
OF SUCCESSIVE DAILY TRANSFER OF CULTURE

P. aeruginosa Strain	Sp. Activity of TSase β*	
	Initially	After 1 Week of Culture
M12 (rsf1010-tryΔ0)	238	11.8
M12 (RSF1010-trpΔ1)	349	7.8
M12 (RSF1010-trpΔ2)	548	13.3
M12 (RSF1010-trpΔ3)	8.1	3.6
PA01	2.2	ND**
M12***	0.5	ND

*A unit of enzyme activity is defined as the utilization of 1
nmol of substrate per min at 37°C. Specific activity is U/mg
protein. **ND, not done. *** *P. aeruginosa* M12 cells were
cultured in glutamate minimal medium containing 50 μg of tryptophan per ml.

characters for vehicles in *Pseudomonas* cells; this includes stability, multicopy, selective drug resistance, and ability of insertional inactivation (4) (Table 2).

CLONING AND EXPRESSION OF LEUCINE GENE FROM THERMOPHILIC
BACTERIA(S)

A high production of thermophilic and stable enzymes in bacteria is desirable for the construction of future biocatalyst systems for the chemical industry. Therefore, a hybrid plasmid was
constructed consisting of *E. coli* plasmid PBR322 and *Thermus
thermophilus* (optimal temperature for growth 75°C) chromosomal
DNA. The pBR322-T.*leu* plasmid was found to have *leu*B and
*leu*C genes from *T. thermophilus* HB27 in intact form, because
the crude extract of *E. coli* cells having this plasmid showed

activity for the conversion of citraconate to α-ketobutyrate at 80°C, for which *leu*B and *leu*C gene products (β-isopropylmalate dehydrogenase and α-isopropylmalate isomerase) were necessary.

The crude extract of *T. thermophilus* HB27 cells showed a clear activity peak of isopropylmalate dehydrogenase at 80°C; whereas that of *E. coli* W3110 wild type cells showed optimal activity at 45°C and no activity above 60°C. On the other hand, the crude extract of *E. coli* cells containing pBR322-T.*leu* hybrid plasmid showed a broader peak at 80°C. After successive daily transfers of the culture in minimal medium containing 20 μg/ml ampicillin for 10 days (approximately 70 generations), the optimal temperature of the activity was decreased by 5° (75°C); but a considerable increase in the level of enzymic activity occurred, giving a level at 37°C equivalent to that of *E. coli* wild type cells. Longer successive transfers of the culture (up to 50 days) did not alter the location of the peak further, resulting only in some decrease of the enzyme activity. These observations suggest the possibility of producing large amounts of the thermophilic enzyme in mesophilic cells, and also show the adaptability of the *T. thermophilus* enzyme (or gene) in *E. coli* cells to the *E. coli* environment.

STREPTOMYCES AND YEASTS

A linear plasmid-like DNA was isolated by agarose gel electrophoresis from a lysate of *Streptomyces* sp. 7434-AN$_4$ which produces the lankacidin group antibiotics. The DNA (pSLA2) with a MW of 11.2×10^6 was cleaved into five and three fragments, respectively, with *Xma*I and *Bam*NI on the definite sites from the end, but not digested by *Eco*RI and *Hind*III. Upon treatment of the strain with ethidium bromide, variants were obtained which had lost the ability to produce the antibiotics. These variants were found to have lost pSLA2. These results suggest that the linear plasmid-like DNA is involved in the production of lankacidin group antibiotics. (6) (Table 3).

In conjunction with the use of a yeast plasmid (2 μm DNA) as a cloning vector, success in the transformation of yeast has led to the establishment of a yeast system for genetic manipulation. Yeasts are eukaryotic microorganisms having many biological features that are structurally and functionally similar to those of higher eukaryotes; and therefore they are promising candidates for hosts in which other yeast and foreign eukaryotic genes may be cloned. On examination of protoplast lysates from various yeast strains by CsCl-ethidium bromide density gradient centrifugagion and by agarose gel electrophoresis, a strain of *Kluyveromyces*

TABLE 3

SUMMARY OF PLASMIDS FROM FOUR STRAINS OF *STREPTOMYCES* (7)

Strain	Antibiotics Produced	Plasmid	S-value	MW ($\times 10^{-6}$)
S. puniceus KCC-S-0406	viomycin	pSPUI	45.8	17.3
		pSPU2	69.0	39.1
Streptomyces sp. 2217-G$_1$	cycloheximide	pSCY1	21.9	3.2
		pSCY2	46.7	18.0
Streptomyces sp. 7068-CC$_1$	neomycin	pSNE1	51.7	22.2
		pSNE2	75.9	47.0
S. hygroscopicus 434	geldanamycin	pSHY1	67.9	37.9

lactis was found to harbor two kinds of DNA plasmids, designated
pGK1-1 and pGK1-2. Electron microscopic analysis revealed that
pGK1-1 and pGK1-2 plasmids are linear DNAs with sizes of 2.7 and
4.1 μm, respectively in striking contrast to the bacterial plasmids,
which are known to be circular DNAs. The molecular weights estimated by gel electrophoresis, assuming the pGK1 plasmids to be
linear DNAs and using *Hin*dIII fragments of λ DNA as internal
markers, were 5.3 and 8.2 x 10^6 daltons, respectively. These values are in agreement with the sizes of linear DNAs determined by
electron microscopic analysis. The possibility exists that the
linear DNA plasmids may have arisen by the cleavage of circular
DNA plasmids existing *in vivo*. In such a case, however, the
cleavage must have occurred at specific sites on the presumed circular DNAs, because it was found that the pGK1 plasmids had specific sites for *Eco*RI or *Bam*HI digestion. In this connection,
it should be remembered that no restriction enzymes have been found
so far in yeasts or in other eukaryotes. Interesting questions
remain as to whether the pGK1 plasmids might be useful as cloning
vectors, what functions they perform *in vivo*, and how these linear plasmids replicate.

These plasmids produce a toxin killing *K. lactis* itself,
Saccharomyces cerevisiae, *K. thermotolerans*, *Candida utilis*,
Hansenula anomala., which provides a useful genetic marker character for genetic engineering. These plasmids are transferrable
into *Saccharomyces cerevisiae*. The transferred yeasts also produce toxin against other yeasts.

PROTOPLAST FUSION

The polyethylene glycol (PEG) method (8) for plant protoplast
fusion has proven to be widely applicable for cell fusion of fungi, yeasts, bacteria and *Streptomyces*, and is thus of great interest to researchers in microbial genetics. This method permits
hybridization or genetic exchange which is impossible in sexual
conjugation; incorporation of organelles, e.g., nuclei, mitochondria and chloroplasts, and of foreign cells or substances; and
transformation of various lower and higher organisms.

In the genus *Brevibacterium* (9) and related genera, genetic
transfer systems are not well established. The development of
genetic systems to produce recombinants is desirable in view of
the industrial importance of this group of bacteria as potent
amino acid producers. *Bacillus subtilis* and *Bacillus megaterium* are readily obtainable as protoplasts, which are recon-
vertible to bacillary form; these are necessary conditions if they
are to be fused and cloned.

Protoplasts of *Brevibacterium flavum* cells were obtained after treatment with penicillin G (0.3 U/ml) during exponential growth, followed by lysozyme treatment in hypertonic medium. Protoplasts of antibiotic-resistant and auxotrophic strains were fused by a modified polyethylene-glycol technique, and allowed to revert to bacillary form on selective media. The clones were analyzed; and recombinants were found that did not segregate. The fusion technique was used on yeast protoplast to achieve hybridization between haploid cells of a fission yeast, *Scizosaccharomyces pombe*, carrying the same mating type (h⁻) which cannot mate sexually. Other yeast protoplast fusions include intraspecific hybridization in sexual yeasts between the identical mating types of *Schizosaccharomyces pombe*, *Saccharomyces cerevisiae* (10), *Saccharomycodes lipolytica*, and *Kluyveromyces lactis* and the asexual yeast, *Candida tropicalis*. Interspecific hybridization was done with *Kluyveromyces lactis* x *K. fragilis*; and intergenetic hybridization was carried out on *Candida tropicalis* x *Saccharomycopsis fibuligera*.

The protoplast preparation is an important part of the fusion experiment. Usually, yeast protoplasts are prepared by treatment of the cells with cell-wall degrading enzymes such as snail gut juice enzymes (e.g., helicase and glusulase) and microbial products from *Streptomyces* sp., *Trichoderma* sp., or *Arthrobacter* sp. The pretreatment of cells with thiol compounds such as mercaptoethanol and dithiothreitol greatly enhances protoplast formation. Aggregation and fusion is induced with the aid of 20-40% PEG (MW 4000-6000). The coexistence of 10-100 mM $CaCl_2$ with PEG is required for effective fusion. The fusion efficiency is much higher when the protoplast preparation is freed from cell walls, as far as possible; but prolonged treatment with a cell-wall degrading enzyme has an adverse effect on the regeneration of protoplasts into viable cells. Embedding of protoplasts in gelatin or agar is desirable for good regeneration. Yeast protoplasts are osmotically stabilized by 0.6-1.0 M sorbitol or KCl. However, since a high concentration of PEG itself serves as a stabilizer, the use of sorbitol or KCl may be omitted during the fusion process. Recently, polyvinyl alcohol (MW 500) was found to be an excellent fusion agent.

In our laboratory, intraspecific transfer of mitochondria into protoplasts of *S. cerevisiae* was attempted. Mitochondria were isolated from an oligomycin-resistant respiring haploid strain $AN^R OR$ 12D carrying the genotype a *his4 leu2 thr4* $C^S E^S O_{II} \rho^+ \omega^+$ Protoplasts were prepared from cells of a neutral petite (respiration-deficient) haploid strain BO60 AF-1 having the genotype *a ade2 arg4 leu2 trp* $C^0 E^0 O^0 \rho^0 \omega^0$, and were mixed with the above isolated mitochondria. The mixture was treated with PEG-$CaCl_2$ and spread on selective plates containing

nutrients necessary for the protoplast regeneration. After incubation at 30°, oligomycin-resistant respiring colonies appeared at
the low frequency of 10^{-8}, while almost all of the protoplasts regenerated into petite colonies. The oligomycin-resistant respiring
colonies were shown to have the same mitochondrial genotype $C^S E^S$
$O_{II}^R \rho^+ \omega^+$ as the donor strain; while the nuclear genotype was a ade2
arg4 leu2 trp, identical with that of BO60 AF-1 (11).

Cells of the nitrogen-fixing bacterium Azotobacter vinelandii
and the unicellular cyanobacterium Anacystis nidulans were introduced into protoplasts of Saccharomyces cerevisiae by the
polyethylene glycol (PEG) method. Factors influencing the uptake
frequency were examined, and experimental conditions were established for maximizing the uptake frequency. Under optimal conditions, each protoplast took up a few bacterial cells. Electron-
microscopic studies showed the localization of integrated bacterial cells in membrane-bound vesicles of the cytoplasm or large vac-
uoles. The protoplasts at the intermediate stages of uptake revealed two major mechanisms of uptake: a) "endocytosis" by a
single protoplast and b) "cell fusion" between two or more protoplasts. Some bacterial cells disintegrated during the subsequent
incubation period through a heterophagy-like process.

Twelve strains of Chlorella which lack the sporopollenin
layer in their cell wall were treated with polysaccharide degrading enzyme mixtures. Osmotically labile protoplasts were obtained
from two of them (C. ellipsoidea C-87 and C. saccharophila C-211).
The absence of the cell wall was demonstrated by the calcofluor
stain and by electron microscopy. Some protoplasts adhered to
each other; it seemed like a cell fusion. Protoplasts of C.
ellipsoidea C-87 were able to regenerate the cell wall and to grow
on a regeneration medium.

A CHLORELLA SPECIES FIXING NITROGEN

A unicellular alga which can grow in the light without a combined nitrogen source was isolated from a hot spring. The cells
were almost spherical, usually 5-10 μm in diameter. Adsorption
spectra of the water-soluble pigments and of the acetone-extracted
ones revealed the existence of chlorophyll a and b and the
absence of phycobilins. Thin sections examined by electron
microscopy revealed an eukaryotic organization with features
typical of the coccoid green algae (the Chlorococcales). Cells
divided by internal cytokinesis with subsequent liberation of
daughter cells from the parental wall in a way similar to
Chlorella. The alga reduced acetylene to ethylene and incorporated $^{15}N_2$ into cell protoplasm when incubated in a low oxygen
atmosphere. Nitrogenase activity was light-dependent, microaero-
philic, and thermophilic. Although the association of symbiotic

nitrogen fixing prokaryotes with the cells may still be possible, any such organisms have not so far been detected.

REFERENCES

1. SAKAGUCHI, K. & OKANISHI, M., ed., "Molecular Breeding and Genetics of Applied Microorganisms," Kodansha--Academic Press, Tokyo, (1980).
2. TANAKA, T. & SKAGUCHI, K. *Molec. Gen. Genet. 165:* 269 (1978).
3. NAGAHARI, K. & SAKAGUCHI, K. *J. Bacteriol. 136:* 312 (1978).
4. NAGAHARI, K. *J. Bacteriol 133:* 1527 (1978).
5. NAGAHARI, K. KOSHIKAWA, T., & SAKAGUCHI, K. *Gene. 10:* 137 (1980).
6. HAYAKAWA, T., TANAKA, T., SAKAGUCHI, K., OTAKE, N. & YONEHARA, H. *J. Gen. Appl. Microbiol. 25:* 255 (1979).
7. HAYAKAWA, T., OTAKE, N., YONEHARA, H., TANAKA, T., & SAKAGUCHI, K. *J. Antibiotics 12:* 1348 (1979).
8. KAO, K. N. & MYCHAYLUK, M. R. *Planta 115:* 355 (1974).
9. KANEKO, H. & SAKAGUCHI, K. *Agric. Biol Chem. 43:* 1007 (1979).
10. GUNGE, N. & TAMARU, A. *Jap. J. Genet 53:* 41 (1978).
11. GUNGE, N. & SAKAGUCHI, K. *Molic. Gen. Genet. 170:* 243 (1979).

ENZYMES ACTIVE ON UNNATURAL SYNTHETIC COMPOUNDS: NYLON

OLIGOMER HYDROLASES CONTROLLED BY A PLASMID AND THEIR CLONING

H. Okada, S. Negoro and S. Kinoshita

Department of Fermentation Technology, Osaka
University Suita-shi, Osaka, Japan

It is generally believed that every organism that inhabits
the earth is highly adapted to its environmental conditions, and
that it must modify its biological activities if it is to adapt
successfully to changes in the selective force of the environment.
Since the middle of this century the rapid progress of chemical
industries has led to the distribution of a wide variety of unnatural sunthetic compounds around the earth as industrial products
and wastes. Where these unnatural synthetic compounds have accumulated, the selective force of nature may have been altered somewhat, which resulted in cleature and selection of new biological
abilities or new enzymes.

One example of such an unnatural synthetic compound is nylon
oligomer. Nylon-6 is produced from ε-caprolactam by ring cleavage
polymerization, and consists of more than 100 residues of 6-amino-
hexanoic acid. During the polymerization, some molecules fail to
polymerize at the oligomer stage, while others undergo head-to-tail
condensation to form cyclic oligomers. These by-products are contained in the waste and discharged from nylon factories.

In this presentation, we will discuss three aspects of the
microbial degradation of this unnatural synthetic nylon oligomer,
a compound which did not exist on the earth before the establishment of industrial production some thirty years ago. First, the
enzymes which degrade nylon oligomer have been created by evolution.
Second, the evolution is considered to have occurred on satellite
DNA structures called plasmids. Third, the genes coding the newly
evolved enzymes have been cloned in *Escherichia coli* and can be
used for breeding useful strains for the utilization of waste com-

pounds. These results are important from the standpoint of the
response of nature to unnatural synthetic compounds.

A MICROORGANISM METABOLIZING 6-AMINOHEXANOIC ACID CYCLIC DIMER
AND THE ENZYMES RESPONSIBLE

From the waste water of a nylon factory we isolated a microorganism which can use 6-aminohexanoic acid cyclic dimer, a waste
product of the factory, as the sole source of carbon and nitrogen.
The isolate was classified as *Flavobacterium* SP, KI72(1).

In purifying the enzymes concerned, a cyclic dimer hydrolyzing activity was saparated from a linear dimer-hydrolyzing activity
by DEAE-Sephadex column chromatography. We concluded, therefore,
that the degradation pathway is as follow (2)

$$
\begin{array}{ccc}
\underset{\displaystyle CO-(CH_2)_5-NH}{\overset{\displaystyle NH-(CH_2)_5-CO}{\mid\qquad\qquad\mid}} \xrightarrow{\;EI\;} &
\underset{\displaystyle CO-(CH_2)_5-NH_2}{\overset{\displaystyle HN-(CH_2)_5-COOH}{\mid\qquad\qquad}} \xrightarrow{\;EII\;} &
\underset{\displaystyle HOOC-(CH_2)_5-NH_2}{\overset{\displaystyle H_2N-(CH_2)_5-COOH}{}}
\end{array}
$$

where EI = 6-aminohexanoic acid cyclic dimer hydrolase and EII =
6-aminohexanoic acid linear oligomer hydrolase.

The EI enzyme was purified to homogenity; Table 1 summarizes
the physico-chemical and enzymatic characteristics of the purified
EI enzyme. We tested more than 100 kinds of natural compounds
with an amide bond, including linear and cyclic amides, diketopi-
peradines, di-, tri-, tetra- and pentapeptides and casein, but
none of them were hydrolyzed. The EII enzyme was also purified
to homogenity (3). Its physicochemical and enzymatic characteristics are summarized in Table 2. It was active toward 6-amino-
hexanoic acid oligomers from the dimer to the hexamer and also the
icosamer, but not the hectamer. The hydrolysis activity toward
the oligomer decreased with increase in the degree of polymerization.

Enzymes active toward unnatural synthetic compounds but not
toward natural compounds had no reason to exist before the creation
of synthetic compounds. Moreover, the observation, described
later, that the enzymes are coded on a less conservative plasmid
strongly suggests that EI and EII enzymes evolved after the establishment of the nylon industry.

TABLE 1

PROPERTIES OF 6-AMINO HEXANOIC ACID CYCLIC DIMER HYDROLASE

Property	Result
MW	100,000 (50,000 x 2)
Optimum temperature	$34^{\circ}C$
Optimum pH	7.4
K_m	5.9 mM
Turnover number	8.0 sec^{-1}
Active on	6-aminohexanoic acid cyclic dimer

TABLE 2

PROPERTIES OF 6-AMINO HEXANOIC ACID LINEAR OLIGOMER HYDROLASE

Property	Result	
MW	84,000 (42,000 x 2)	
Optimum temp.	$40^{\circ}C$	
Optimum pH	7.2	
	Ahx-oligomer*	Relative Activity (%)
Active on	Ahx_2	100
	Ahx_3	43
	Ahx_4	26
	Ahx_5	14
	Ahx_6	8
	Ahx_{20}	0.3
	Ahx_{100}	0

*Ahx = 6-aminohexanoic acid.

DEPENDENCE OF EI AND EII ENZYMES UPON PLASMID, pOAD2 (4)

If the enzymes responsible for the metabolism of the cyclic
dimer are coded on the plasmid, the metabolic activity would be
eliminated in high frequency by treatment with curing agents such
as mitomycin C, ethidium bromide, and sodium dodecyl sulfate.
After treatment, more than 80% of the clones had lost the metabolic
activities toward both the cyclic and the linear dimer.

If the genes of nylon oligomer metabolism are located on a
plasmid, the metabolic activities should be restored by transformation of the cured strain with the plasmid DNA of the wild strain.
Though in low frequency, transformants were obtained when KI723,
a cured strain, was used as the recipient.

The metabolic activities toward the cyclic dimer and the
linear dimer were eliminated by curing and restored by transformation. The EI and EII enzyme activities after these treatments were
studied. Compared to the wild-type strain the cured strains
possessed less than 0.5%, the lower limit of detection, of both
EI and EII activities. , while the transformants had about 80% of
EI and EII activities. These results indicate that the inability
of cured strains to grow on the cyclic and linear dimers arises
from their lack of EI and EII enzymes.

The above results were confirmed immunologically. Fig. 1A
shows a double diffusion experiment. The anti-EI serum in the

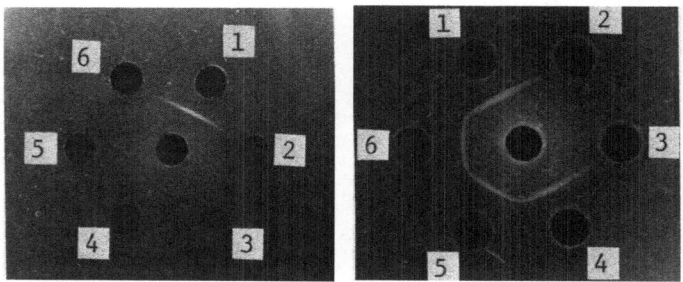

Fig. 1. Double diffusion tests between anti-EI serum and cell
         extracts of wild, cured, and transformant strains. 1A:
         center well anti-EI serum; well 1 cell extract of wild
         strain KI72; wells 2-6 cell extracts of cured strains
         KI722, KI723, KI724, KI7212, and KI7212, respectively.
         1B: center well anti-EI serum; wells 1 and 5 cell extract
         of wild strain KI72; wells 2 and 3 cell extracts of cured
         strains KI722 and KI723; and wells 4 and 6 cell extracts
         of transformants KI723T1 and KI723T2, respectively.

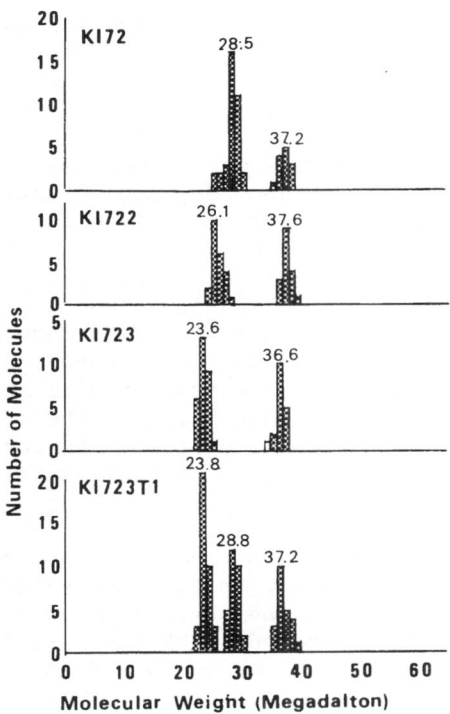

Fig. 2. Distributions of MW of plasmids harbored in wild (KI72),
cured (KI722 and KI723), and transformant (KI723T1)
strains.

center well produced a clear precipitin line with the cell extract
of the wild strain but not with those of cured strains. Fig. 1B
shows the precipitin lines between the cell extract of the transformants and anti-EI serum fused with that of the wild strain at
each corner, indicating that the EI proteins produced by the wild
and the transformant strains are immunologically identical.

Another feature of curing is the loss of the plasmid harbored
in the wild strain. Plasmid samples were prepared and observed
electronmicroscopically. Plasmids of two different lengths were
observed in preparations from the wild strain and the cured
strains. So we measured the length of 50 to 60 plasmid DNA molecules from each strain against the internal marker. The results
are shown in Fig. 2 as a frequency distribution histogram. The
plasmids harbored in the wild strain fall into two size groups with
mean MWs of 37 and 29 Mdal. As will be discussed later, the popu-
lation of smaller plasmids was in fact a mixture of two kinds of
plasmid with molecular sizes of 29 and 26 Mdal, but these were in-
distinguishable in terms of length. The transformant KI723T1,

which was derived from KI723 by transformation harbored three
kinds of plasmids, two of them 37, and 23 Mdal, being the same as
those of the recipient strain KI723, and the EI and EII genes are
coded on the middle-sized plasmid, that of MW of 28.8 Mdal.

   Electrophoresis on agarose gel allows the separation of plasmids, as their migration distance depends on their molecular size.
The resulting DNA bands were individually cut from the agarose gel
electrophoregram, extracted, and subjected to a restriction endo-
nuclease HindIII Hydrolysis. Then the resulting fragments were
subjected to slab gel electrophoresis. The results are shown in
Fig. 3. Lane T T[2] contained the HindIII fragments of the
middle-sized plasmid of the transformant gave six bands, and the
small-sized plasmid of KI722 gave hour HindIII fragments, which
can be seen in lane 2. The broad plasmid band of the wild strain
gave ten HindIII fragment bands, which can be totally accounted
for as the sum of those obtained from T(2) and KI722. Moreover,
the sum of the MW of the restriction fragments should give the MW
of the plasmid from which they were derived. However, in the case
of the wild strain, W, the sum of the MW of the ten HindIII
fragments was 55 Mdal, much larger than the estimated MW of 29
Mdal. This is evidence that the smaller plasmid in the wild
strain, which had been estimated at 29 MDAL, is in fact a mixture
of two kinds of plasmids with MWs of 29 and 26 Mdal.

   The conclusions drawn from the above experimental results are
summarized in Fig. 4. The wild strain, KI72, has three kinds of
plasmids, which were named pOAD1, pOAD2, and pOAD3 in order of increasing size. Curing readily eliminates pOAD2, accompanied by

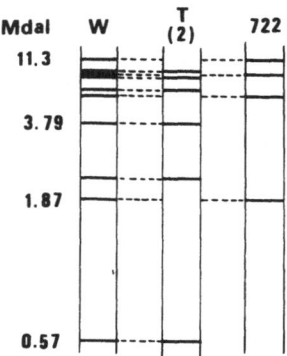

Fig. 3.   Illustration of an agarose gel electrophoregram of
          HindIII fragments generated from the small size
          plasmid of wild strain KI72 (lane W), middle size
          plasmid of a transformant KI723T1 (lane T(2)), and
          small size plasmid of a cured strain KI722 (lane 722).

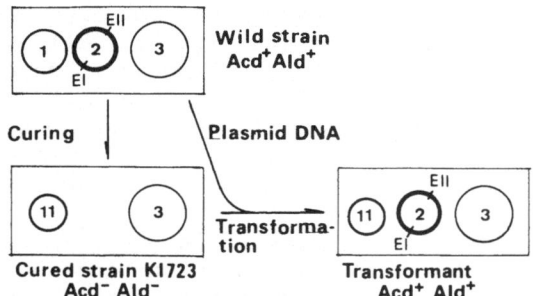

Fig. 4. Changes of plasmids harbored in wild, cured, and
transformant strains.

the loss of EI and EII enzyme activities. On curing, pOAD1 may
be conserved intact, as was the case with KI722; or it may be
deleted, as was the case with KI723. On transformation with the
plasmid DNA of the wild strain, KI723 received the pOAD2 plasmid
and recovered its EI and EII enzyme productivities.

CONSTRUCTION OF PHYSICAL MAP OF pOAD2 and pOAD21

To obtain more detailed information on the pOAD2 plasmid, we
constructed a physical or restriction map of it. To construct a
restriction map of pOAD2, a new mutant of KI72 was introduced into
the experimental system. This strain, KI725, was isolated on a
mutant unable to grow on the cyclic dimer but still able to grow
on the linear dimer. It harbored a partially defective plasmid
with a MW of 6 Mdal less than that of pOAD2. The defective plasmid
was named pOAD21. Since, KI725 cannot produce the EI enzyme, the
gene controlling EI production is believed to be located on the
deleted region of the original plasmid. We first constructed the
restriction map of pOAD21. The restriction map of pOAD2 should
have the same structure as that of pOAD21 except for the insertion
of the 6 Mdal fragment on which the EI gene is located.

CONSTRUCTION OF HYBRID PLASMIDS AND THEIR EXPRESSION

To determine more precisely the location of EI and EII genes
on pOAD2, we cloned these genes on pBR322 and transferred to *E.
coli*. *Hind*III fragments of pOAD2 were inserted to the vector
plasmid at the   dIII restriction site, and the resulting hybrid
plasmids were transferred to *E. coli* C600 cells. Transformants
harboring the hybrid plasmid were selected as ampicillin resistant
and tetracyclin sensitive clones. Cell extract of transformant
strains were incubated with the cyclic dimer or the linear dimer;

and the reaction products were subjected to paper chromatography
to detect the EI and EII enzyme activities. Cell extracts from
transformants Nos. 5, 18, 23, 27, 33, and 34 produced the linear
dimer from the cyclic dimer, which means the EI gene was cloned
in these cells. Cell extracts of transformants Nos. 14, 16, 28
and 29 showed EII enzyme activity. No transformants producing
both EI and EII enzymes in the same cell were obtained.

The hybrid plasmid DNAs harbored in the transformants were
analyzed with respect of HindIII digestion patterns. The results
showed that all EI-positive transformants harbored hybrid plasmids
consisting of the vector DNA and the HindIII-C-fragment of pOAD2,
and all EII-positive transformants harbored a plasmid consisting
of the vector DNA and the HindIII-A-fragment only or with the
additional fragments. These results reveal that the structureal
genes of EI and EII are located on the HindIII-C and A fragments,
respectively.

CONSTRUCTION OF MINIPLASMIDS OF pNDH5

The hybrid plasmid which enables a transformant to produce
EI enzyme was nominated pNDH5. A restriction map of this plasmid
was constructed, as shown in Fig. 5 in a linearized form. Both
orientation of the HindIII-C fragment inserted into pBR322 were
active for EI synthesis, indicating the promoter gene of EI located
within the E-fragment. By partial or complete digestion of pDNH5
with a restriction enzyme, the ligation by DNA ligase, several
mini-plasmids were constructed. The smallest and EI enzyme produc-

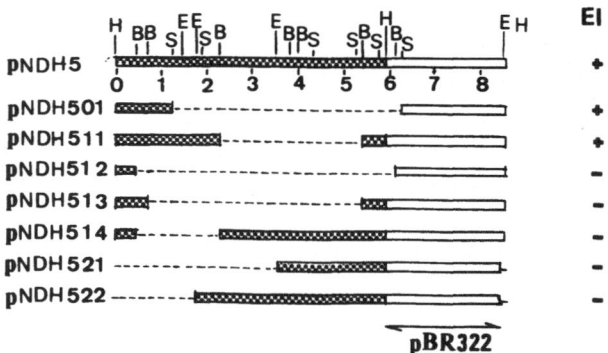

Fig. 5. Restriction maps of pNDH5 and miniplasmids derived from
         it and their productivity of EI enzyme. Open bar indi-
         cates the part of DNA derived from the vector pBR322 and
         spotted bar the part of DNA derived from pOAD2. H:
         HindIII, E: EcoRI, B: BamHI, S: SmaI.

ing plasmid, pNDH501 contained 1.3 Mdal DNA originated from pOAD2, which is almost the same size as that calculated from the base of the MW (50,000) of the subunit of the EI enzyme.

To study the evolutional origin of the EI gene, a survey of DNA fragments having a homologous base sequence as that of the EI gene was made according to Southern brotting technique using pNDH501 as the survey probe. Among DNA fragments from the chromosome, pOAD1, pOAD2 and pOAD3 no fragment had the homologous base sequence as pNDH501 except for *Hind*III-C fragment of pOAD2 on which EI gene is coded.

EII GENE AND ITS TRANSCRIPTIONAL AND EVOLUTIONAL ORIGINS

Plasmid pNDH29 is a hybrid plasmid of the *Hind*III-A fragment of pOAD2 and the vector DNA. By a similar procedure to that used for EI gene analysis, deletion plasmids were prepared from pNDH29. As shown in Fig. 6 it is evident that the A_3 region is not necessary for the expression of the EII gene while both the A_1 and A_2 regions are. To test the possibility that A_1 (A_2) region covers a structural gene of EII and that transceiption starts from the A_2 (A_1) region, we constructed new hybrid plasmids in which A_1 or A_2 region was replaced by *lacUV5* promotor. About half of the plasmids in which the A_1 gene was joined to the *lac* promotor gene expressed EII synthesis at 10 times the level of pNDH29. Thus we can conclude that the structural gene of EII is present in the A_1 region.

To study the origin of the EII gene, the DNA regions of the chromosome and plasmid of *Flavobacterium* sp. KI72 able to form heteroduplex with pNDH2912 were surveyed. By the method of

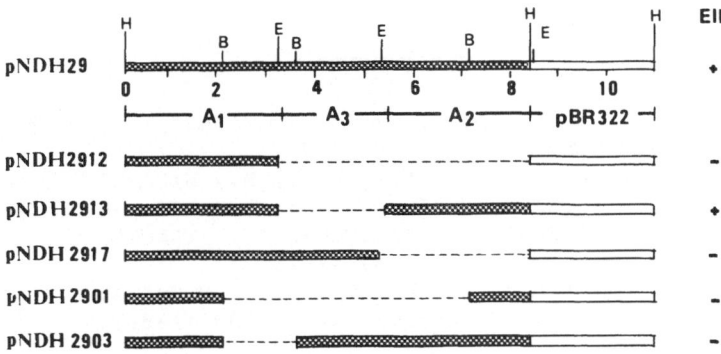

Fig. 6. Restriction maps of pNDH29 and miniplasmids derived from it, and their EII enzyme productivity.

Southern brotting, none of the DNA fragments derived from the
chromosome pOAD1 and pOAD3 formed hetero-duplex with pNDH2912 in
which HindIII-A₁ fragment is cloned. Some HindIII fragments
derived from pOAD2, fragments B, C, and E in addition to the A
fragment, produced heteroduplex with pNDH2912 DNA.

CONCLUSION

There are two possible reasons for an enzyme to be active on
an unnatural substrate. One is that an unnatural compound could
be decomposed by an enzyme if it were an analogue of that enzyme's
physiological substrate. The other is that an unnatural substrate
could be decomposed by a newly evolved enzyme. The data presented
here show that both EI and EII enzymes have no activity on any
physiological substrates, including the linear and cyclic amides
and peptides tested.

In addition to the lack of activity on natural compounds and
the enzyme's low specific activity, we found that the structural
genes of EI and EII were coded on a plasmid, pOAD2, in *Flavobacterium* sp. KI72. These evidences support the possibility that
the EI and EII enzymes were newly evolved on a plasmid by adapta-
tion to newly distributed waste products.

The fact that the structural genes of the two newly evolved
enzymes, EI and EII, were coded at distant loci of the same plasmid
and that an anti-serum for the EI enzyme did not react with the
EII enzyme suggests the possibility that the two enzyme evolutions
occurred independently. In addition to the above evidence, EII
enzyme activity was observed in another bacterial strain, *Corynebacterium aurantiacum*, indicating that such an enzyme evolution
of adapting to unnatural synthetic compounds such as nylon oligomers is not a very rare phenomenon in nature.

REFERENCES

1. KINOSHITA, S., KAGEYAMA, S., IBA, K., YAMADA, Y., & OKADA, H.
       Agric. Biol. Chem. 39: 1219 (1975).
2. KINOSHITA, S., NEGORO, S., MURAMATSU, M., BISARIA, V. S.,
       SAWADA, S., & OKADA, H. *Eur. J. Biochem. 80:* 489 (1977).
3. KINOSHITA, S., TERADA, T., TANIGUCHI, T., TAKENE, Y., & OKADA,
       H. *Eur. J. Biochem. 116:* 547 (1981).
4. NEGORO, S., SHINAGAWA, H., NAKATA, A., KINOSHITA, S., HATOZAKI,
       T., & OKADA, H. *J. Bacteriol. 143:* 238 (1980).

CLOSING SESSION
Chairman: G. Manecke

RECENT DEVELOPMENTS AND FUTURE ASPECTS OF ENZYME ENGINEERING

E. Katchalski-Katzir

The Weizmann Institute of Science, Rehovot and
Tel-Aviv University
Tel-Aviv, Israel

The opening chapter in this volume cites some of the directions and highlights of the first ten years of these International Enzyme Engineering conferences. Therefore, in concluding this volume it seems appropriate to critically evaluate our achievements and failures, to survey the new vistas opened up, and to identify some of the new trends worth pursuing. Enzyme engineering has reached maturity; and our sessions here at Kashikojima have renewed our confidence in the importance of this relatively new field. Future developments in enzyme engineering should be expected because of the outstanding characteristics of biocatalysts: a) their ability to catalyze the synthesis, degradation, and modification of chemical compounds under mild conditions; b) their stereospecificity; c) their ability to enhance complex sequential chemical reactions; d) the vast number of different enzymes available in nature; e) the increasing information concerning enzyme structure and function, as well as their mode of action within living cells; f) the development of sophisticated chemical engineering techniques in the design and construction of enzyme- and cell-reactors; g) the urgent need to utilize biomass by its transformation into useful chemicals as well as gas and liquid fuels; and h) the continued demand for specific catalysts in industry, in the clinic, and in research laboratories.

It should be noted, however, that progress will be made only if: a) cheap methods for large scale production and purification of biocatalysts become available; b) new techniques for enzyme and cell stabilization are established; c) methods for cofactor stabilization and recycling are worked out; d) the techniques for enzyme and cell immobilization are improved; e) the use of biocatalysts in organic synthesis and in industry are extended; f)

efficient biological techniques for biomass conversion and energy
production are developed, and g) new and reliable enzyme analytical
assays are designed.

The various methods of enzyme and cell immobilization have
been accepted during the last decade as useful techniques in the
laboratory and in industry; and they have promoted the industrial
utilization of several new enzyme-catalyzed processes in the United
States, United Kingdom, USSR, Germany, France, and particularly
Japan. Immobilized glucose isomerase is being used for the preparation of high fructose syrup from glucose, immobilized aminoacylase to prepare optically active amino acids from appropriate
synthetic racemates, immobilized penicillin acylase to catalyze
the production of 6-aminopenicillanic acid from penicillin G, immobilized beta-galactosidase to hydrolyze lactose, immobilized
fumarase to catalyze the production of malic acid from fumarate,
and immobilized aspartase to produce aspartic acid from ammonium
fumarate.

The ideas and findings presented by the speakers, as well
as the data contained in the various posters, suggest that we are
moving in the right direction. Allow me, therefore, to summarize
some of the important findings presented at this conference.

We were astounded to hear that the total annual production
of the fermentation industry in Japan amounts to 2-2 1/2% of their
Gross National Product.

The need for novel efficient methods of isolation and purification of extra- and intra-cellular enzymes on a large scale is
rather obvious. Thus it is gratifying to hear that further progress has been attained in scaling up protein purification by the
liquid-liquid extraction technique, that the instrumentation for
batch treatment of 50 kg cells has been constructed, and that
intracellular enzymes such as fumarase and aspartase are obtained
in high yield. Further development of this and other techniques
for the large scale separation of biopolymers based on their different physical properties, or on their specific biological characteristics will be followed with keen interest. Separation methods
based on affinity chromatography should be scaled up; and specific
reagents, such as monoclonal antibodies binding reversibly with
proteins, should be utilized for final purification.

The remarkable progress attained in recent years in the development of recombinant DNA techniques is well known. The potential of the method was illustrated by explaining the molecular
genetic procedures that have been adopted for the production of
human leucocyte interferon and different vaccines in *E. coli*.
Future progress in genetic engineering will undoubtedly facilitate
the production of enzymes on a large scale and at a relatively

low price. Moreover, it is plausible to assume that molecular
geneticists will not be satisfied with their ability to synthesize
known enzymes but will try to prepare new biocatalysts, which will
provide the chemical engineer and the biotechnologist with novel
enzyme homologues and analogues possessing desirable chemical,
physical and biological properties. As a matter of fact data on
the use of plasmids and lamda phage as vectors for the insertion
of genes coding for given enzymes into some procaryotes are already
available in the literature. In a few cases gene expression has
led to rather high yields of the desired enzyme. Of particular
interest, however, are the findings that in various strains of *E.
coli* one can induce the production of some of the enzymes which
are being used extensively by the genetic engineer himself, such
as restriction enzymes, DNA-ligases, polymerases, and exonucleases.
It was especially interesting to hear from the success in transferring an *E. coli* plasmid (plasmid RSF 1010) into the cells
of *Pseudomonas*. The inserted plasmids multiplied rapidly within their new host cells and induced an almost 300-fold increase
in the amount of tryptophan synthetase originally present. Transfer of a leucine-synthesizing gene derived from the extremely
thermophilic *Thermus thermophilus* into *E. coli* cells also was
exciting to hear. Gene transfer in this case led to the production of a thermophilic enzyme within the mesophilic cells of *E.
coli*. The detection in *Flavobacterium* sp. of new hydrolases
capable of hydrolyzing nylon-oligomers and the procedure for their
cloning in *E. coli* by means of an artificially assembled plasmid
is in its early stages; but this work suggests that microorganism
selection in conjunction with suitable recombinant-DNA procedures
might lead to large-scale production of biocatalysts capable of
degrading highly resistant polymer waste products.

On the whole I feel that cooperation between enzyme engineers
and genetic engineers is highly recommended. A joint effort is
required to get a better understanding of the factors determining
enzyme specificity, catalysis, and stability; to devise new techniques for the large-scale production of enzymes and for the development of novel engineering systems for enzyme utilization in
industry, medicine, and research.

In spite of the considerable number of methods available for
enzyme and cell encapsulation and immobilization by coupling to
different inorganic or organic carriers, new ingenious immobilization techniques are being continuously developed. We heard about
the coupling of proteins to polysaccharides by means of the bifunctional trifluoro chlorosulfonyl reagent, $ClSO_2CH_2-CF_3$, and
by means of p-(β-sulfatoethylsulfonyl)aniline plus diazotization.
A novel method for the entrapment of biocatalysts by means of
photo-crosslinkable resin prepolymers was described, that yielded
gel-entrapped biocatalysts of different degrees of hydrophobicity;
this approach could be used successfully to catalyze the conversion

of different hydrophobic compounds in suitable organic solvents.
In the organic solvents employed, steroids and terpenoids, for
example, underwent characteristic bioconversion by means of specific microbial cells whose stability was markedly improved by
the immobilization technique employed. Lipases first adsorbed
on Celite and then entrapped in a hydrophobic gel showed high
activity and good stability, and thus could be used to catalyze
ester-exchange in different triglycerides. The use of synthetic
and native polyelectrolytes also was suggested for cell entrapment. A new technique for enzyme and cell entrapment recently
has been developed in my laboratory at Tel-Aviv University. It
is based on preparation of a linear polyacrylamide of desired chain
length, transformation of some of the amide side chains into the
corresponding hydrazides, admixture with the desired enzyme, bacteria or yeast, and crosslinking with a dialdehyde such as glyoxal
or glutaraldehyde. The method was found particularly suitable
for the preparation of gels in which viable cells continue to
metabolize freely and even multiply with little interference.

At some of our previous meetings we discussed the observation
that many enzymes were stabilized by attachment to external carriers, by intramolecular crosslinking, or by tight encapsulation.
As a plausible explanation for this unexpected phenomenon it was
suggested that enzyme immobilization leads to "freezing" of the
native, active conformation of the enzyme; thus preventing conformational fluctuations which might eventually lead to thermal,
irreversible denaturation. Some of the speakers at this meeting
have indicated that bacteria, yeast, plant and animal cells also
can, in some cases, be markedly stabilized by encapsulation or
anchorage to specific microcarriers. The reason for this behavior
is still unknown. One should bear in mind, however, that within
the immobilized systems under consideration close contacts between
cells are established which might lead to the activation of latent
chemical or physical communication systems, and even to cell surface modification. Furthermore, the high density of cells attained
might enable specific metabolites, which markedly effect cell
growth as well as cell behavior, to accumulate.

Classical biochemists are aware of a great number of chemical
reactions catalyzed by enzymes. I am thus surprised that organic
chemists and chemical engineers are still reluctant to use immobilized enzymes and immobilized cells as specific heterogeneous
catalysts in some of the chemical reactions in which they are
interested. I do hope, however, that sessions such as the one
on the applications of biocatalysts for new reactions and organic
synthesis, and the one devoted to the industrial applications of
immobilized biomaterials will, in due course, enhance collaboration
between experts who have specialized in related fields.

The industrial scale enzymatic synthesis of semisynthetic penicillins and cephalosporins continues to be an important topic. Penicillin acylases from different microorganisms were used to remove the acyl side chains of the penicillins and cephalosporins produced by direct fermentation, and the 6-aminopenicillanic acid (6APA) and 7-aminocephalosporanic acid (7ACA) and 7-aminodeacetoxy-cephalosporanic acid (7ADCA) obtained were reacylated enzymatically to yield the corresponding semisynthetic penicillins and cephalo-sporins which contain the new acyl groups required. The processes worked out, for the enzymatic synthesis of ampicillin from 6APA and D-phenylglycine methyl ester and for the synthesis of cepha-lexin 'from 7ADCA and D-phenylglycine methyl ester, suggest that a search for other enzyme catalyzed synthetic reactions which can yield synthetic or semisynthetic drugs might be rewarding from the economic point of view.

An ingenious enzymic method for the synthesis of L and D-amino acids was based on multifunctional pyridoxal enzymes in the synthesis of L-amino acids, such as L-tyrosine, L-tryptophan and 3,4-L-Dopa and on microbial dihydropyrimidinase in the synthesis of D-amino acids via the corresponding hydantoin derivatives. The combined approaches of an organic chemist and an enzymologist were required to complete the synthetic reactions described; and it is just this combined approach which will ascertain further progress in the utilization of enzymes and microorganisms by the chemical and pharmaceutical industry.

The work on ester exchange reactions in triglycerides, using lipase adsorbed on Celite containing a small amount of water or glycerol as the heterogeneous catalyst, impressed me in particular. This procedure enabled the conversion of olive oil into cocoa butter-like fat by the replacement of the oleic acid moiety with palmitic or stearic acid.

At this point it is worth mentioning the immobilized enzyme process for the production of fructose and alkene oxides. In this process D-glucose is converted into D-fructose, in practically quantitative yield, by a procedure consisting of two steps, the first enzymic and the second chemical. During the enzymic step D-glucose is oxidized to D-glucosone by immobilized pyranose-2-oxidase derived from *Polyporus obtusus*; in the second step fruc-tose is obtained by reduction of glucosone with hydrogen gas, using palladium as catalyst. The hydrogen peroxide appearing as a by-product in the enzymic oxidation of glucose, is utilized to trans-form ethylene or propylene into the corresponding alkene oxides by the consecutive action of two enzymes: haloperoxidase acting in the presence of chlorine or bromine ions to yield the corres-ponding chloro- or bromo-hydrins, and halohydrin epoxidase, iso-lated from suitable microorganisms, which catalyzes the transfor-mation of halohydrins into alkene oxides. If this process proves

economically feasible, it will provide a novel technique for the
enzymic conversion of glucose into fructose and also introduce
immobilized enzyme technology into the petrochemical industry.

Bioengineers interested in scaling up enzyme-catalyzed reactions naturally want to know the mode of action, efficiency, stability, and cost for various enzyme reactors in order to facilitate
the choice of the type of reactor to be constructed. Enzyme membrane reactors, in particular, should be evaluated further. In
one type of membrane reactor the enzymes are retained in solution;
and the low MW products are removed continuously via an ultrafiltration membrane. In another enzyme membrane reactor the enzymes
are immobilized by entrapment in the micropores of the membrane;
and biotransformation occurs as a result of the flow of substrate
under pressure via the enzyme loaded membrane. L-Alanine was prepared in the first type of membrane reactor by three different
enzymic routes: racemic resolution of N-acetyl-DL-alanine, production of L-aspartic acid from fumaric acid followed by β-decar-
boxylation, and reductive amination of pyruvate. In the third
reaction a water-soluble NADH-polyethyleneglycol conjugate was
used as a high MW nondializable cofactor, which could be regenerated by formate dehydrogenase with formate as cosubstrate. In
evaluating the usefulness of this continuous membrane reactor,
one should bear in mind that the efficiency of this type of reactor will be determined to a large extent by the stability in solution of the enzymes, as well as by the efficiency and stability
of the ultrafiltration membrane. The main advantage of the second
type of enzyme-membrane reactor stems from the fact that practically all types of mass-transfer limitations are eliminated; this
type should be a useful system for carrying out fast enzyme reactions with high enzyme turnover numbers.

In the above, I have focused my attention on immobilized enzyme reactors; however, one should not ignore the progress attained
during the past two years in the design, construction, and utilization of immobilized cell reactors. New techniques have been developed for immobilization by encapsulation, gel entrapment, or
anchorage to suitable surfaces of bacteria, yeast, fungi, and cells
of higher plants and animals; and the reactor efficiency with living or inactivated cells has been studied. It has been found that
cell reactors can be of particular use in the preparation of compounds resulting from a sequence of complex enzymic reactions,
such as antibiotics, alkaloids, steroids, different fermentation
products, peptides, and proteins. Furthermore, since cofactors
are continuously regenerated within living cells, their supply
from the outside is not required. In an efficient immobilized
cell reactor for the continuous production of L-tryptophan from
indole and serine by *E. coli*, the bacterial cells were entrapped
in polyacrylamide and chitosan beads, placed in a continuous flow

stirred tank reactor, and used to produce impressive yields of tryptophan. In another immobilized cell reactor *Saccharomyces cerevisiae* and *Zymomonas mobilis* were entrapped in beads of calcium alginate and karrageenan, respectively, packed in column reactors, and used for the continuous production of ethanol by fermentation for long periods of time without deterioration.

Starch and sucrose, which can be transformed readily into fermentable hexoses, have been used extensively as basic raw materials for the production of alcohol and other fermentable products. Large scale fermentation plants have been established to produce large amounts of high grade ethanol, which can be used directly as a liquid fuel or in admixture with petrol as gasohol. As the price of oil has dropped markedly during the last year, it is still premature to predict whether the biotechnological processes worked out for alcohol production are economically sound. Cellulose is obviously the cheapest and most abundant raw material for the production of glucose and its various fermentation products. Considerable efforts are being made to work out efficient methods for the isolation of cellulose from wood and agricultural waste products and to design suitable processes for its enzymic degradation. Unfortunately, in addition to cellulose biomass contains lignin and hemicellulose; and these have to be removed or modified before making any attempt to hydrolyse cellulose by the various known exo- and endo-cellulases or by suitable microorganisms containing these enzymes.

No thorough discussion on biomass conversion and energy production took place at the conference. However, the kinetics of the degradation of cellulose, methods for increasing the yields of glucose from biomass, and methods for transformation of yeast non-fermentable pentoses such as xylose into fermentable ones were of interest. Xylose was obtained by hydrolysis of hemicellulose, and then converted into xylulose by xylose isomerase to yield a pentose which was readily fermented by yeast. The insertion into yeast of the microbial gene coding for xylose isomerase yielded yeast which readily fermented hexoses as well as pentoses.

In the analytical area enzyme columns, enzyme membranes, enzyme electrodes, and enzyme thermistors represent the most promising analytical enzyme devices designed so far; but their use in the clinic, the laboratory, and the chemical industry is still at its beginnings. However, it is worth recalling some words of caution that the potential market for analytical applications of immobilized enzymes will require mainly instruments which have been designed for continuous monitoring, bed analyses, or measurement of special parameters. The microbial sensor if of considerable practical value and illustrates a new vista in the analytical field as a result of work with different cell sensor systems.

The possible medical application of enzyme technology is rather appealing, as immobilized enzymes might be used to remove or alter specifically harmful compounds that have accumulated accidentally within the blood system. However, many difficulties have to be overcome; so that it is not surprising that no success has been attained so far in the direct application of enzyme engineering to medicine. Still, the data presented give sufficient indication that efforts to find medical applications are being continued.

We can look back at what has been achieved in enzyme engineering during the past ten years with a feeling of satisfaction. Enzyme engineering has reached maturity and its further progress has been assured. Scientists and engineers working in the various disciplines of biochemistry, organic chemistry, microbiology, genetics, chemical engineering, and biotechnology have contributed to the progress attained; and we are fortunate in being able to work in a field which is of both theoretical and practical interest. It is obvious, however, that many of our aims have not yet been achieved. The new enzyme technologies developed are still not widely used; the different areas in which enzyme engineering might contribute have so far been given only limited interpretation; the engineering aspects of the new discipline have still to be clarified; and the economic evaluation of the various enzyme processes has still to be carried out.

As a result of the work being done, problems of a basic nature have arisen concerning the behavior of biopolymers and cells under the particular conditions employed. One might therefore expect that future developments in molecular biology and genetic engineering will initiate further progress in the different areas of enzyme engineering.

CONFERENCE PARTICIPANTS

ADO, YUTAKA
    Kyowa Hakko Kogyo Co, Ltd.,
    3-6-6 Asahimachi, Machidashi, Tokyo 194, JAPAN.
AIZAWA, MASUO
    Associate Professor, University of Tsukuba,
    Materials Sci. Inst., Sukura-mura, Ibaraki, 305, JAPAN
AMON, JR., WILLIAM F.,
    Vice President, Cetus Corporation
    600 Bancroft Way, Berkeley, CA 94710, USA.
ANDERSON, ARTHUR W., DR.
    Vice President Research & Development, C.M. Mushrooms, Inc.
    20408 Boulevard, Hayward CA 94545, USA
ARAKI, KEIICHI
    Supervisor of R & D, Chiyoda Chem. Eng. & Const.
    221, 3-13 Moriya-cho, Kanagawa-ku, Yokohama, JAPAN
ARIMA, KEI
    Professor Emeritus, Famiiru Hisakata, 5-24-21 Koishikawa,
    Rm. 1102, Bunkyo-ku, Tokyo 113, JAPAN
ARMSTRONG, DAVID J.
    Scientist, Ralston Purina Company
    Checkerboard Square, St. Louis, MO 63188, USA
AUCHINCLOSS, LESLIE
    Managing Director, Biocon Ltd.
    Kilnagleary, Carrigaline, Country Cork, IRELAND
BAILEY, JAMES E.
    Professor, California Inst. of Tech.,
    Chem. Eng. Dept., 206-41, Pasadena, CA 91125, USA
BAKKER, HARRY G.
    DSM, Corporate Dev., Post Box 65, Heerlen 6400 AB, NETHERLANDS
BARET, JEAN LUC
    Project Leader, Corning Europe, Inc.
    Boite Postte #3, 77211 Avon Codex, FRANCE
BARTOLI, FRANCESCO
    Doctor in Chemistry, ASSORENI Via. E. Ramarini,
    32, P.O. Box 15, Monterotondo 00015, ITALY

BELL, KEITH
    Director, Southern Cross Lab. Ltd.
    P.O. Box 2, Dural, NSW, 2158, AUSTRALIA
BEPPU, TERUHIKO
    Professor, University of Tokyo Agricultural Chem. Dept.,
    Yayoi 1-1-1, Bunkyo-ku, Tokyo, 113, JAPAN
BEREZIN, ILIA V.
    Professor and Dir., A.N. Bash Inst. of Biochem.,
    Leninskly Prospekt 33, Moscow, 117071, USSR
BLANCH, HARVEY W.
    Assoc. Prof. University of California, Dept. of Chem. Eng.
    Berkeley, CA 94720, USA
BRANNER-JORGENSEN, SVEN
    Group Leader, Novo Industri A/S, Novo Alle,
    Bagsvaerd DK-2880, DENMARK
BRODELIUS, PETER
    Univ. of Lund, Pure and Applied Biochem., Chemical Center,
    P.O. Box 740, S-22007 Lund, SWEDEN
BROUN, GEORGES B.
    Professor, University of Technology
    BP 233, 10206 Compiegne, FRANCE
CASHION, PETER J.
    Associate Professor, Univ. of New Brunswick, Dept. of Biology,
    Fredericton NB, E3B 5A3, CANADA
CHAMBERS, ROBERT P.
    Professor and Dept. Head, Auburn University
    Chem. Engineering Dept., Auburn, AL 36849, USA
CHANG, THOMAS M.
    Professor, McGill University, Art. Cells/Organs Research Center
    3655 Drummond Street, Montreal PQ H3G 1Y6, CANADA
CHIBATA, ICHIRO
    Director, Tanabe Seiyaku Co., Ltd. 16-89,
    Kashima-3-chome, Yodogawa-ku, Osaka 532, JAPAN
CHUN, MOONJIN
    Professor, Korea University, Dept. of Agri. Chem.,
    Anam-Dong 5-1, Seoul 132, KOREA
COLE, SANDFORD S.
    Director of Conferences, Engineering Foundation
    345 E. 47th Street, New York, NY 10017, USA
COMERER, HAROLD A.
    Assistant Director, Engineering Foundation
    345 East 47th Street, New York, NY 10017, USA
COULET, PIERRE R.
    Charge Recherche CNRS, Universite Claude Bernard de Lyon,
    Lab. Biol. Memb., CNRS, 43 Blvd 11 Novembre 1918,
    Villeurbanne 69622, FRANCE
DAHLMANS, JACQUES J.
    Research Manager, D.S.M. Central Lab. P.O.B. 18,
    Geleen 6160 MD, NETHERLANDS

DANIEL, ROY MCIVER
    University of Waikato, Hamilton, NEW ZEALAND
DANIELL, E.
    University of California Molecular Biology Dept.
    Berkeley CA 94720, USA
DANIELSSON, BENGT
    University of Lund, Chemical Center, Pure & Applied Biochemistry
    P.O. Box 740, S-22007 Lund, SWEDEN
DAVIS, FRANK F.
    Prof. of Biochemistry, Rutgers University, Nelson Biological Lsbs,
    Busch Campus, New Brunswick, NJ 08903, USA
DAY, DONAL F.
    Louisana State Univ., Audubon Sugar Institute,
    Baton Rouge, LA 70803, USA
DOHAN, LUC
    Department Manager, Corning Europe Inc.
    Biote Postte #3, 77211 Avon Codex, FRANCE
DRIOLI, ENRICO
    Professor, University of Naples, Inst. Chem. Eng.,
    Piazzale Techhio, I-80125 Napoli, ITALY
DUARTE, JOSE M.C.
    Chemical Engineer, Quatrum, Empresa Nac. Quim. Organ.,
    Av. Joao XXI-10-6-DTO, 1000 Lisboa, PORTUGAL
DUNNE, CHARLES P.
    Research Chemist, U.S. Army Natick R & D Labs., DRDNA-YEB,
    Environ. Sci. & Eng. Div., Natick, MA 01760, USA
EGOROV, N.S.
    Moscow University, Moscow 117230, USSR
ENDO, ISAO
    Inst. of Phy & Chem Res., Chem Eng Lab,
    Hirosawa 2-1, Wako-shi, Saitama 351, JAPAN
ENGASSER, JEAN-MARC
    Institut National Polytec. de Lorraine, ENSAIA,
    1 rue Grandville, Nancy 54042, FRANCE
FINK, DAVID J.
    Project Manager, Battelle Columbus Labs.
    505 King Avenue, Columbus, OH 43201, USA
FLASCHEL, ERWIN
    Swiss Fed. Inst. of Tech., Inst. Chem. Eng.
    Lausanne 1015, SWITZERLAND
FLEMING, SAMUEL M.
    Mngr., Technology Dev., Fluor Corporation
    P.O. Box C 11944, Santa Anna, CA 92711, USA
FUKUI, SABURO
    Professor, Kyoto University, Lab. of Indus. Biochem.,
    Dept. of Ind. Chem., Eng. Faculty, Yoshida,
    Sakyo-ku, Kyoto 606, JAPAN
FUKUSHIMA, SUSUMU, PROFESSOR
    Kansai University, Chem. Eng. Dept., Senriyama, Suita,
    Osaka, 564, JAPAN

FURUHASHI, KEIZO
      Bio Research Ctr Co. Ltd., 17-35, 3-chome,
      Niizo-Minami, Toda Saitama 335, JAPAN
FURUSAKI, SHINTARO
      Associate Professor, University of Tokyo, Dept. of Chem. Eng.
      Bunkyo-ku, Tokyo 113, JAPAN
GAMBACORTA, AGATA
      Rieereatore Eng., Inst. di Chimica
      Via Tolano 2, Arco Fezice, Napoli, ITALY
GAUTHERON, DANIELE C.
      Professor, Universite Claude Bernard de Lyon, Lab. Biol. Tech.
      Memb., CNRS, 43 Blvd, 11 Novembre 1918, Villeurbanne 69622,
      FRANCE
GELFAND, D.H.
      Cetus Corporation, 600 Bancroft Way, Berkeley, CA 94710, USA
GELLF, GERARD
      Research Associate, Universite de Compiegne
      Lab. de Tech. Enzymatique, BP 233, Compiegne 60206, FRANCE
GESTRELIUS, STINA M.
      Section Head, Novo Industri A/S Novo Alle,
      Bagsvaerd DK-2880, DENMARK
GEYER, HANS U.
      Commercial Director, Miles Kali-Chemie GmbH, P.O. Box 690307
      D-3000 Hannover, F.R. GERMANY
 GLOGER, MANFRED
      Dipl. Chemiker, Boehringer Mannheim GmbH,
      Biochemica Werk Tutzing, Bahnhofstr. 9-15,
      D-8132 Tutzing, Obb., F.R. Germany
GONDO, SHINICHIRO,
      Professor, Fukuoka Institute of Tech., Wijiro-machi,
      Higashiku, Fukuoka 811-02, JAPAN
GRAVES, DAVID J.
      Associate Professor, Univ. of Pennsylvania,
      Dept. of Chem. Eng., Towne Bldg. D3, Philadelphia, PA 19104,
      USA
GRECO, GUIDO,
      Professor, University of Naples, Inst. Chem. Eng.,
      Piazzale Tecchio I-80125, Naples, ITALY
GUILBAULT, GEORGE G.,
      Professor, University of New Orleans, Dept. of Chemistry,
      Lake Front, New Orleans, LA 70122, USA
HAMSHER, JAMES J.
      Assistant Director, Pfizer, Inc., Central Research,
      Eastern Point Road, Groton, CT 06340, USA
HAN, MOON H.
      Manager, KIST, Biotechnology Res. Dept.,
      P.O. Box 131 Dongdae-Mun, Seoul, KOREA
HARANO, YOSHIO,
      Professor, Osaka City University, Faculty of Engineering,
      459 Sugimoto 3, Sumiyoshi-ku, Osaka 558, JAPAN

HASHIMOTO, KENJI
    Professor, Kyoto University, Dept. of Chem. Eng.,
    Yoshida, Sakyo-ku, Kyoto 606, JAPAN
HASHIMOTO, YUKIO
    Fuji Oil Co., Ltd.
    1-Sumiyoshi-cho, Izumisano-shi, Osaka 598, JAPAN
HATAKEYAMA, HIROYUKI
    Takara Shuzo Co., Ltd., Shijodori Higashinotoin,
    Shimogyo-ku, Kyoto 600-91, JAPAN
HAYASHI, HIDECHIKA,
    Toyo Soda Manufacturing Co., Ltd.,
    2743-1 Hayakawa, Ayase-shi, Kanagawa, JAPAN
HIDAKA, HIDEMASA
    Director, Meiji Seika Kaisha Ltd. 580 Horikawa-cho,
    Saiwai-ku, Kawasaki, Kanagawa, 210, JAPAN
HIKUMA, MOTOHIKO
    Ajinomoto Co Inc., Central Research Labs., 1-1 Suzuki-cho,
    Kawasaki-ku, Kawasaki, 210, JAPAN
HIROHARA, HIDEO
    Sumitomo Chemical Company, Biol. Sci. Lab., 40 2-chome,
    Tsukahara, Takatsuki City, Osaka 569, JAPAN
HUANG, SHIH YOW
    Professor, National Taiwan Univ., Dept. of Chem. Eng.,
    Taipei, Taiwan 107, CHINA
HUITRON, CARLOS
    Inst. de Invest. Biomed., Apartado Postal 70228,
    Ciudad Universitaria 20, D.F., 04510 Mexico, D.F., MEXICO
ICHIJO, HISAO
    Res. Inst. for Polymers & Textiles, 1-1-4 Yatabe-higashi,
    Tsukuba, Ibaraki, 305, JAPAN
IKEDA, YONOSUKE
    Professor, Tokyo Univ. of Agriculture, 1-1-1 Sakuraga-oka,
    Setagaya-ku, Tokyo, 156, JAPAN
ISE, NORIO
    Professor, Kyoto University, Department of Polymer Chemistry
    Yoshida Hommachi, Sakyo-ku, Kyoto 606, JAPAN
JENSEN, VILLY J.
    Manager, Novo Industri Novo Alle,
    Bagsvaerd, Copenhagen 2880, DENMARK
JONES, J. BRYAN
    Professor, University of Toronto, Dept. of Chemistry,
    80 St. George Street, Toronto M5S 1A1, CANADA
KAETSU, ISAO
    Chief of Laboratory, Japan Atomic Energy Res. Inst.,
    Watanuki-machi, Takasaki, Gunma-ken, 370, JAPAN
KARUBE, ISAO
    Associate Professor, Tokyo Institute of Tech.,
    Res. Lab. of Res. Util., 4259 Nagatsuta-cho, Midori-ku,
    Yokohama 227, JAPAN

KATCHALSKI-KATZIR, EPHRAIM
     Professor, Weizmann Inst. of Science
     Rehovot 76100, ISRAEL
KATOH, SHIGEO
     Assistant Professor, Kyoto University Chem. Eng. Dept.,
     Yoshida Hon-machi, Sakyo-ku, Kyoto 606, JAPAN
KAZANSKAYA, NOVELLA F.
     Doctor of Science, Moscow State University, Chem. Dept.,
     Lenin Hills, Moscow 117234, USSR
KIM, CHONG YOL,
     Sr. Research Scientist, Miles Laboratories, Inc.
     1127 Myrtle Street, P.O. Box 932, Elkhart, IN 46515, USA
KIMURA, KAZUO
     Manager, Kyowa Hakko Kogyo Co.
     Hofu Plant, Hofu-shi, Yamaguchi, Yamaguchi-ken, JAPAN
KLEIN, JOACHIM
     Professor & Dir., Institut fur Chem. Tech., Tech. Univ.
     Braunschweig, D-3300, Braunschweig, F.R. GERMANY
KLIBANOV, ALEXANDER M.
     Assistant Professor, Dept. of Nutrition & Food Sci., M.I.T.,
     Room 16-209, Cambridge, MA 02139, USA
KNUDSEN, STEN L.
     Tech. Service Manager, Novo Industri Japan, Ltd., Sakura Bldg.
     6F, 3-3, Uchikanda, 1-chome, Chiyoda-ku, 101, JAPAN
KOBAYASHI, TAKESKI
     Associate Professor, Dept. of Chemical Eng., Nagoya University,
     Chikusa-ku, Nagoya 464, JAPAN
KOMAKI, TOSHIAKI
     Nagase & Co., Ltd., 1-17 Shinmachi 1-chome,
     Nishi-ku, Osaka 550-91, JAPAN
KONECNY, JAN O.
     Senior Scientist, Ciba-Geigy, Ltd.
     Werk Klybeck K-121.403, Basel CH-4002, SWITZERLAND
KOSHCHEYENKO, KIIRA A.
     Doctor of Biology, USSR Academy of Sciences, Inst. of
     Biochem. & Physiology of Microorganisms,
     Pushchino, Moscow Region, 142292, USSR
KRAEMER, DIETER M.
     ROHM GMBH Pharmaceutical Laboratory,
     Postfach 4242, Darmstadt, D-61, F.R. GERMANY
KULA, MARIA-REGINA
     Gesellschaft fur Biotech., Forschung mbH,
     Mascheroder Weg 1, Braunschweig-Stockheim, D3300 F.R. GERMANY
KUSUNOKI, KOICHIRO
     Professor, Kyushu University, Dept. of Chem. Eng.
     6-10-1 Hakozaki, Higashi-ku, Fukuoka 812, JAPAN
LAI, CHING-LIANG
     Taiwan Sugar Research Institute 54,
     Sheng Chan Road, Tainan, Taiwan 700, CHINA

LANGER, ROBERT S.
    Associate Professor, Mass. Inst. of Tech., Dept. Nut. Food Sci.,
    Bldg. 16, Room 309, Cambridge, MA 02139, USA
LASKIN, ALLEN I.
    Head, Biosciences Res., Exxon Res. & Eng. Co.
    Exxon Research Center, P.O. Box 45, Linden, NJ 07036, USA
LEACH, ROBERT E.
    Marketing Manager, Corning Glass Works, Hain Plant,
    50-3, Walnut Street, Corning, NY 14870, USA
LEE, YOON Y.
    Associate Professor, Auburn University, Chemical Eng. Dept.
    Auburn, AL 36849, USA
LEGOY, MARIE D.
    Lecturer, Univ. Technologie Compeigne, Lab. Technol.
    Enzymatique, BP 233, Compiegne 60206, FRANCE
LEHTONEN, PAAVO OLAVI
    Production Manager, Finnish Sugar Co., Ltd., Hanko Plant,
    P.O. Box 38, SF-10901 Hanko, FINLAND
LINKO, PEKKA
    Professor/Director, Dept. of Chemistry, Helsinki Univ. of
    Technology, Otaniemi, Otakaari I, SF-02150 Espoo 15, FINLAND
LINKO, YU-YEN
    Senior Scientist, Lab. of Biochem & Food Tech.,
    Helsinki Univ. of Tech., SF-02150 Espoo 15, FINLAND
LIU, SHU-HUANG
    Shanghai Inst. of Biochemistry, Academia Sinica,
    320 Yo-Yang Road, Shanghai, CHINA
MAEDA, HIDEKATSU
    Vice Director, Fermentation Res. Inst.,
    Agency Ind. Sci. Tech., Higashi 1-1-3,
    Yatabemachi Ibaragi 300-21 JAPAN
MANECKE, GEORG
    Professor, Free Universitat Berlin, Inst. f Organische Chemie,
    Takustr. 3, 1000 Berlin 33, F.R. GERMANY
MARCINOWSKI, STEFAN
    BASF Aktiengesellschaft D-ZH Tagungen B9, Ludwigshafen,
    Rhein, 6700, F.R. GERMANY
MARKKANEN, PERTTI
    Associate Professor, Helsinki Univ. of Tech.
    Otakaari 1, Kemistintie 1, 02150 Espoo 15, FINLAND
MATSUNO RYUICHI
    Assoc. Prof., Kyoto University, Dept. Food Sci. & Tech.,
    Faculty of Agric., Sakyo-ku, Kyoto 606, JAPAN
MATTIASSON, BO
    Associate Professor, University of Lund, Pure & Appl.
    Biochem., Chemical Center, P.O. Box 740, S-22007 Lund, SWEDEN
MAY, SHELDON W.
    Professor of Chemistry, Georgia Inst. of Tech., Dept. of
    Chemistry Atlanta, GA 30332, USA

MAZZOLA, GIORGIO
Laboratorio di Chemica Degli Ormoni, CNR, Via Mario Bianco 9,
Milano 20131, ITALY
MESSING, RALPH A.
Senior Research Associate, Corning Glass Works, Sullivan Park,
Corning, NY 14831, USA
MICHAELS, ALAN S.
Professor, Stanford University, Chemical Engrg. Dept.,
341 Seeley Mudd Bldg., Stanford, CA 94305, USA
MIRABEL, B.
Dr., Rhone Poulenc Recherches, 182 Avenue Aristide Briand
92160 Antony, FRANCE
MIURA, YOSHIHARU
Professor, Faculty of Pharmaceutical Sciences, Osaka Univers-
ity 1-6 Yamadaoka, Suita, Osaka, JAPAN
MONSAN, PIERRE F.
Professor, Inst. Nat. Sci. Appl., Dep. Genie Biochim.,
Avenue de Rangueil, Toulouse 31077, FRANCE
MORIMOTO, KEIICHI
Mgr of Tech Info Section, Kiren Brewery Co., Ltd.
Research Labs Miyahara, Takasaki-chi, Miyahara 370-12, JAPAN
MOSBACH, KLAUS
Professor, University of Lund, Biochemistry 2, Chemical
Center P.O. Box B740, Lund 7, S-22007, SWEDEN
MURACHI, TAKASHI
Professor, Clinical Science Dept., Kyoto Univ. Hospital
Sakyo-ku, Kyoto 606, JAPAN
MURAKAMI, KAZUO
Professor, Inst. of Appl. Biochem., University of Tsukuba
Ibaraki-ken 305, JAPAN
NAKAMURA, KOZO
Associate Professor, University of Tokyo, Dept. of Agric.
Chem., Bunkyo-ku, Tokyo 113, JAPAN
NAKATA, NORIO
Technologist, Shell Kagaku, Kasumigaseki Bldg., 3-2-5,
Kasumigaseki, Chiyoda-ku, Tokyo 100, JAPAN
NEIDLEMAN, SAUL L.
Director, Cetus Corporation, New Venture Research,
600 Bancroft Way, Berkeley, CA 94710, USA
NICOLAUS, BARBARA
Ricercatore CNR, Napoli C.N.R. via Toiano 2,
Arco Felice, Napoli, ITALY
NISHIURA, MASARU
Daiichi Radioisotope Labs. 10-5, Nihombashi,
3-chome Chuo-ku, Tokyo, 103, JAPAN
OHLSON, STEN
Gambro, AB, F-Avd. Sambro AB, S-22010, Lund, SWEDEN
OHSHIRO, TAKESHI
Assistant Professor, Osaka Univ. Med. School, Sec. Dept. of
Surgery 1-1-50, Fukushima-ku, Osaka 553, JAPAN

OHTA, TAKAHISA
    Professor, University of Tokyo, Dept. of Agricultural
    Chemistry Bunkyo-ku, Tokyo 113, JAPAN
OKACHI, RYO
    Pharmaceutical Res. Lab., 1188, Shimotogari, Nagaizumicho,
    Suntogun, Shizuokaken, 411 JAPAN
OKADA, HIROSUKE
    Professor, Osaka University, Dept. of Fermentation Tech.
    Yamada-kami, Suita-shi, Osaka, 565, JAPAN
PANSOLLI, PAOLO
    Dr. of Chemistry, Assoreni, Via Ramarini,
    Lab. Processi Micro., Monterotondo, 00015, ITALY
PEITERSEN, NICOLAI
    Head of Research Lab, Christian Hansen's Lab A/S
    Masnedogade 22, Copenhagen DK-2100, DENMARK
POULSEN, POUL BORGE
    Product Manager, Novo Industri A/S Novo Alle,
    Bagsvaerd DK 2880, DENMARK
PRENOSIL, JIRI E.
    Chem. Eng. Dept. Eid. Tech. Hochsch., CH8092 Zurich,
    SWITZERLAND
RAMMESKOW, NEILS
    President, Novo Industri Japan, Ltd.
    Sakura Bldg. 6F, 3-3 Uchikanda, 1-chome, Tokyo, 101, JAPAN
RENKEN, ALBERT
    Professor, Swiss Fed. Inst. of Tech.,
    Institute of Chem. Eng., CH-1015 Lausanne, SWITZERLAND
RENN, DONALD W.
    Research Fellow, FMC Corporation, 5 Maple Street
    Rockland, ME 04841, USA
ROEDER, ALBERT
    Boehringer Manheim M Gmbh, Biochemic A Werk,
    Bahnh of Strasse 9-15, Tutzing 8132, F.R. GERMANY
ROMETTE, JEAN
    Lab. Tech. Enzymnatique, Univ. Tech. Compeigne,
    BP 233, Compiegne 60206, FRANCE
ROSEVEAR, ALAN
    UK Atomic Energy Authority, Biochemistry Group, Bldg 353,
    Harwell, Oxfordshire OX11 ORA, ENGLAND
ROYER, GARFIELD P.
    Professor, Ohio State University, Dept. of Biochem.,
    484 W 12th Avenue, Columbus, OH 43210, USA
RYU, DEWEY D.Y.
    Professor, Dept. of Chemical Engineering, Univ. of
    California at Davis, Davis, CA 95616, USA
SAKAGUCHI, KENJI
    Chief, Lab. Micro. Chem., Mitsubishi-Kasei Institute Life
    Sci., 11 Minamiooya, Machida-shi, Tokyo, 194, JAPAN

SAMEJIMA, HIROTOSHI
     Manager, Technical Div., Kyowa Hakko Kogyo Co., Ltd.
     Ohtemachi 1-6-1, Chiyoda-ku, Tokyo 100, JAPAN
SANTOS, NELSON NEBEL
     Division Head, Fundacao de Tecnologia Ind.,
     Av. Venezuela 82/304, Rio de Janeiro, 20081, BRAZIL
SATO, TADASHI
     Senior Scientist, Tanabe Seiyaku Co., Ltd.
     16-89, Kashima-3-chome, Yodogawa-ku, Osaka, 532, JAPAN
SAUBER, KLAUS
     Hoechst AG, H 780, Postfach 800320, 6230 Frankfurt-Hochst,
     F.R. GERMANY
SCHMID, ROLF D.
     Manager, Henkel KGaA, Henkelstrasse 67, D-4000,
     Duesseldorf 1, F.R. GERMANY
SCHMIDT-KASTNER, GUNTER
     Dipl. Chemiker, Bayer AG, VE-Biochemie Dept.
     Friedrich Ebert. Str. 217, D5600 Wuppertal 1, F.R. GERMANY
SCOTT, DON
     President, Fermco Biochemics Inc.
     2638 Delta Lane, Elk Grove Village, IL 60007, USA
SCOUTEN, WILLIAM H.
     Associate Professor, Bucknell University,
     Chemistry Dept., Lewisburg, PA 17837, USA
SEIDMAN, MARTIN
     Manager, A.E. Staley Mfg. Co., Research Center,
     Decatur, IL 62525, USA
SERIS, LOUIS JEAN
     S.N. Elf. Aquitaine, Centre de Recherches Lacq,
     Artix 64170, FRANCE
SFAT, MICHAEL R.
     President, Bio-Technical Resources,
     7th and Marshall, Manitowoc, WI 54220, USA
SHIMIZU, SAKAYU
     Research Associate, Kyoto University,
     Dept. of Agriculture Chemistry, Sakyo-ku, Kyoto 606, JAPAN
SHIMIZU, SHOICHI
     Professor, Nagoya University, Dept. Food Science & Tech.,
     Furo-cho, Chikusa-ku, Nagoya 464, JAPAN
SHINKE, RYU
     Associate Professor, Kobe University Agriculture Faculty,
     1-Rokkodai-cho Nada-ku, Kobe 657, JAPAN
SILVER, RICHARD S.
     Staff Engineer, Gulf R & D Company, P.O. Drawer 2038,
     Pittsburgh, PA 15230, USA
SINSKEY, ANTHONY J.
     Professor, M.I.T., Nutrition & Fd Dept., Room 56-121
     Cambridge, MA 02139, USA

SMINK, D.A.
    Manager, Product Dev., Gist Brocades,
    Wateringseweg 1, 2600 MA Delft, NETHERLANDS
SODA, KENJI
    Professor, Kyoto University, Chemical Research Inst.,
    Gokasho, Uji 611, JAPAN
STRAMONDO, JAMES G.
    Senior Scientist, Genetech, Inc., 460 Pt. San Bruno Blvd.,
    South San Francisco, CA 94080, USA
SUNDARAM, P.V.
    Professor, 160 Eldams Road, Madras-600018, INDIA
SUZUKI, SHUICHI
    Professor, Tokyo Inst. of Technology, Res. Lab. of Res. Util.,
    Nagatsuta-cho 4259, Midori-ku, Yokohama 227, JAPAN
TAKASAWA, SEIGO
    Research Manager, Kyowa Hakko Kogyo Co.,
    3-6-6 Asahimach, Machida-shi, Tokyo 194, JAPAN
TAKINAMI, KOICHI
    Chief Biochemist, Ajinomoto Company, Inc.
    1-1 Suzuki-cho, Kawasaki-ku, Kanagawa-ken, Kawasaki, 210,
    JAPAN
TAMURA, NORIYOSHI
    Mitsubishi Petrochem Co., Central Research Labs,
    1315 Ami-cho, Inashiki-gun, Ibaraki-ken 300-03, JAPAN
TANAKA, ATSUO
    Lecturer, Kyoto University, Lab. Indus. Biochem.,
    Dept. of Ind. Chem., Eng. Faculty, Yoshida, Sakyo-ku,
    Kyoto 606, JAPAN
THOMAS, DANIEL
    Professor, University of Compiegne, Labortorie Enz. Tech.,
    BP 233, Compiegne 60206, FRANCE
TOMODA, KATSUMI
    Chief, Section of Enzyme, Takeda Chemical Ind., LTD. 17-85,
    Jusohomachi 2-chome, Yodogawa-ku, 532, Osaka, 532, JAPAN
TORCHILIN, VLADIMIR P.
    Lab. of Enzyme Eng., Cardiology Research Center of USSR,
    Academy of Medical Sciences, Petroverigskly Lane 10,
    Moscow 101837, USSR
TOSA, TETSUYA
    Senior Scientist, Tanabe Seiyaku Co., Ltd.
    16-89, Kashima-3-chome, Yodogawa-ku, Osaka 532, JAPAN
TSAO, GEORGE T.
    Professor, Purdue University, Lorre,
    West Lafayette, IN 47907, USA
TSUDA, YOSHIHISA
    Research Biochemist, UOP Inc. Research Center,
    10 UOP Plaza, Des Plaines, IL 60016, USA
UCHIDA, KEIICHI
    Senior Engineer, Asahi Glass Co., Ltd. 2-1,
    Marunouchi 2-chome, Chiyoda-ku, Tokyo 100, JAPAN

UHLIG, HELMUT
     Dipl. Chemiker, Rohm GmbH, Postfach 4242,
     Chemische Fabrik, 6100 Darmstadt I, F.R. GERMANY
UMEZAWA, HAMA
     Professor Emeritus, Director, Institute of Microbial Chem-
     istry, 14-23, Kamiosaki, 3-chome, Shinagawa-ku, Tokyo 141,
     JAPAN
URABE, ITARU
     Osaka University, Fermentation Tech., Dept. of Faculty of
     Engineering, Yamada-kami, Suita-shi, Osaka 565, JAPAN
YABUKI, AKIRA
     Ajinomoto Company Inc., 1-1 Suzuki-cho,
     Kawasaki-ku, Kawasaki-shi, Kanagawa, 210, JAPAN
VAN BEYNUM, GERHARD
     Gist Brocades N.V., Postbus 1, Delft, 2600 MA, NETHERLANDS
VENKATSUBRAMANIAN, K.
     Asst. Corp. Dir., H.J. Heinz Company, P.O. Box 57,
     Pittsburgh, PA 15230, USA
VERONESE, FRANCESCO M.
     Professor, Universita di Padova, Instituto di Chimica Organica
     Farmaceutica, Via Marzola I, I-35100 Padova, ITALY
VIETH, WOLF R.
     Professor, Rutgers University, Dept. of Chem. and Biochem. Eng.,
     Busch Campus, Piscataway, NJ 08854, USA
WAGNER, FRITZ
     Professor, Lab of Biochem. & Biotech., Technischen Universitat
     Braunschweig, Mascheroder Weg 1, 3300 Braunschweig,
     Stockheim, F.R. GERMANY
WANDREY, CHRISTIAN
     Professor, Institute of Biotechnology II der Kernforschung-
     sanlage, Postfach 1913, D-5170 Juelich 1, F.R. GERMANY
WANG, ZHEN-XIANG
     Academia Sinica, Inst. of Microbiology, Beijing, CHINA
WATANABE, SATOSHI
     Lecturer, Kyoto University, Dept. Thoracic Surgery,
     Chest Disease Res. Inst., Sakyo-ku, Kyoto 606, JAPAN
WEAVER, JAMES C.
     Lecturer, M.I.T.-Harvard, Biomed. Engr. Ctr. for Clinical
     Instru., Rm. 20A-128, MIT, Cambridge, MA 02139, USA
WEETALL, HOWARD H.
     Manager, Biosciences Res., Corning Glass Works,
     Sullivan Science Park, Corning, NY 14830, USA
WEIBEL, MICHAEL K.
     Weibel Assoc., Inc., P.O. Box 108,
     Gallows Hill Road, West Redding, CT 06896, USA
WEIGAND, WILLIAM A.f
     Program Director, National Science Found.,
     1800 "G" Street, N.W., Washington, DC 20550, USA

WEISSMANN, CHARLES
        Professor, University of Zurich, Inst. fur Molekularbiologie,
        Hoenggerberg, CH-8093 Zurich, SWITZERLAND
WICHMANN, ROLF
        Dipl. Chem., Institute of Biotechnology 1,
        II der Kernforschungsanlage, Postfach 1913,
        D-5170 Juelich 1, F.R. GERMANY
WINGARD JR., LEMUEL B.
        Professor, University of Pittsburgh, Dept. of Pharmacology,
        School of Medicine, Pittsburgh, PA 15261, USA
YABUKI, AKIRA
        Ajinomoto Company Inc., 1-1 Suzuki-cho,
        Kawasaki-ku, Kawasaki-shi, Kanagawa, 210, JAPAN
YAMADA, HIDEAKI
        Professor, Kyoto University, Dept. of Agricultural Chemistry,
        Sakyo-ku, Kyoto 606, JAPAN
YAMANAKA, SHIGERU
        Chief Biochemist, Ajinomoto Company, Inc.
        1-1 Suzuki-cho, Kawasaki-ku, Kawasaki-shi, Kanagawa-ken,
        210, JAPAN
YAMANE, TSUNEO
        Asst. Professor, Kansai University, Dept. of Chemical Eng.,
        Suita-shi, Osaka 564, JAPAN
YATA, NAOKI
        Managing Director, BioRes Center Co., Ltd. 3-17-35,
        Niizo Minami, Toda-shi, Saitama-ken, Toda-shi 335, JAPAN
YODA, KENTARO
        Chief, Enzyme Tech. Lab., Toyobo Co., Ltd.
        Katata Research Center, 1300-1 Honkatata, Otsu 520-02, JAPAN
YOKOZEKI, KENZO
        Biochemist, Ajinomoto Company, Inc.
        1-1, Suzuki-cho, Kawasaki-ku, Kawasaki, 210, JAPAN
YU, YAO TING
        McGill University, Cells and Organs Res. Centre,
        3655 Drummond St., McIntyre, Montreal H3G 1Y6, CANADA
        Present address: Department of Chemistry, Nankai University,
                        Tienjin, CHINA
YUAN, ZHONG-YI
        Shanghai Inst. of Biochemistry, Academia Sinica,
        320 Yo-Yang Road, Shanghai, CHINA
ZAFFARONI, Pasquale
        Fermentation Dept. Mgr., Assoreni,
        P.O. Box 15, Monterotondo, Rome I-00015, ITALY
ZHANG, SHU-ZHENG
        Professor, Academia Sinica, Institute of Microbiology,
        Beijing, CHINA

Acetyl cellulose, 47
Acid phosphatase, 229
  immobilized in membrane,
  229
Aclacinomycin A, 19
Acyl CoA oxidase, 371, 467
Acyl-CoA synthetase, 467
Acyl enzyme intermediate, 92
Acyl exchange, 152
Acyl transfer,
  enzymatic, 91-96
  kinetics, 93
Acylation, pH effect on rate,
  94
Acylation, reduction of en-
  zyme stability, 287
Adenine arabinoside, 197
  synthesis of, 198
Adenosine, kinase, 351
Adenylate kinase, 351
Adriamycin, 18-21
Adsorption, moving-bed, 273
Affinity chromatography, 145,
  436
  milk clotting enzymes, 75
Affinity supports, 76
Agar, 268
Agarose
  cyanate root, density of,
  223
  tritylated, 219
  gel electrophoresis, of
  plasmids, 496
Ajmalicine, 138, 203

L-Alanine, production of, 271
Alanine racemase, inactivation,
  272
Alcaligenes eutropha, 143
  immobilized in alginate, 323
  immobilized, in carrageenan,
  323
Alcohol dehydrogenase, 107-115,
  163, 451
  immobilized in microcapsules,
  453
Alcohol, secondary, oxidation,
  111
Aldehyde dehydrogenase, immobilized on nylon, 473
Aldehyde reduction, by alcohol
  dehydrogenase, 110
  stereospecific, 110
Alginate, 182, 336, 347
  diffusion through, 188
  pressure stability, 247
Alkaloid production, 137
Alkaline phosphatase, immobilized on ABSE-agar, 266
  immobilized on trityl agarose,
  220
Alkaline protease, immobilized
  on Dowex resin, 289
Alumina, 174
L-Amino acids, assays for, 396
Amino acids,
  optically active, synthesis of,
  97-105
  reductive amination, 63

Amino acylase, immobilized on
  DEAE Sephadex, 265
α-Aminobutyric acid, production
  of, 265
7-Aminodeacetoxy cephalo-
  sporanic acid, 82
  production of, 291-292
Aminoglycoside antibiotics,
  26-27
6-Aminohexanoic acid hydrolase,
  cyclic dimer enzyme, 492
  linear oligomer enzyme, 492
Aminohexanoic acid hydrolase,
  gene restriction map, 499
Aminohexylated cellulose, 259
6-Aminopenicillanic acid, 91
Ammonia,
  conversion to glutamate, 164
  electrode, 388
  fumarate, 272
  gas, microbial sensor for,
  387
Amorphous cellulose, 331
Amoxicillin, synthesis, 89
Amperometric sensor, 413
Ampicillin, formation, 83
α-Amylase pretreated starch,
  275
β-Amylase, immobilized on
  phenol formaldehyde resin,
  275
Androsterene dione, 194
Angiotensin-I, 75
Antibiotics, 11-33
  production economics, 42
Antibodies
  in assays, 470
  immobilized, 397
  immobilized in liposomes,
  462
  partitioning of, 360
Antigens, immobilized, 397
Anthracyclines, 18-21
  daunosamine, 21
  trisaccharide, 20
Antithrombin III-heparin,
  immobilized on sepharose
  457
Antithrombogenic polymer,
  443, 457

Aromatic amines, removal from
  water, 319
Arthrobacter globiformis, immobilized in polyacrylamide
  gel, 123
Arthrobacter simplex, immobilized on,
  modified polyvinyl alcohol,
  182
  modified polyethylenimine,
  182
L-Asparaginase, immobilized in
  hollow fibers, 465
Aspartase, 269
Aspergillus oryzae, 81
Aspergillus phoenicis, immobilized in,
  albumin, 130
  alginate, 130
  gelatin, 130
  k-carrageenan, 130
  polyurethane, 130
ATP, regeneration of, mathematical model, 352
Aureobasidium, cellulase
  activity, 353-354
Avicel, 349
B. ammoniagenes, immobilized
  in polyacrylamide, 237
B. subtilis, 417, 479
B. subtilis, immobilized on
  acetele cellulose, 418
Bacillus stearothermophilus,
  233, 240
Barley starch, treatment with
  enzymes, 276
Bestatin, 16
Bestatin kinetics, 17
Binding assays, 359
  with cells, 363
Biochemical oxygen demand, 419
Biomass conversion, 509
Birch sulfite liquor, fermenta-
  tion to ethanol, 338
Bleomycin, 21-25
  biosynthesis, 22
  copper complex, 23
  DNA interaction, 24
Bleomycinic acid, 22

Brevibacterium flavum, 237
  immobilized on carrageenan,
  237
  immobilized on polyethylene
  imine, 237
Brevibacterium flavin, protoplast of, 487
Brick, 178
Bridged bicyclics, reduction
  of, 111
Caldariella acidophila, 209
cAMP, in enzyme biosynthesis,
  46
cAMP, intracellular, glucose
  regulation of, 49
Candida biodinii, 70
Cane molasses, fermentation
  to ethanol, 340
Candida rugosa, immobilized
  in polyacrylamide gel, 268
Carbodiimide, 119, 373, 443
Carbon dioxide electrode, 413,
  419
Carbon dioxide, from immobilized cells, 337
Carboxylation, of enzyme
  support, 295
Carboxypeptidase-Y, 117–120
  immobilized on polyethylene
  glycol, 119
Carrageenan, 336
α–and β– Casein, 289
Catharanthus roseus, immobilized in polyacrylamide
  gel, 137
Catharanthus roseus, immo-
  bilized in gels, 203
Catheters, enzyme treated, 444
Catheters, fibrinolytic activity, 445
Cells, artificial, 452
Cell fusion, 488
Cell to carrier ratio, immobilized cells, 228
Cellobiose, 354
  hydrolysis of, 231
Cellulase
  adsorption on cellulose,
  326

adsorption, temperature
  effect, 328
effect on cellulose structural parameters, 327
extracellular from *Aureobasidium fungus*, 353
hydrolysis of cellulose, 349
from *Trichoderma reesei*,
  355
Cellulose acetate beads, 217
Cellulose
  beads, 218
  conversion to methane, 179
  crystallinity, 326
  degree of polymerization, 329
  hydrolysis of, rate expression,
  349–350
  hydrolysis of, enzymatic,
  325–333, 349
  hydrolysis of, kinetic model,
  330–332
  specific surface area, 326
  x-ray diffraction, during
  hydrolysis of, 325
Cephalosporins, synthesis, 81–
  90, 91–96
CHEMFETs, 378
Chemiluminescence, 370, 423
Chiral compounds, synthesis of,
  113
Chlorella, addition of cell wall
  by protoplasts, 488
Chlorella, nitrogen fixing, 488
cChloroplasts, immobilized on
  methacrylates, 226
Cholesterol, assay for, 370
Cholesterol, conversion of, 157
Cholesterol ester, assay for,
  370
Cholesterol oxidase, 157, 423
Chymostatin, 12
Chymotrypsin, 77
α–Chymotrypsin, immobilized in
  lyposomes, 462
α–Chymotrypsin, photoimmobilization of, 150
cis-Cinnamoyl-trypsin, activation by light, 145
Citrobacter freundii, 165

Cleavable spacers, 211
Clinical analysis, 474
  automated, 369
Clinical chemistry, immobilized
  enzymes, 379
  major cost factors, 382-383
Clinical tests, costs of, 406
Cloning, from thermophilic
  bacteria, 483
Cloning, nylon degradation
  genes, 491
Clostridium acetobutylicum,
  immobilized in alginate,
  154
  in two phase system, 153
Coenzyme regeneration, 311
Coenzyme, economics, 66
Cofactor regeneration, 129
Collagen membranes, 409, 411
Compaction, in immobilized
  enzyme reactor, 315
Computer controlled mass
  spectrometer, 429
Concentration polarization,
  229
Concanavalin A, 362
Cordierite, 174, 284
Corn stover, 305
Cortisol, 182
Corynebacterium, 127
Creatine, 475
Creatine amidinohydrolase,
  immobilized on cellulose
  acetate, 475
Creatinine, 475
Creatinine, amidinohydrolase,
  immobilized on cellulose
  acetate, 475
Crystallinity, of hydrolyzed
  cellulose, 326
Curvularia lunata, immobilized
  in polyethylene glycol, 131
Cyanogen bromide, 211, 223
L-Cysteine, enzymatic synthe-
  sis of, 97
Cypridina, 423
6-Deacetyl-7-aminocephalo-
  sporanic acid, acylation of,
  93

cis-2-Decalone, stereospecific
  reduction, 114
Δ-1-Dehydrogenase, 183
Dextran, 359
Dextran-NAD, 373, 452
Dextrose, 37
1,8-Diamino-4-aminomethyl
  octane, 421
Diffusion resistance, 184, 205,
  214, 224, 285
Diffusivity, effective, 186
Digoxin, 361
Dihydropyrimidinase, 103
Dihydropyrimidinase, substrate
  specificity, 104
Diketocoriolin-B, 15
3,3'-Dimethoxybenzidine, 320
Dimethyl sulfoxide, 197
Diphiobis (succinimidyl propio-
  nate), 212
Dithioglycol, 235
Dithiothreitol, 235
DNA, foreign, barriers to, 480
Dopastin, 14
Drug release, 226
Drug targeting, 461
E. aerogenes, 197
E. coli, 45, 271, 479
  immobilized on,
    agar, 268
    agar beads, 291
    alginate, 252
    cellulose beads, 217
    chitosan, 252
    epoxy, 252
    eudragit, 252
    gelatin, 215
    polyacrylamide gel, 252, 271
    tannin, 217
Economics,
  in analysis, 405
  clinical laboratory, 383
  of clinical tests, 406
  of immobilized enzymes, 381
Effectiveness factor, 186, 282
Elastatinal, 12
Electrochemical potential, 48
Electrolyte analyzers, 397

α,β-Elimination reactions,
  135
α,λ-Elimination reactions,
  135
Endocytosis, 488
Enriched fructose corn
  syrup, 37, 40
Enzymatic hydrolysis of
  cellulose, kinetic model,
  330-332
Enzymatic synthesis, penicillins, cephalosporins, 81-90
Enzyme,
  applications, future needs,
  503
  deactivation, kinetics of,
  229, 287
  electrodes, 396, 409, 402-
  404, 411, 475
  heat stable, 161
  immobilization, in situ,
  201
  immunoassay, 469
  induction, 46
  induction, math model, 45-
  56
  inhibitors, 11-33
  purification, 75
  purification, extraction,
  72-74
  purification, parametric
  pumping, 77
  reactors, 305
  regeneration, 202
  re-immobilization, 202
  stability, 65, 218, 230,
  233, 235, 237-238, 239,
  254, 267, 279, 296, 305,
  416, 476
  stability, modification of,
  by chemicals, 287
  stabilization, 506
Enzymes,
  activity on non-natural
  compounds, 411-501
  in analysis, 509
  in analysis, immobilized,
  377

immobilized in lipid polyamide microcapsules, 164
immobilized by radiation, 226
in medicine, 510
milk clotting, 75
multi, 61-67
in organic solvents, 141, 151,
  191-200
in organic synthesis, 107-
  115, 117-120, 167-170, 506
photoimmobilization, 149
photosensitive, 149
Epichlorohydrin, activation of
  cellulose, 259
Epoxides, from olefins, 169
Epoxy resins, 184
Erythrocytes, immobilized on
  methacrylates, 226
Ethanol, production of, 337,
  343, 347
Ethanol, production of, cost
  estimation, 348
Extracorporeal perfusion, 434
Extraction, liquid-liquid, 69-74
Fatty acids, assay for, 371,
  467, 468
Fermentation,
  ethanol, with immobilized
  cells, 335-342
  of lactic acid, 299
  of molasses, 343, 347-348
  of wine, 245
  to produce heparinase, 435
  of E. coli, 251
  mass spectrometer monitoring
  of, 429
Fibrinolytic activity, 443
Field affect transistors, 378,
  425
Flavobacterium, 500
Flavobacterium heparinum, 435
Flow stabilizing devices, in
  fluidized bed, 309
Fluidized bed reactor, 347
Fluorescein labelled glucoamylase, 214
Food analysis, immobilized
  enzymes, 380
Formate dehydrogenase, purification, 70

Fortimicin, 29
Fructose corn syrup, 273
Fructose, production eco-
    nomics, 40
Fructose, purification, 38
Fumarase, 238, 269
Fumarase, inactivation, 272
Fusaric acid, 14
Galactose electrode, 410
Galactose oxidase, 410
α-Galactosidase, 216
β-Galactosidase, 217, 339
    biosynthesis model, 45-49
    immobilized on,
        cellulose beads, 217
        glutaraldehyde, 339
        ion exchange resin, 201
        membranes, 210
        tannin, 217
    lactose induction, 45
Gamma-irradiation, 225
Gas analysis, 387
Gas production, with immo-
    bilized cells, 173-179
Gel electrophoresis, 437
Gene transfer, 505
Genes, structural, 500
Genetic engineering, 479-
    489, 505
Glass, porous, surface area,
    279
Glucoamylase, 411
    immobilized on,
        carbon, 213
        collagen fibrils, 281
        DEAE Sephadex, 265
        phenol formaldehyde
        resin, 275
        porous glass, 279
        silanized zircon and
        cordierite, 284
Gluconic acid, production,
    188
Glucose adsorption, 274
Glucose,
    assay for, 370
    electrode, 410, 413
    fructose, production,
    37-42

fructose, production eco-
    nomics, 40
isomerase, 52, 205, 216,
    227, 273, 275
isomerase, aminoethylation
    of, 206
isomerase, immobilized on
    anion exchange resin, 266
isomerase, immobilized on
    polyacrylamide gel, 205
oxidase, 410, 413, 415, 422,
    423
production from soluble
    starch, 282
sensor, 381, 415
β-Glucosidase, immobilized in
    membrane, 230
Glutamate, assay for, 370
Glutamate dehydrogenase, 163
    immobilized in microcapsules,
    454
Glutaraldehyde, 77, 146, 205,
    215, 227, 241, 268, 275, 289,
    291, 295, 339, 370, 415, 421,
    453
Gram-positive bacteria, 479
Grape juice, fermentation to
    ethanol, 338
Hapten, assay for, 360
Heparin, 433
Heparinase, 435
    immobilized on sepharose, 437
    inhibitors, 436
    production of, 436
    purification, 436
Hesperiginase, recovery of, with
    tannin, 262
Hexamethylene diamine, 227
High fructose corn syrup, 37-41
Horseradish peroxidase, in water
    purification, 320
Hybrid plasmids, 498
Hydrogen dehydrogenase, 426
Hydrogen peroxide, 475
Hydrogen, semiconductor device,
    425
Hydrogen, semiconductor enzyme
    assay, 426
Hydrogenase, 143

Hydrogenation catalyst, 120
Hydrophilic gel, 193
Hydrophobic
  compounds, biocatalysis
  of, 506
  gel, 193
  environment, 191
  supports, 219, 220
2-Hydroxydecalin stereoisomers, synthesis of, 115
Hydroxylation, stereo-
  chemical, 168
D-p-Hydroxyphenylglycine,
  synthesis of, 103
11-α-Hydroxyprogesterone,
  130
3-α-Hydroxysteroid dehydrogenase, 129
17-β-Hydroxysteroid dehydrogenase, 124
Hyphomicrobium, 103
IgG, assay for, 421
Immobilization of enzymes,
  flow rate, effect of, 214
Immobilized cells, 42-43
Immobilized cells, effect of
  coagulants, 227
Immobilized enzymes,
  advantages in analysis,
  377
  in analysis, 395
  in automated analyses, 369
  in clinical chemistry, 379
  economics, 381
  in food analysis, 380
Immobilized heparinase, in
  vivo testing, 439
  immunoassay, 469
  immuno-modifiers, 15-18
  immunosensors, 421
Indole, effect on enzyme
  stability, 255
Inhibitors, effect on immobilized enzymes, 396
Inhibitor, of heparinase,
  436
Invertase, immobilized on,
  corn stover, 305
  polyacrylamide gel, 315

polyvinyl alcohol, 208
Ionotropic gelation, 181
Iron, removal from water, 263
Isocyanate, 193
Kanamycin, 28
α-Keto acids, 62
Ketonization, 168
3-Ketosteroid-Δ-dehydrogenase,
  123-124, 269
Kinetic model, cellulose hydrolysis, 330-332
Kinetic parameters, starch
  hydrolysis, 283
Kluyvera citrophila, 81
Kluyveromyces fragilis,
  immobilized in alginate, 336
Kluyveromyces lactis, plasmids
  from, 486
β-Lactam,
  N-acylations, 91-96
  α-amino antibiotics, synthesis
  of, 85
  antibiotics, 26-27
  antibiotics, synthesis, 82
  resistance, 27-28
β-Lactamase,
  activity, 86
  deficient mutants, 87
  induction, 88
Lactase,
  high stability, 297
  immobilized on,
    amphoteric ion exchange
    resin, 295
    chitosan coated silica gel,
    309
    plexazym, 293
Lactate, assay for, 370
L-Lactate dehydrogenase, 161,
  373
L-Lactate, production of, 239
Lactic acid, production of, 301
Lactobacilli, 245
Lactobacillus bulgaricus, immobilized in alginate, 299
Lactobacillus delbrueckii,
  immobilized in alginate, 299
Lactones, chiral, synthesis of,
  113

Lactose, 297
    efflux, pH effect, 50
    induction, 45-49
    translocation, 45-46
    transport, 47
Lamda phage, 505
Lankacidin antibiotics, 484
L-Leucine dihydrogenase, 63
L-Leucine, production of,
    312
Leuconostoc oenos, immobi-
    lized in calcium alginate,
    246
Leukemia, treatment with im-
    mobilized enzyme, 465
Leupeptin, 11
Linoleic acid, air oxidation
    of, 141
Lipase,
    ester exchange with tri-
    glyceride, 197
    immobilized on,
        celite, 196
        gels, 152
        polyethylene glycol, 152
    in organic solvent, 151
    stability of, 152
Lipid, polyamide membrane
    microcapsules, 163
Lipid polyamide microcapsules,
    454
Liposomes, 461
Lipoxygenase, 141
Luciferin analog, 423
Luminol, 424
Lymph drainage, in surgery,
    446
L-Lysine electrode, 413
Lysozyme, immobilized on
    collagen-polypropylene,
    459
Lysozyme, immobilized, implanted
    in vivo, 459
Lysyl amine residues, conversion
    to guanidine analogs, 212
Malate humerate, equilibrium
    constant, 303
L-Malic acid, fermentation of,
    in wine, 245-250

L-Malic acid, production of,
    238, 303
Malic dehydrogenase, 451
    immobilized in microcapsules,
    453
Malolactic catalysts, important
    factors, 247
Malolactic fermentation, 245
Malolactic treatment of wine,
    pilot plant, 250
Maltose, assay for, 411
Mass spectrometer, in fermenta-
    tion, 429
Mass transfer coefficient, 282
Medical applications, of en-
    zymes, 510
Membrane, asymmetric, 475
Membrane reactors, 61-67, 313
1-Menthol, production of, 196
Meso-diols, stereospecific oxi-
    dation of, 113
Metal ions, absorption of, on
    tannin, 261
Methane gas, microbial sensor
    for, 390
Methane production, with im-
    mobilized cells, 177
Methanobacter, immobilized on
    ceramics, 179
p-Methoxy phenylglycine, pro-
    duction of, 265
Methyl α-D-mannopyranoside, 362
Methyl perhydroindanone pro-
    pionic acid, 125
Methyl succinate, hydrolysis of,
    stearoselective, 196
Methylomonas flagellata, im-
    mobilized in acetele cellu-
    lose, 390
Microbial cell,
    growth in supports, 254
    immobilized, diffusion re-
    sistance, 186
    in organic solvents, 191-200
Microbial degradation, of syn-
    thetic nylon, 491
Microbial sensors, 387
Microcapsules, membrane permea-
    bility, 453

Microcapsules, nylon-polyethylenimine, 452
Micrococcus lysodeikticus,
459
Microencapsulation, 163
Milk,
curd, β-casein content
of, 290
curd, tension of, 290
curdling of, with immobilized
enzymes, 289-290
skim, treatment of, 295
Mitomycin, 418
Mitotic index, plant cells,
204
Molasses, conversion to
ethanol, 347
Mononucleotides, production
of, 308
Mucor miehi, thermal stability, 287
Multichannel analyzer, 370
Multienzyme systems, 61-67,
275-277
immobilization of, 453
Multiple enzymes, immobilized,
372
Mutagens, microbial sensor
for, 417
Mutation, recE, 480
NAD, 129, 163, 311, 373, 452,
473
analogs of, 239
polyethylene glycol adduct,
311
immobilized on polyethylene
glycol
Membranes, polarization, 62
Nitrate, assay for, 420
Nitrite, assay for, 420
Nitrite, oxidation of, 392
Nitrifying bacteria, 388
Nitrogen dioxide, microbial
sensor for, 391
Nitrogen fixing cells, proto-
plasts from, 488
Nitrogenase, 488
Nitrobacter, 388
Nitrosomonas, 388

Nocardia corallina, immobilized
in polyacrylamide gel, 139
Nocardia,
in hollow fiber reactor, 126
immobilized, stability, 126
IMMOimmobilized in polyacrylamide
gel, 125
Nocardia rhodocrous, 157, 194
Nocardia, steroid activity, 125
Nozzle Separater, 72
Nuclease, immobilized on agarose,
223
Nuclease pl, immobilized on
ABSE-cellulose, 266
Nylon polyethylenimine micro-
capsules, 453
Nylon-6, substrate for enzymes,
491
β-Oestradiol, assay for, 414
Optically active amino acids,
enzymatic synthesis of, 97-105
Organoleptic properties, of
wine, 246
Oudenone, 14
Ovomucoid, immobilized on chito-
san, 77
Oxirane groups, 293
Oxygen concentration, 416
Oxygen, mass transfer, 157
Oxygenation reactions, enzymatic,
167
Palladium poly(ethylenimide), 121
Palladium, semiconductors, 425
Papain, 12
Particle size, of supports, 213
Partition affinity assay, 359
Partition coefficient, 71, 73
Penicillin acylase, 81-90, 216,
269
microbial sources, 82
Penicillin amidohydrolase, 92
immobilized on Eupergit C, 235
Penicillin G, hydrolysis,
kinetics, 93
Penicillin, synthesis, 81-90,
91-96
Penicillinase, inactivation, 84
Penicillium chrysogenum, 81
Pepleomycin, 22

Pepsinostreptin, 76
Pepstatin, 13
Pepstatin A, immobilized on
  aminohexal-agarose, 75
Peptide synthesis, deblocking,
  117
Peroxidase, 362, 423
Peroxidase catalyzed oxidation, 320
Peroxyacid resins, 211
Petrochemical industry, immobilized enzymes in, 508
Phenols, removal from water,
  319
Phosphatidyl ethanol amine,
  461
Phosphatidylinositol, 462
6-Phosphogluconate dehydrogenase, 233
Photo cross linkable resin,
  131, 192
Photoprinting, with enzymes,
  146
Photosynthesis, artificial,
  143
pH parametric pumping, 77
Pilot plant,
  immobilized enzymes, 293,
  297, 305
  immobilized cells, 249, 268,
  292, 300, 340, 347
  immobilized nuclease p1,
  266
Plant cells, immobilized, 203
Plasmids
  for amino hexanoic acid hydrolases, 494
  from B. subtilis, 480
  composite, instability, 482
  pLS28, 480
  separation by electrophoresis,
  496
  from Streptomyces, 485
  pTL12, 480
  in wild and transformed
  strains, 495
Platelet aggregation, 457
Platinum catalysts, alterna-
  tives for, 323

Polyacrylamide gel, 268
Polycondensation of oligomers,
  184
Polyethylene glycol, 193, 311,
  359
  coupled to carboxypeptidase,
  119
  NAD analogs, 62, 239
  in protoplast fusion, 486
  salt system, 71
Polyethylenimine, 443
Polymyxin B, 459
Polynucleotide kinase, 220
Polynucleotide ligase, immo-
  bilized on trityl agarose,
  220
Polynucleotide phosphorolase,
  immobilized on ABSE-agarose,
  266
Polypropylene glycol, 193
Polyurethanes, 184
Polyvinyl alcohol, ABSE deriva-
  tives, 268
Polyvinyl alcohol fibers, 207
Polyvinylsulfate, immobilized on
  cepharose, 437
Pore size, of support, 224
Pore diameter, for immobilized
  cells, 173
Potentiometric sensor, 415
Prednisolone, 124
Printed circuit board, 146
Prochiral diols, selective
  oxidation of, 112
Product inhibition, 153
Progesterone, hydroxylation of,
  133
Propylene oxide, production,
  139
Protamine, 434
Proteases, inhibitors, 11
Protease, from Trichoderma
  reesei, 355
Protein modifiers, 211
Protoplast fusion, 486
Pseudomonas, in genetic en-
  gineering, 482
Pseudomonas dacunhae, immo-
  bilized in carrageenan, 271

Pseudomonas melanogenum, 81
Pseudomonas oleovorans,
  hydroxylation with, 169
Pseudomonas putida, 104, 135
Pullulanase, immobilized on
  phenol formaldehyde resin,
  275
Purification,
  cellulases, 354
  economics, 42
  enzymes, 75-76
  of heparinase, 436
  lactase, solvent precipitation, 295
  methods, scale up, 504
  nozzle separater, 72
  parametric pumping, 77
  of proteases, 355
  proteins, 69-74
  recovery, 267
  scale-up, 69-74
  two-phase system, 69
Pyridoxal enzymes, 97, 507
Radiation polymerization,
  225
Radioimmunoassay, 361, 471
Reactor,
  batch, 276, 312
  bubble tower, 140
  with cells, batch, 184
  column, 202, 216, 228, 265,
  275, 292, 299, 304, 370
  column, alginate, 248
  columns, for immobilized
  cells, 175
Reactors, ethanol fermentation,
  337
  fermentation, batch, 251
  flow rate, effect of, 255
  fluidized bed, 309-310,
  347, 438
  hollow fiber, 126, 465
  immobilized enzymes, 508
  immobilized cells, 508
  membrane, 61-67, 313
  membrane rotary disc, 281
  moving-bed adsorber, 274
  nylon tube, 405
  packed bed, 279, 295, 305,

315, 337, 344
  packed bed, pressure drop,
  293
  packed bed, void fraction,
  316
  pilot plant, packed bed,
  249
  plug flow, 351
  recycle, 374
  residence time effects, 337
  stirred tank, 129, 255, 378
  tubular, 473
  ultrafiltration membrane,
  209, 229, 239
  void volume, 341
Recombinant systems, in *B.*
  subtilis, 479
Redox dye, immobilized, 144
Regeneration of coenzymes, 311
Regio specificity, synthesis,
  enzymatic, 169
Reichstein's Substance S,
  hydroxylation of, 131-132
Renin, 75
Rennet,
  excess thermal stability, 287
  immobilized on Dowex resin,
  289
  modified by oxidation, 288
Replication, barrier to, 480
Residence time, 177
Residence time distribution,
  309
Restriction map,
  pOAD2, 497
  pOAD21, 497
  pNDH5, 498
Rhizopus delemar,
Rhodotorula minuta, immobilized
  in gels, 195
Rhizopus stolonifer, immo-
  bilized in polyethylene
  glycol, 133
3'-RNase, immobilized on ABSE
  Sephadex, 266
Rotary disc reactor, 281
Routine analysis, with enzymes,
  405
Saccharification, 276

Saccharomyces cerevisiae,
  123
  binding assay of, 364
  immobilized in alginate, 336
  immobilized on agar, 343-345
  immobilized in polyacrylamide gel, 123
Sake, clarified with immobilized tannin, 262
Sarcosine oxidase, immobilized
  on cellulose acetate, 475
Selenium amino acids, synthesis
  of, 135
Semiconductor, metal oxide
  type, 425
Serpentine, 138
Sewage cells, immobilized on
  alumina, 174
Sewage cells, immobilized on
  cordierite, 174
S. Formosensis, immobilized
  in alginate, 347
S. Formosensis, immobilized
  on methacrylates, 226
S-octopamine, 169
Spergualin, 25
Stability, of enzymes, 215
Staphylococcus aureus, 363
Starch
  barley, hydrolyzed by immo-
  bilized cells, 336
  hydrolysis of, 279
  hydrolysis, math model, 285
  katayma, 284
  potato, 284
  soluble, hydrolysis
  kinetics, 283
Steroid dehydrogenase iso-
  enzymes, 127
Steroid dehydrogenation, 194
Steroid hydroxylation, 198
Steroid modifications, 129
Stirred tank reactor, 129, 255
Streptococci, binding assay
  of, 366
Streptomyces, 12
Streptomyces, gene engineer-
  ing of, 484
Streptomyces griseus, 145

Streptomyces, immobilized on
  soy bean protein, 227
Streptomyces phaeochromogenes,
  205
Streptomyces roseofulus,
  immobilized,
    in cross-linked gelatin,
    216
    with glutaraldehyde, 268
Succrose hydrolysis, 306
p-(β-Sulfatoethylsulfonyl)-
  aniline, coupling reagent,
  266
Sulfoxidation, stereochemical,
  168
Sulfur amino acids, synthesis
  of, 135
Sulfur dioxide, enzyme inhibition, 248
Sulphato ethylsulphonyl aniline,
  307
Supports, aminohexal-agarose, 75
Supports, chitosan
Supports, for cells,
  acetyl cellulose, 418
  agar, 268, 291
  albumin, 130
  aliginate, 130, 154, 252, 299,
  347, 323, 336
  alumina, 174
  anion exchange resin, 303
  artificial membranes, 209
  brick, 178
  calcium alginate, 181, 246
  κ-carageenan, 130, 237, 271,
  323, 336
  cellulose beads, 217
  cellulose acetate butyrate
  capsules, 304
  ceramics, 173
  chitosan, 182, 252
  cordierite, 174
  epoxy carriers, 185, 252
  eudragit, 252
  gels, 195, 203
  gelatin, 130
  glutaraldehyde, 268
  methacrylates, 226

modified polyethylenimine,
182
modified polyvinyl alcohol,
182
polyacrylamide gel, 123, 125,
137, 139, 238, 252, 268, 271
polyethylene glycol, 131-133
polyethylene imine, 237
polyurethane, 130
soy bean protein, 227
tannin, 217
Supports, for chemicals, liposomes, 462
Supports, for chemicals,
sepharose, 437
Supports, for compounds, aminohexylated cellulose, 259
Supports, for enzymes,
ABSE-agar, 266
ABSE-agarose, 266
ABSE-Sephadex, 266
anion exchange resin, 266
carbon, 213
celite, 152, 192
cellulose acetate, 475
chitosan on silica gel, 309
collagen, 409
collagen-polypropylene, 459
DEAE Sephadex, 265
Dowex resin, 289
Eupergit C, 235
hollow fibers, 465
ion exchange resin, 201
lipid-polyamide, 454
maleic anhydride-methyl vinyl
ether, 443
nylon, 444, 473
nylon-polyethylenimine, 453
phenol formaldehyde resin,
275
phenolic amphoteric ion exchange resin, 295
platinum, 415
polyacrylamide gel, 315
polyethylene glycol, 152
polyvinyl alcohol fibers,
207
porous glass, 280
sepharose, 437, 457

silica beads, 241
trityl agarose, 220
zircon and cordierite, 284
Supports, for enzymes, in
organic solvents, 142
Supports, polyacrylamide, 123,
144
Supports, polyethylene glycol,
62, 119
Suture material, enzyme treated,
448
Synzymes, 117
Tannin, benzoguinone derivative,
218
Tannin, immobilized,
aminohexylated cellulose, 259
reaction with metals, 261
specificity of, 260
Targeting, of drugs, 461
Testosterone, 195
Thermolysin, immobilized on
Dowex resin, 289
Thermophile, 290
Thermophilic bacteria, 216,
303, 483
Thermophilic bacteria, gene of,
in plasmid, 483
Thermophilic enzymes, 233, 239-
240
Thermophilic enzymes, stability
to chymotrypsin, 234
Thermus caldophilus, 161
Thermus rubens, immobilized in
cellulose acetate butyrate
capsules, 304
Thermus rubens, immobilized on
anion exchange resin, 303
Thermus thermophilus, 240
2-Thiapyranones, reduction of,
stereospecific, 112
Thiele modulus, 187
Thionine dye, immobilized on
polyacrylamide, 144
Thoracic surgery, enzyme
treated tubes in, 446
Thrombi reduction, with enzyme
treated filaments, 447
Thrombus formation, 457
Thyroxine, assay for, 469

Tissue paper, hydrolysis of, 349

Transcription, barrier to, 482

Transducer, with enzymes, 378

Trichoderma reesei, proteases and cellulase, 355

Triglyceride, ester exchange, 151

Triglyceride esterification, 194

Triiodothyronine, 361

Tritium, removal from water, 322

Tritylated agarose, 219

Trypsin, 77

Trypsin, immobilized on Dowex resin, 289

Trypsin, immobilized on ion exchange resin, 201

L-Tryptophan, enzymatic synthesis of, 97

L-Tryptophan, production of, 251-257
dilution rate, 256

Tryptophan synthetase, 251
genetic engineered, 483

Tryptophan synthetase, in immobilized cells, 253

Tryptophanase, 97

Tubular recycle membrane reactor, 313

Tumor antibiotics, 18-25

Two phase system, 157, 159, 359

β-Tyrosinase, 97

L-Tyrosine, enzymatic synthesis of, 97

Tyrosine phenol lyase, 165

Tyrosine, synthesis of, 165

Ultrafiltration membranes, 209, 229

Urea, assay for, 370

Urethane prepolymer, 193

Uric acid, assay for, 370

Urokinase, immobilized,
on maleic anhydride-methyl vinyl ether copolymer, 443
on nylon, 444
on sepharose, 457

Vectors for gene coding, 505

Void volume, in reactors, 341

Waste water, analysis of, 419

Water, purification of, 319

Whey, treatment of, 295

Whey ultrafiltrate, fermentation to ethanol, 339

Xanthine oxidase, 423

Yeast cells, immobilized on epoxy carriers, 185

Yeast cells, polyethylene glycol modified, 365

Yeast plasmid, 484

Zircon, 284

Zymomonas mobilis, immobilized in carrageenan, 336